INTERMEDIATE QUANTUM MECHANICS

THIRD EDITION

HANS A. BETHE
Cornell University
Ithaca, New York

ROMAN JACKIW
Massachusetts Institute of Technology
Cambridge, Massachusetts

CRC Press
Taylor & Francis Group
Boca Raton London New York

CRC Press is an imprint of the
Taylor & Francis Group, an informa business

First published 1986 by Westview Press

Published 2018 by CRC Press
Taylor & Francis Group
6000 Broken Sound Parkway NW, Suite 300
Boca Raton, FL 33487-2742

CRC Press is an imprint of the Taylor & Francis Group, an informa business

Visit the Taylor & Francis Web site at
http://www.taylorandfrancis.com

and the CRC Press Web site at
http://www.crcpress.com

ISBN 13: 978-0-201-32831-8 (pbk)

Cover design by Suzanne Heiser

Advanced Book Classics

Negele/Orland: Quantum Many-Particle Systems, ABC ppbk, ISBN 0-7382-0052-2

Nozieres/Pines: Theory of Quantum Liquids, ABC ppbk, ISBN 0-7382-0229-0

Nozieres: Theory of Interacting Fermi Systems, ABC ppbk, ISBN 0-201-32824-0

Parisi: Statistical Field Theory, ABC ppbk, ISBN 0-7382-0051-4

Pines: Elementary Excitations in Solids, ABC ppbk, ISBN 0-7382-0115-4

Pines: The Many-Body Problem, ABC ppbk, ISBN 0-201-32834-8

Quigg: Gauge Theories of the Strong, Weak, and Electromagnetic Interactions, ABC ppbk, ISBN 0-201-32832-1

Richardson: Experimental Techniques in Condensed Matter Physics at Low Temperatures, ABC ppbk ISBN 0-201-36078-0

Rohrlich: Classical Charges Particles, ABC ppbk ISBN 0-201-48300-9

Schrieffer: Theory of Superconductivity, ABC ppbk ISBN 0-7382-0120-0

Schwinger: Particles, Sources, and Fields Vol. 1, ABC ppbk ISBN 0-7382-0053-0

Schwinger: Particles, Sources, and Fields Vol. 2, ABC ppbk ISBN 0-7382-0054-9

Schwinger: Particles, Sources, and Fields Vol. 3, ABC ppbk ISBN 0-7382-0055-7

Schwinger: Quantum Kinematics and Dynamics, ABC ppbk, ISBN 0-7382-0303-3

Thom: Structural Stability and Morphogenesis, ABC ppbk, ISBN 0-201-40685-3

Wyld: Mathematical Methods for Physics, ABC ppbk, ISBN 0-7382-0125-1

Advanced Book Classics

Anderson: Basic Notions of Condensed Matter Physics, ABC ppbk, ISBN 0-201-32830-5

Atiyah: K-Theory, ABC ppbk, ISBN 0-201-40792-2

Bethe: Intermediate Quantum Mechanics, ABC ppbk, ISBN 0-201-32831-3

Clemmow: Electrodynamics of Particles and Plasmas, ABC ppbk, ISBN 0-20147986-9

Davidson: Physics of Nonneutral Plasmas, ABC ppbk ISBN 0-201-57830-1

DeGennes: Superconductivity of Metals and Alloys, ABC ppbk, ISBN 0-7382-0101-4

d'Espagnat: Conceptual Foundations Quantum Mechanics, ABC ppbk, ISBN 0-7382-0104-9

Feynman: Photon-Hadron Interactions, ABC ppbk, ISBN 0-201-36074-8

Feynman: Quantum Electrodynamics, ABC ppbk, ISBN 0-201-36075-4

Feynman: Statistical Mechanics, ABC ppbk, ISBN 0-201-36076-4

Feynman: Theory of Fundamental Processes, ABC ppbk, ISBN 0-201-36077-2

Forster: Hydrodynamic Fluctuations, Broken Symmetry, and Correlation Functions, ABC ppbk, ISBN 0-201-41049-4

Gell-Mann/Ne'eman: The Eightfold Way, ABC ppbk, ISBN 0-7382-0299-1

Gottfried: Quantum Mechanics, ABC ppbk, ISBN 0-201-40633-0

Kadanoff/Baym: Quantum Statistical Mechanics, ABC ppbk, ISBN 0-201-41046-X

Khalatnikov: An Intro to the Theory of Superfluidity, ABC ppbk, ISBN 0-7382-0300-9

Ma: Modern Theory of Critical Phenomena, ABC ppbk, ISBN 0-7382-0301-7

Migdal: Qualitative Methods in Quantum Theory, ABC ppbk, ISBN 0-7382-0302-5

CONTENTS

Editor's Foreword

Addison-Wesley's *Frontiers in Physics* series has, since 1961, made it possible for leading physicists to communicate in coherent fashion their views of recent developments in the most exciting and active fields of physics—without having to devote the time and energy required to prepare a formal review or monograph. Indeed, throughout its nearly forty-year existence, the series has emphasized informality in both style and content, as well as pedagogical clarity. Over time, it was expected that these informal accounts would be replaced by more formal counterparts—textbooks or monographs—as the cutting-edge topics they treated gradually became integrated into the body of physics knowledge and reader interest dwindled. However, this has not proven to be the case for a number of the volumes in the series: Many works have remained in print on an on-demand basis, while others have such intrinsic value that the physics community has urged us to extend their life span.

The *Advanced Book Classics* series has been designed to meet this demand. It will keep in print those volumes in *Frontiers in Physics* or its sister series, *Lecture Notes and Supplements in Physics*, that continue to provide a unique account of a topic of lasting interest. And through a sizable printing, these classics will be made available at a comparatively modest cost to the reader.

The lecture-note volume *Intermediate Quantum Mechanics*, written by Nobel laureate Hans A. Bethe in collaboration with his distinguished colleague, Roman Jackiw, introduces the reader to the intricacies and applications of quantum mechanics. It is written in the lucid style for which both authors are justly famous, and remains as fresh today as when it first appeared over twenty-five years ago. It continues to be one of the best accounts of this sine-qua-non material that every graduate student in physics seeks to master, and I am pleased to welcome its authors to the *Advanced Book Classics* series.

David Pines
Urbana, Illinois
October 1997

PREFACE TO THE FIRST EDITION

This book is intended to serve as a text for a second course in quantum mechanics, for graduate students in both theoretical and experimental physics. It is assumed that the student has a knowledge of the principles of quantum mechanics, equivalent to the first eight chapters of Schiff's *Quantum Mechanics,* or the entire book of Merzbacher. I believe that the general exposition of the theory as given in these books should be followed by a discussion of the applications to problems in which the basic physics is essentially known and well understood, notably the structure of atoms and the theory of atomic collisions, so that the solidity of the theory becomes apparent. After this, the student will be better prepared to study nuclear physics, where the forces are unknown, or solid-state physics, where the physical approximations are often tentative. I have stressed the connection with experimental information and with the physical picture rather than the formal development of the theory. Some recent books have stressed all too much the formal side.

Good books are available for study at this level, e.g., Condon and Shortley, *Theory of Atomic Spectra;* Slater, *Quantum Theory of Atomic Structure;* Mott and Massey, *Theory of Atomic Collisions;* Heitler, *Quantum Theory of Radiation.* However these books are mainly intended for the expert, or at least the student who wishes to specialize in one particular field of quantum mechanics. The present book is intended to give to the graduate student in physics enough knowledge in at least one of the fields, namely, atomic structure, so that he can then intelligently follow discussions on various coupling schemes in atoms, nuclei, and fundamental particles. It will give him a working knowledge of Clebsch-Gordan coefficients. It also gives a detailed treatment of optical transition probabilities, including quantitative calculation.

H. A. BETHE

Ithaca, New York
January 1964

PREFACE TO THE SECOND EDITION

In preparing this Second Edition, the chapters on theory of atomic structure have essentially been retained from the First Edition, although we hope to have made some improvements in presentation. We have added to the book a fairly extensive section on collision theory. Some of this material cannot be found in other textbooks. For a more extensive treatment, the reader is referred to the classic book by Mott and Massey, *Theory of Atomic Collisions.*

We wish to point out that the present edition, like the First Edition, is not intended to be used as the sole textbook discussing the fundamental theory in a second course of quantum mechanics. Rather, it is intended as a supplement to a text stressing fundamental theory, such as Messiah's *Quantum Mechanics,* Volume II. A few chapters on field theory which were in the First Edition have been omitted because this subject is covered adequately in other books.

H. A. BETHE
R. W. JACKIW

April 1968

PREFACE TO THE THIRD EDITION

In our Third Edition we have made revisions, which were suggested during the almost quarter century that *Intermediate Quantum Mechanics* has been in print and in use. They are mainly of pedagogical nature, reflecting our own teaching experience and the many useful comments we have received from colleagues, for which we are very grateful. Also a brief description of recent progress in the treatment of electron correlations is included. These new developments, based on the Bruckner, Bethe, Goldstone method in nuclear physics, help substantially in calculating accurate atomic energy levels.

H.A. BETHE
R. JACKIW

August, 1985; Aspen, Colorado

Part I

THEORY OF ATOMIC STRUCTURE

There are at least three good reasons for undertaking a careful study of atomic structure. First, on the basis of the quantum theory every known feature of the electronic structure of atoms can be explained. Knowledge of this structure is important for chemistry, solid state physics, spectroscopic determination of nuclear properties (hyperfine structure, etc.), and many other applications. The quantitative validity of these explanations is limited only by computational difficulties. Second, the excellent agreement between theoretical and experimental results in the tremendously wide range of atomic phenomena provides a crucial test for the validity of quantum mechanics. Finally, the theory of atomic structure is a convenient "theoretical laboratory" in which one can become acquainted with many physical ideas and mathematical tools which are relevant to other branches of physics. Some aspects of the theory of nuclear structure, for example, parallel quite closely atomic theory.

Chapter 1

MISCELLANEOUS RESULTS FROM ELEMENTARY QUANTUM MECHANICS

SCHRÖDINGER EQUATION

This chapter will serve to summarize the results of elementary quantum mechanics and to set the notation that will be employed in this book. For a more complete treatment of these topics, the reader is referred to any textbook on quantum mechanics.

The starting point for a nonrelativistic quantum mechanical description of the electronic configuration of an N-electron atom with nuclear charge Ze is the *Schrödinger equation*:

$$i\hbar \frac{\partial}{\partial t} \Psi = H\Psi \tag{1-1}$$

For the present, we concern ourselves only with the stationary properties of this electronic configuration. Thus, the Hamiltonian is time independent, and the usual separation occurs as

$$\Psi = \psi e^{-iEt/\hbar} \tag{1-2a}$$

$$H\psi = E\psi \tag{1-2b}$$

where E is the energy eigenvalue. If we neglect all the spin interactions of the electrons and all nuclear effects (e.g., finite size and mass), the Hamiltonian in (1-2b) has the form

$$H = \sum_j \frac{p_j^2}{2m} - Ze^2 \sum_j \frac{1}{r_j} + \sum_{i>j} \frac{e^2}{r_{ij}} \tag{1-3}$$

where m is the electronic mass, e the absolute value of the electronic charge, r_j the absolute value of the position vector of the jth

electron, and $r_{ij} = |\mathbf{r}_i - \mathbf{r}_j|$. The summations in the first and second terms extend over all the N electrons. The summation in the last term is over all pairs ($i \neq j$), each pair being counted once; i.e.,

$$\sum_{i>j} = \sum_{i=2}^{i=N} \sum_{j=1}^{j=i-1}$$

The first term in (1-3) represents the kinetic energy of the electrons; the second is the Coulomb interaction of the electrons with the nucleus; the last term is the interelectron Coulomb interaction.

In the calculation of atomic structure it is convenient to use Hartree's *atomic units* (au), which we shall employ in Chapters 1 to 4. The unit of mass is the rest mass of the electron; of charge, the magnitude of the electronic charge; of length, the first Bohr radius of the hydrogen atom $\hbar^2/me^2$. In these units. (1-2b) becomes

$$\left[\sum_j \nabla_j^2 + 2\left(E + Z\sum_j \frac{1}{r_j} - \sum_{i>j} \frac{1}{r_{ij}}\right)\right]\psi = 0 \qquad (1\text{-}4)$$

For one electron, (1-4) can be solved exactly. The solution corresponding to bound states is

$$\psi(\mathbf{r}) = R_{n\ell}(r)Y_{\ell m}(\Omega)$$

$$Y_{\ell m}(\Omega) = \sqrt{\frac{(\ell - |m|)!}{(\ell + |m|)!}\,\frac{2\ell + 1}{4\pi}}\;(-1)^{(m+|m|)/2} \qquad (1\text{-}5)$$

$$\times\, P_\ell^{|m|}(\cos\theta)\, e^{im\varphi}$$

$$R_{n\ell}(r) = \frac{2}{n^2}\left\{\frac{(n-\ell-1)!\,Z^3}{[(n+\ell)!]^3}\right\}^{\frac{1}{2}}\left(\frac{2Zr}{n}\right)^{\ell} e^{-Zr/n}\, L_{n+\ell}^{2\ell+1}\left(\frac{2Zr}{n}\right)$$

where $P_\ell^{|m|}$ is the associated Legendre function, and L_q^p is the generalized Laguerre polynomial. The energy levels are

$$E_n = -\frac{1}{2}\frac{Z^2}{n^2} \qquad (1\text{-}6)$$

The quantum numbers n, ℓ, m describing the solutions are called the principal, azimuthal, and magnetic quantum numbers, respectively. They take on the values

$$n = 1, 2, \ldots$$

$$\ell = 0, 1, \ldots, n-1$$

$$m = -\ell, -\ell+1, \ldots, \ell-1, \ell \qquad (1\text{-}7)$$

For a detailed discussion of these *hydrogen atom* solutions the reader is referred to Bethe and Salpeter,[1] pp. 4-33.

The separation of the one-body Schrödinger equation into radial and angular variables is always possible when the potential is spherically symmetric. The wave function is then a product of a radial function $R_{n\ell} \equiv \mathcal{R}_{n\ell}/r$ and a spherical harmonic $Y_{\ell m}$, as in (1-5). The energy eigenvalues will in general depend on the principal and azimuthal quantum numbers $E_{n\ell}$, and states are labeled by these quantum numbers, with the letters s, p, d, ... frequently replacing the ℓ-values 0, 1, 2... . The principal quantum number n is defined by the statement that $\mathcal{R}_{n\ell}$ has $n-\ell-1$ nodes, not counting the nodes at $r=0$ and $r=\infty$. This also gives an ordering for the eigenvalues, i.e., $E_{n\ell}$ increases with increasing n. It is a special symmetry property of the Coulomb potential which renders the energy eigenvalues (1.6) ℓ-independent (Schiff,[2] pp. 234-239).

For more than one electron, no exact solutions of (1-4) have been found. For two electrons, very accurate approximation methods exist; for multielectron systems, only less accurate methods are available.

APPROXIMATE METHODS OF SOLUTION

In the approximate treatment of (1-4), we need to make use of the various well-known approximation techniques: *bound state perturbation theory, time-dependent perturbation theory, Born approximation, variation methods*, and the *WKB approximation*. We summarize here the principal results of these.

In *bound state perturbation theory*, the Hamiltonian is split into an unperturbed part and a small perturbing part.

$$H = H^0 + \lambda H^1$$

Here λ is some convenient parameter, which is assumed to be small and which will distinguish orders of the perturbation. The solutions of H^0 are supposed known.

$$H^0 u_n = E_n u_n$$

---

[1] H. A. Bethe and E. E. Salpeter, *Quantum Mechanics of One- and Two-Electron Atoms*, Plenum, New York, 1977.

[2] L. I. Schiff, *Quantum Mechanics*, 3rd ed., McGraw-Hill, New York, 1968.

We assume that the unperturbed system is in a definite state u_n with energy E_n and the effect of the perturbation λH^1 is such that E, the exact eigenvalue, is much closer to E_n than to $E_{n\pm 1}$. If u_n is non-degenerate, we can write

$$E \approx E^{(0)} + \lambda E^{(1)} + \lambda^2 E^{(2)}$$

$$E^{(0)} = E_n$$

$$E^{(1)} = H^1_{nn}$$

$$E^{(2)} = \sum_m{}' \frac{|H^1_{nm}|^2}{E_n - E_m}$$

$$H^1_{nm} = \int d\tau\, u_n^* H^1 u_m \tag{1-8}$$

(The symbol $\int d\tau$ indicates integration and summation over all the variables on which the wave functions depend.)
The prime on the sum indicates that the term $m = n$ is omitted. If the unperturbed state is degenerate, we must find the proper linear combination of the degenerate unperturbed eigenfunctions such that the perturbing Hamiltonian is diagonal.

The *time-dependent perturbation theory* proceeds by replacing the time-dependent Schrödinger equation

$$i\hbar \frac{\partial \Psi}{\partial t} = H\Psi \tag{1-1}$$

by

$$i\hbar \frac{\partial a_k}{\partial t} = \lambda \sum_n H^1_{kn} e^{i\omega_{kn} t} a_n \tag{1-9}$$

where

$$H = H^0 + \lambda H^1$$

$$H^0 u_n = E_n u_n$$

$$\Psi = \sum_n a_n u_n e^{-iE_n t/\hbar} \tag{1-10}$$

$$\omega_{kn} = \frac{E_k - E_n}{\hbar}$$

The set of equations (1-9) for all k is entirely equivalent to (1-1).

This set is obtained by substituting for Ψ in (1-1) the expression for Ψ as given in (1-10). The set (1-9) can be solved by successive approximation, i.e., by substituting

$$a_n = \sum_{i=0}^{\infty} \lambda^i a_n^{(i)} \tag{1-11}$$

in (1-9) and equating equal powers of λ. If it is assumed that initially the system is in the state m, we put

$$a_k^{(0)} = \delta_{km}$$

$$a_k^{(1)} = (i\hbar)^{-1} \int_{-\infty}^{t} H_{km}^1 (t') e^{i\omega_{km} t'} \, dt' \tag{1-12}$$

If H^1 is independent of time, except for being turned "on" and "off" at times 0 and t, respectively, an integration of (1-12) yields

$$a_k^{(1)} = \frac{-H_{km}^1}{\hbar} \frac{e^{i\omega_{km} t} - 1}{\omega_{km}} \tag{1-13a}$$

$$|a_k^{(1)}(t)|^2 = 4\,|H_{km}^1|^2 \frac{\sin^2 \frac{1}{2}\omega_{km} t}{\hbar^2 \omega_{km}^2} \tag{1-13b}$$

Equation (1-13b) is the first-order probability of transition from state m to state $k \neq m$. The *transition probability per unit time* will then be

$$w = \frac{2\pi}{\hbar} |H_{km}^1|^2 \frac{1}{\hbar} \left(\frac{2 \sin^2 \frac{1}{2}\omega_{km} t}{\pi \omega_{km}^2 \, t} \right) \tag{1-14a}$$

The term in parentheses may be recognized as a representation of the Dirac delta function in the limit $t \to \infty$. (We assume the perturbing Hamiltonian has been "turned on" for a long time—as we must if both the states k and m are defined with no uncertainty.) We may then rewrite (1-14a) as

$$w = \frac{2\pi}{\hbar} |H_{km}^1|^2 \frac{1}{\hbar} \delta(\omega_{km}) \tag{1-14b}$$

$$w = \frac{2\pi}{\hbar} |H_{km}^1|^2 \delta(E_k - E_m) \tag{1-14c}$$

This explicitly exhibits the fact that energy is conserved in first-

order transitions. If the transition is to a continuum (or quasi-continuum) of states about state k, we describe the density of final states by $\rho(E_k)$, the number of states per unit energy, and replace the delta function by the density function, thus obtaining the well-known *Fermi golden rule*:

$$w = \frac{2\pi}{\hbar} |H^1_{km}|^2 \rho(E_k) \qquad (1\text{-}15)$$

This formula may also be derived by the *Born approximation* in scattering theory. Recall that the exact transition probability per unit time is given by (Schiff,[2] pp. 312-324)

$$w = \frac{2\pi}{\hbar} |\langle k|H^1|m^{(+)}\rangle|^2 \delta(E_k - E_m) \qquad (1\text{-}16)$$

Here $|k\rangle$ represents the unperturbed state, while $|m^{(+)}\rangle$ is an eigenstate of the exact Hamiltonian with boundary conditions which are conveniently given by an integral equation.

$$|m^{(+)}\rangle = |m\rangle + \left(\frac{1}{E_m - H^0 + i\varepsilon}\right) \lambda H^1 |m^{(+)}\rangle \qquad (1\text{-}17)$$

(It is understood that ε is set to zero from above after the equation is solved; ε serves to define the inverse to the operator $E_m - H^0$, which without some regularization has no inverse since it possesses a zero eigenvalue $(E_m - H^0)|m\rangle = 0$.) The Born approximation consists of solving (1-17) in powers of λ. When only the lowest order is kept, $|m^{(+)}\rangle \approx |m\rangle$ in (1-16), thus (1-14c) and (1-15) are regained.

When using (1-15), it should be remembered that the density of final states per unit energy $\rho(E)$ satisfies the following relations. The total number of free-particle states in the momentum interval between $\mathbf{p}$ and $\mathbf{p} + d^3\mathbf{p}$ is

$$\rho(E)\,dE = \frac{d^3\mathbf{p}}{(2\pi\hbar)^3} \qquad (1\text{-}18a)$$

per quantum degree of freedom. An equivalent formula is

$$\rho(E) = \frac{d\Omega p^2}{(2\pi\hbar)^3} \frac{dp}{dE} = \frac{d\Omega p^2}{(2\pi\hbar)^3 v} \qquad (1\text{-}18b)$$

where the velocity v satisfies $\frac{dE}{dp} = v$, which is also true relativistically.

The *variation principle* consists in selecting a function ψ, varying the quantity $\langle\psi|H|\psi\rangle$ in some arbitrary fashion subject to the condition $\langle\psi|\psi\rangle = 1$ and thus obtaining a stationary value for $\langle\psi|H|\psi\rangle$. If the trial function ψ and the method of variation are completely arbitrary, the variation principle obtains Schrödinger's time-independent equation

$$\delta \int \psi^* H \psi \, d\tau = 0 \qquad \int \psi^* \psi \, d\tau = 1 \tag{1-19a}$$

The subsidiary normalization condition is introduced by a real Lagrange multiplier E, i.e.,

$$\delta\left[\int \psi^* H\psi \, d\tau - E \int \psi^* \psi \, d\tau\right] = 0$$

$$\delta \int \psi^* (H - E)\psi \, d\tau = 0$$

$$0 = \int \delta\psi^* (H - E)\psi \, d\tau + \int [(H - E)\psi]^* \delta\psi \, d\tau \tag{1-19b}$$

The Hermiticity of H − E was used in equation (1-19). Considering the variation of ψ^* and ψ to be independently arbitrary yields

$$(H - E)\psi = 0 \tag{1-20}$$

An *approximate* application of the variation principle consists in choosing the trial function to be of some specific form, and carrying out the variation in some prescribed fashion; i.e., neither the choice of trial function nor the mode of variation is completely arbitrary. An example is the *Ritz method* where the trial function depends on several parameters and the variation is effected by varying these parameters. We shall use the Ritz method in variation calculations for the helium atom in Chapter 3. More general variation calculations will be employed in deriving the Hartree-Fock equations, Chapter 4, and in the Thomas-Fermi Dirac-model, Chapter 5. Whenever a special form of trial function or a particular method of variation is used, the stationary value of $\langle H\rangle$ is no longer the exact eigenvalue of H. The following set of equations shows that a variational estimate for $\langle H\rangle$ is always an upper bound for the lowest energy eigenvalue.

$$\psi = \sum_n a_n u_n$$

$$\langle\psi|\psi\rangle = 1 = \sum_n |a_n|^2$$

$$Hu_n = E_n u_n$$

$$H\psi = \sum_n a_n E_n u_n$$

$$\langle\psi|H|\psi\rangle = \sum_n |a_n|^2 E_n \geq E_0 \sum_n |a_n|^2 = E_0 \quad \text{(1-21a)}$$

E_0 is the lowest energy state. Hence,

$$\langle H\rangle \geq E_0 \quad \text{(1-21b)}$$

One can also obtain upper bounds for the higher levels if the trial function is orthogonal to the exact eigenfunctions of all the lower states, see p. 48.

The *WKB approximation* is useful in solving a one-dimensional Schrödinger equation for highly excited states. For example, the radial wave function for a particle moving in a spherically symmetric potential V(r) satisfies such a one-dimensional equation:

$$\frac{d^2\mathcal{R}(r)}{dr^2} + 2\left[E - V(r) - \frac{\ell(\ell+1)}{2r^2}\right]\mathcal{R}(r) = 0 \quad \text{(1-22)}$$

Here we have decomposed the total wave function $\psi(\mathbf{r})$ into $[\mathcal{R}(r)/r]Y_{\ell m}(\Omega)$, and we shall be interested in solutions in the region where the quantity

$$\Phi(r) = 2\left[E - V(r) - \frac{\ell(\ell+1)}{2r^2}\right]$$

is positive. In this region, the WKB solution is

$$\mathcal{R}(r) = A\Phi^{-1/4}\cos\left[\int_a^r \Phi^{1/2}\,dr' - \frac{\pi}{4}\right] \quad \text{(1-23)}$$

where A is a normalization constant, and a is the classical turning point; viz., it is the zero of $\Phi(r)$ (see Figure 1-1). (We assume

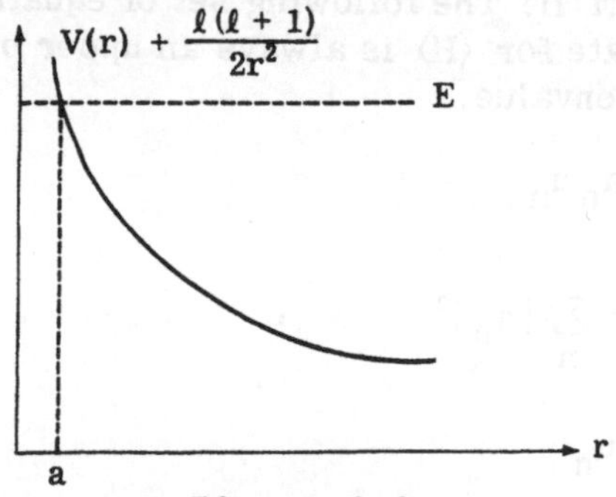

Figure 1-1

For r > a, Φ is positive and WKB solution (1-23) holds.

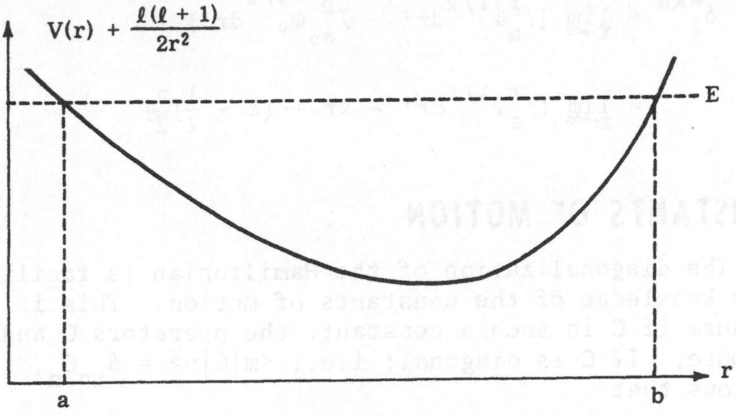

Figure 1-2
For $a < r < b$, Φ is positive and the WKB solution (1-23) holds.

there is only one turning point.) Frequently, we shall replace the centrifugal potential $\ell(\ell + 1)/2r^2$, occurring in Φ, by $(\ell + \frac{1}{2})^2/2r^2$. This improves the WKB solution.

When there are two turning points a and b, as in Figure 1-2, the WKB solution holds in the region between them where $\Phi(r) > 0$. The phase

$$\alpha(r) = \int_a^r \Phi^{1/2}\, dr' - \frac{\pi}{4} \tag{1-24a}$$

must satisfy the quantization condition

$$\alpha(b) = (n + \tfrac{1}{4})\pi \tag{1-24b}$$

When $\Phi(r)$ remains positive for $r \to \infty$, we are dealing with a scattering solution, and the WKB approximation to the scattering phase shift is obtained by comparing (1-23) with its non-interacting counterpart.

$$\mathcal{R}_0(r) = A\Phi_0^{-1/4} \cos[\int_{a_0}^r \Phi_0^{1/2} dr' - \frac{\pi}{4}] \tag{1-25a}$$

Here a_0 is the positive zero of $\Phi_0 \equiv k^2 - \frac{(\ell+1/2)^2}{r^2}$; $2E \equiv k^2$.

For large r, we may ignore the difference between $\Phi^{-1/4}$ and $\Phi_0^{-1/4}$ and the WKB phase shift is

$$\delta_\ell^{WKB} = \lim_{r\to\infty} [\int_a^r \Phi^{1/2} dr' - \int_{a_0}^r \Phi_0^{1/2} dr']$$

$$= \lim_{r\to\infty} [\int_a^r \Phi^{1/2} dr' - kr + (\ell + \tfrac{1}{2})\tfrac{\pi}{2}] \qquad (1\text{-}25b)$$

CONSTANTS OF MOTION

The diagonalization of the Hamiltonian is facilitated by a knowledge of the constants of motion. This is so, because if C is such a constant, the operators C and H commute. If C is diagonal; i.e., $\langle m|C|n\rangle = \delta_{mn} C_n$, it follows that

$$0 = \langle m|[C, H]|n\rangle$$
$$= (\langle m|C|m\rangle - \langle n|C|n\rangle)\langle m|H|n\rangle \qquad (1\text{-}26)$$

In deriving (1-22) the completeness relation

$$\sum_\ell |\ell\rangle\langle\ell| = I$$

has been used. From (1-26) one sees that in a representation that diagonalizes C, H has nonzero matrix elements only between states with the same eigenvalue of C. That is, H breaks up into nonvanishing submatrices only between states of equal C. Since it is frequently possible to find a diagonal representation for C from the symmetries of the problem, the secular problem of diagonalizing H is vastly simplified.

Parity will be such a constant if H is invariant under inversion, which is certainly the case for spherically symmetric potentials, and can be true for other more general potentials also.

For a large class of Hamiltonians with which we shall be concerned, the total angular momentum **J** will be a constant of the motion. The commutation relations between the components of **J** can be obtained from an investigation of infinitesimal rotations (Schiff,[2] pp. 194-212). These commutation relations are ($\hbar = 1$ in atomic units).

$$[J_a, J_b] = i J_c \qquad (1\text{-}27a)$$

with a, b and c comprising a cyclic permutation of x, y, z. This may also be written as

$$[J_a, J_b] = i\epsilon_{abc} J_c \qquad (1\text{-}27b)$$

where $\epsilon_{abc} = 0$ when the three indices are not distinct, and $\epsilon_{abc} = 1$ (-1) when the three indices are an even (odd) permutation of x, y, z. From (1-27) it follows that the components of J commute with the square of the total angular momentum

$$[J_a, J^2] = 0$$

$$J^2 = J_x^2 + J_y^2 + J_z^2 \tag{1-28}$$

Hence we may diagonalize simultaneously J^2 and one component of **J**.

In a representation in which J^2 and J_z are diagonal, if we label the rows and columns by a pair of symbols j and m, we find that the eigenvalues of J_z for a fixed j are m, where m ranges in unit steps from $-j$ to j. The eigenvalues of J^2 for any m are $j(j+1)$, where 2j is either a positive integer or zero.

The elements of J_x and J_y can be obtained from the matrix J_+ defined by

$$J_+ = J_x + iJ_y \tag{1-29a}$$

The only nonzero elements of J_+ are

$$\langle j, m+1 | J_+ | j, m\rangle = \sqrt{(j-m)(j+m+1)} \tag{1-29b}$$

so that J_+ has the property that for any normalized state $|j, m\rangle$

$$J_+ | j, m\rangle = \sqrt{(j-m)(j+m+1)}\, | j, m+1\rangle \tag{1-29c}$$

The adjoint $J_+^\dagger \equiv J_-$ has the property

$$J_- | j, m\rangle = \sqrt{(j+m)(j-m+1)}\, | j, m-1\rangle \tag{1-29d}$$

J_+ and J_- are called the *raising* and *lowering operators*, respectively.

The representations of J_x, J_y, and J_z corresponding to integral j can arise from the orbital angular momentum **L** defined by

$$\mathbf{L} = \sum_i \mathbf{r}_i \times \mathbf{p}_i \tag{1-30}$$

where p_i is the linear momentum of the ith particle. L satisfies the commutation relations (1-27) and for one particle the eigenfunctions are the spherical harmonics $Y_{\ell m}(\Omega)$.

SPIN

A particle or any other quantum system may possess intrinsic angular momentum which cannot be expressed in terms of the classical position and momentum coordinates. The components of this angular momentum can be half-integers, since J_z no longer has the representation $-i\partial/\partial\varphi$. In dealing with spin alone we shall replace **J** by **S**, j by s, and m by m_s. For a system with definite spin s, we have

$$S^2 = s(s + 1)I$$

$$S_z = m_s I \tag{1-31}$$

where 2s is either zero or a positive integer; m_s ranges in unit steps from $-s$ to s; and I is the identity operator in spin space.

While the commutation relations (1-27) and the Hermiticity of **J** imply that spin can take on only integer or half-integer values, no present physical theory can predict the spin of individual fundamental particles. Thus like charge and mass, spin is an experimentally determined attribute of fundamental particles. Electrons, nucleons, and μ mesons have spin $\frac{1}{2}$; π- and K-mesons have spin 0; photons, to the extent that they can be considered particles, have spin 1.

The most natural quantum mechanical description of spin is accomplished with the help of relativistic generalizations of Schrödinger's equation. Spin 0 particles are described by the Klein-Gordon equation, and spin $\frac{1}{2}$ particles, by the Dirac equation. We shall discuss these equations in Part IV. For the present we confine ourselves to nonrelativistic descriptions of particles with spin.

Let us for a moment revert to ordinary units retaining $\hbar$ explicitly so that the eigenvalue of S^2 is $\hbar^2 s(s + 1)$. Since the spin is fixed for every particle, the total spin s of a quantum system cannot grow arbitrarily large and $\hbar\sqrt{s(s + 1)} \to 0$ as $\hbar \to 0$. This shows that spin has no classical analog. In particular, early models of the "spinning electron" are meaningless.

The existence of spin indicates that quantum mechanical operators corresponding to classical dynamical quantities cannot always be obtained by the elementary rule: "replace the classical quantities **r** and **p** by quantum mechanical operators and symmetrize." Evidently, in addition to the above prescription, we must stipulate that the quantum mechanical operator may contain terms which vanish in the classical limit $\hbar \to 0$. (We now return to atomic units.)

The state function of a particle must depend on the 2s + 1 components of the spin. In general, if the Hamiltonian couples strongly space motion and spin motion, 2s + 1 space-spin functions of the form $\psi(\mathbf{r}, m_s)$ will be required to specify that particle. If the coupling can be ignored, the wave function separates into

$$\psi(\mathbf{r}, m_s) = \psi(\mathbf{r})\,|\,m_s\rangle \tag{1-32}$$

Since spin space consists only of 2s + 1 points, $|m_s\rangle$ is completely specified by 2s + 1 numbers. An obvious choice for these spin functions is the set of orthonormal eigenvectors of the diagonalized S^2 and S_z matrices. For example, for the case $s = \frac{3}{2}$

$$|\tfrac{3}{2}\rangle = \begin{pmatrix}1\\0\\0\\0\end{pmatrix} \quad |\tfrac{1}{2}\rangle = \begin{pmatrix}0\\1\\0\\0\end{pmatrix} \quad |-\tfrac{1}{2}\rangle = \begin{pmatrix}0\\0\\1\\0\end{pmatrix} \quad |-\tfrac{3}{2}\rangle = \begin{pmatrix}0\\0\\0\\1\end{pmatrix}$$

The total angular momentum of a particle possessing both spin and orbital angular momentum is

$$\mathbf{J} = \mathbf{L} + \mathbf{S} \tag{1-33}$$

L and **S** each satisfy the commutation relations (1-27) and since they operate on different variables, they commute. Hence **J** also satisfies the commutation relations (1-27). For an isolated system, e.g., an atom, **J** will always be a constant of the motion since **J** generates infinitesimal rotations in space, and space is isotropic. However, **L** need not be conserved, since it generates rotations only in the position variables under which the isolated system need not be invariant. If **L** happens to be a constant, this is a consequence of a specific physical feature of the Hamiltonian and not of general geometrical properties of the space. Specifically, for a radially symmetric Hamiltonian with no spin-orbit coupling **L** is conserved. Of course, when **J** and **L** are conserved, so is **S**.

DESCRIPTION OF SPIN $\frac{1}{2}$ PARTICLES

The spin $\frac{1}{2}$ functions and spin matrices shall be our main concern since they describe a single electron. Writing $\mathbf{S} = \frac{1}{2}\sigma$ we find from the general expression (1-29) for J_+, J_- that

$$\sigma_x = \begin{pmatrix}0 & 1\\1 & 0\end{pmatrix} \quad \sigma_y = \begin{pmatrix}0 & -i\\i & 0\end{pmatrix} \quad \sigma_z = \begin{pmatrix}1 & 0\\0 & -1\end{pmatrix} \tag{1-34}$$

$$|\tfrac{1}{2}\rangle \equiv \alpha = \begin{pmatrix}1\\0\end{pmatrix} \qquad |-\tfrac{1}{2}\rangle \equiv \beta = \begin{pmatrix}0\\1\end{pmatrix} \tag{1-35}$$

σ_x, σ_y, σ_z are called the *Pauli spin matrices*. The following relations for the Pauli spin matrices can be verified:

$$\sigma_a \sigma_b = i\epsilon_{abc}\sigma_c + \delta_{ab}$$

$$a, b = x, y, z \tag{1-36a}$$

$$(\sigma\cdot\mathbf{A})(\sigma\cdot\mathbf{B}) = \mathbf{A}\cdot\mathbf{B} + i\sigma\cdot\mathbf{A}\times\mathbf{B} \tag{1-36b}$$

where **A** and **B** are vector operators which commute with σ.

Since the spin operators of two electrons commute, $\mathbf{S} = \mathbf{S}_1 + \mathbf{S}_2$ is a spin operator again. Four orthonormal spin functions for a pair of electrons can be chosen to be $\alpha_1\alpha_2$, $\beta_1\beta_2$, $\alpha_1\beta_2$, $\beta_1\alpha_2$. In such products of spin functions, $\mathbf{S}_1$ operates only on the first spin function, $\mathbf{S}_2$ only on the second. The inner product between two such product functions is the product of the individual inner products. For example

$$\alpha_1\alpha_2\cdot\beta_1\beta_2 = (\alpha_1\cdot\beta_1)(\alpha_2\cdot\beta_2)$$

The four spin functions enumerated above correspond to total m_S: 1, −1, 0, 0, respectively. Since **S** is an angular momentum it should be possible to find a representation in which S_z and S^2 are diagonal. It can be verified that $\alpha_1\alpha_2$, $\beta_1\beta_2$ are eigenvectors of S^2. To construct the remaining eigenvectors from $\alpha_1\beta_2$ and $\alpha_2\beta_1$ we observe that $\alpha_1\alpha_2$ is symmetric and corresponds to $s = 1$, $m_S = 1$. Therefore the lowering operator $S_- = S_{1-} + S_{2-}$ will generate the wave function corresponding to $s = 1$, $m_S = 0$, and since $\alpha_1\alpha_2$ is symmetric, the resulting wave function will also be symmetric. Explicitly,

$$
\begin{aligned}
S_-\alpha_1\alpha_2 &= 2^{1/2}|s = 1, m_S = 0\rangle \\
&= (S_{1-} + S_{2-})\alpha_1\alpha_2 = [\beta_1\alpha_2 + \alpha_1\beta_2]
\end{aligned}
$$

$$|s = 1, m_S = 0\rangle = 2^{-1/2}[\alpha_1\beta_2 + \beta_1\alpha_2] \tag{1-37a}$$

The fourth eigenvector is obtained by observing that the only normalized linear combination of $\alpha_1\beta_2$ and $\beta_1\alpha_2$ which is orthogonal to (1-37a) is the antisymmetric

$$|s = 0, m_S = 0\rangle = 2^{-1/2}[\alpha_1\beta_2 - \beta_1\alpha_2] \tag{1-37b}$$

Table 1-1 lists the four basis vectors for a diagonal representation of S^2 and S_z, together with the eigenvalues of S^2, the total spin s, and the total z component m_S. The first three states, corresponding to $s = 1$, are described by symmetric spin functions and are called a *triplet*. The last one, with $s = 0$, is antisymmetric and is called a *singlet*. (In adding two equal, nonzero, angular momenta $J = J_1 + J_2$, the simultaneous eigenfunctions of J^2 and J_z will always be either symmetric or antisymmetric. However, for general addition of angular momenta this need not be so. See Chapter 6.)

Table 1-1

	S^2	s	m_s
$(\alpha_1\alpha_2)$	2	1	1
$2^{-1/2}[(\alpha_1\beta_2) + (\beta_1\alpha_2)]$	2	1	0
$(\beta_1\beta_2)$	2	1	−1
$2^{-1/2}[(\alpha_1\beta_2) - (\beta_1\alpha_2)]$	0	0	0

PROBLEMS

1. Give the unit of energy, momentum, and angular momentum in Hartree's atomic units.
2. Show that (1-9) follows from (1-1) using the notation of (1-10).
3. Prove that

$$\lim_{t \to \infty} \frac{2 \sin^2 \frac{1}{2}\omega_{km} t}{\pi\, \omega_{km}^2\, t} = \delta(\omega_{km})$$

This verifies (1-14b).

4. Show that $|m^{(+)}>$ defined in (1.17) satisfies $(H-E_m)|m^{(+)}> = 0$.
5. Show that a variational normalized wave function ψ can always be decomposed as

$$\psi = \sqrt{1-\eta^2}\,\psi_0 + \eta f$$

where ψ_0 is the exact wave function and f is normalized and orthogonal to ψ_0. Verify that $\langle H\rangle$ differs from the exact eigenvalue by terms of order η^2.

6. Verify (1-36).
7. An atom has a nucleus of charge Z and one electron. The nucleus has a radius R, inside which the protons are uniformly distributed. Calculate the effect of the finite size of the nucleus

on the energy levels of the electron, as a function of n and ℓ using perturbation theory. Assume the nuclear radius to be very small compared with the Bohr radius, and approximate the radial electron wave functions accordingly. Discuss the result as a function of n and ℓ. Assuming $R = r_0 A^{1/3}$, $r_0 = 1.2$ fermi, calculate the level shift in cm^{-1} for the 1s and 2p states of H^1 and Pb^{208}. A is the atomic weight.

8. A hydrogen atom in its ground state is placed between the plates of a condenser. A voltage pulse is applied to the condenser so as to produce a homogeneous electric field that has the time dependence:

$$\mathcal{E} = 0 \qquad t < 0$$

$$\mathcal{E} = \mathcal{E}_0 e^{-t/\tau} \qquad t > 0$$

Find the first-order probability that, after a long time, the atom is in the 2s or in one of the 2p states. State the selection rules. Is it realistic to assume that such a voltage pulse can be produced, and if not, why not?

9. The Schrödinger equation for an $\ell = 0$ state with the exponential potential

$$V(r) = -\frac{a^2}{8} e^{-r/r_0} \tag{1}$$

can be solved analytically and has the solution $\psi = u(r)/r$ with

$$u(r) = J_n\left(ar_0\, e^{-r/2r_0}\right) \tag{2}$$

where J_n is the Bessel function of order n, and

$$E = -\frac{1}{8r_0^2} n^2 \tag{3}$$

Thus n may be regarded as the eigenvalue parameter; it is to be determined from the condition

$$u = 0 \qquad \text{at} \qquad r = 0 \tag{4}$$

Prove that (2) solves Schrödinger's equation, by direct substitution. From (4) find the values of a which make $n = 0, \frac{1}{2}$, and 1, respectively; call them a(n). For a = a(1), find the variational energy $E(\Phi)$ using the trial functions $\Phi = u(r)/r$ with

$$u_1(r) = re^{-\lambda r}$$

$$u_2(r) = re^{-\lambda r}(1 + c_2 r)$$

$$u_3(r) = re^{-\lambda r}(1 + c_3 r^2) \qquad (5)$$

Compare the results with each other and with the exact eigenvalue.

10. In deriving the time-independent Schrödinger equation from the variation principle, equations (1-19) to (1-20), we considered $\delta\psi^*$ and $\delta\psi$ to be independent. By splitting ψ into its real and imaginary parts, rederive equation (1-20) and discuss the independence of $\delta\psi^*$ and $\delta\psi$.
11. (a) Express J^2 in terms of J_+, J_-, and J_z.
    (b) Verify that the states enumerated in Table 1-1 are eigenvectors of S^2 and S_z with the proper eigenvalues.

Chapter 2

IDENTICAL PARTICLES AND SYMMETRY

The Hamiltonian for a system of n particles that are identical, that is, particles that can be substituted for each other with no physical change, must be completely symmetric under any interchange of its arguments. Let $\Psi(1, 2, \ldots, n)$ be any solution of the Schrödinger equation, depending on the coordinates (spatial and spin) of the n identical particles 1 to n, and let P be any permutation of the n numbers 1 to n. Then $P\Psi(1, 2, \ldots, n) \equiv \Psi(P1, P2, \ldots, Pn)$ is a function which depends on the coordinates of particle Pi in the same manner as the original function Ψ depended on the coordinates of particle i. Then using the symmetry of H, we can show that the operator P commutes with the Hamiltonian

$$H(P\Psi) = P(H\Psi) \tag{2-1}$$

Thus if Ψ is a solution of the Schrödinger equation, then $P\Psi$ is also a solution (belonging to the same eigenvalue, if we are considering a stationary solution). There are n! different permutations of n objects, so in this manner we get n! wave functions. Some of these may be (and in general are) linear combinations of others, but *in general* we find several linearly independent solutions by our procedure. Thus most of the eigenvalues of a Hamiltonian which is symmetric in n particles will be degenerate. This phenomenon is called *exchange degeneracy* and will be discussed in more detail later.

Whichever of the n! wave functions describes an assembly of n identical particles at some initial time t_0, this same wave function will describe the assembly for all time. This follows immediately from the fact that P commutes with the Hamiltonian H. In other words, $H\Psi$ has the same symmetry with respect to any permutation P as Ψ itself; thus $\partial\Psi/\partial t$ has the same symmetry, and by integrating the Schrödinger equation in small steps of time we see that Ψ will retain its initial symmetry.

Among the various energy levels of H some will be nondegenerate. For such nondegenerate eigenvalue, the function $P\Psi$ must be simply a multiple of Ψ, since Ψ is by assumption the only solution for the given eigenvalue. Moreover, since $P\Psi$ and Ψ have the same normalization,

$$P\Psi = e^{i\alpha}\Psi \tag{2-2}$$

where α is real. Consider now an especially simple permutation, viz., the simple interchange P_{ij} of particles i and j,

$$P_{ij}\Psi(1, \ldots, i, \ldots, j, \ldots, n) = \Psi(1, \ldots, j, \ldots, i, \ldots, n) \tag{2-3}$$

Then if this interchange is applied twice we get back to the original; therefore, using (2-2), $e^{2i\alpha} = 1$ and

$$P_{ij}\Psi = \pm\Psi \tag{2-4}$$

Now it is a fact of nature, established by many observations, that all actual wave functions in physics obey (2-4), either with the plus or the minus sign. In other words, from the great multitude of mathematical solutions of $H\Psi = E\Psi$, nature has selected only the nondegenerate ones. The character of the solution depends on the sign to be taken in (2-4); the plus sign leads to a wave function which is symmetric in all particles, the minus sign corresponds to an antisymmetric wave function. Which symmetry applies depends on the type of identical particles involved; experiment shows that Ψ is antisymmetric for electrons, protons, neutrons, μ-mesons, hyperons; symmetric for π-mesons, K-mesons, and photons. A system may of course contain several types of particles, e.g., protons, neutrons, and π-mesons. Then its wave function will change sign if we interchange the coordinates of any two protons, or of any two neutrons; it will remain unchanged if we interchange two π-mesons; finally if we interchange the coordinates of two particles of different type, e.g., one proton and one neutron, the resulting wave function will in general have no simple relation to the original one.

ARGUMENTS FOR SIMPLE SYMMETRY

As we have pointed out, the special symmetry of the wave function, e.g., the antisymmetry in the coordinates of all electrons in the system, does not follow from the symmetry of the Hamiltonian but is an additional requirement selecting the physically meaningful solutions from among the much larger number of mathematically possible solutions of the Schrödinger equation. We shall now give some arguments which make it reasonable that the physical solutions have such a simple symmetry.

First, we simply postulate that all the physical properties which we may derive from the wave function should remain unchanged if we interchange two identical particles. This reasonable interpretation of the meaning of "identity" requires that $|P_{ij}\Psi|^2 = |\Psi|^2$, which is equivalent to (2-2) and hence leads to (2-4).

To develop the second argument, we must first investigate the other possible types of symmetry of the wave function, aside from complete symmetry and antisymmetry. For this purpose, it is simplest to consider an approximate Hamiltonian which will also be very useful to us throughout our study of atoms. This is,

$$H_0 = H_0'(1) + H_0'(2) + \cdots + H_0'(n) \tag{2-5}$$

i.e., a Hamiltonian which may be written as a sum of terms, each referring to an individual particle. Let ϕ_a, ϕ_b, etc., be normalized eigenfunctions of H_0', thus

$$H_0'(1)\,\phi_a(1) = E_a\,\phi_a(1) \tag{2-6}$$

Then a possible eigenfunction of H_0 is

$$\Psi = \phi_a(1)\,\phi_b(2)\cdots\phi_k(n) \tag{2-7}$$

with the eigenvalue

$$E = E_a + E_b + \cdots + E_k \tag{2-8}$$

This is because $H_0'(1)$ commutes with all functions of coordinates other than those of particle 1. Along with (2-7), the function

$$P\Psi = \phi_a(P1)\,\phi_b(P2)\cdots\phi_k(Pn) \tag{2-9}$$

is an eigenfunction of H_0 with the same eigenvalue (2-8), where P is an arbitrary permutation of the n numbers 1, ..., n. (We could have permuted the functions ϕ_a, etc., instead of their arguments, and would have obtained the same result.) There are n! permutations P, hence (2-9) gives n! linearly independent eigenfunctions if the one-particle functions $\phi_a, \ldots, \phi_k$ are all different. These one-particle functions are called *orbitals*.

All the n! functions $P\Psi$ are degenerate, a phenomenon called *exchange degeneracy*. They are unsymmetric, and we can form more symmetric functions by linear combination. However, the totally symmetric and the antisymmetric functions account only for two of these. The remaining $n! - 2$ linear combinations must then have lesser symmetry. Only in the special, but important, case $n = 2$ do the symmetric and antisymmetric functions exhaust all linearly independent combinations.

To illustrate the other possible symmetries, it suffices to consider the case of three particles, with two orbitals being alike. Then we have the three functions

$$\Psi_1 = \phi_b(1)\,\phi_a(2)\,\phi_a(3)$$

$$\Psi_2 = \phi_a(1)\,\phi_b(2)\,\phi_a(3)$$

$$\Psi_3 = \phi_a(1)\,\phi_a(2)\,\phi_b(3) \tag{2-10}$$

The totally symmetric normalized function is

$$\Psi_S = 3^{-1/2}(\Psi_1 + \Psi_2 + \Psi_3) \tag{2-11a}$$

There is no totally antisymmetric function. Two other linear combinations, orthonormal to Ψ_S, are

$$\Psi_4 = 6^{-1/2}(2\Psi_1 - \Psi_2 - \Psi_3)$$

$$\Psi_5 = 2^{-1/2}(\Psi_2 - \Psi_3) \tag{2-11b}$$

Evidently, Ψ_4 is symmetric, Ψ_5 antisymmetric, in particles 2 and 3. If we apply the permutation P_{12} to Ψ_4 we find

$$P_{12}\Psi_4 = -\tfrac{1}{2}\Psi_4 + \frac{\sqrt{3}}{2}\Psi_5 \tag{2-12}$$

and similarly for $P_{13}\Psi_4$, etc. Thus the two functions Ψ_4 and Ψ_5 obviously belong together: The permutations P_{12} and P_{13} transform these into linear combinations of each other and are represented by 2×2 matrices in the "subspace" of Ψ_4, Ψ_5. The symmetric function Ψ_S stands by itself; it transforms into itself under any permutation P, so that all permutations are represented by 1×1 matrices, viz., unity, in the subspace of Ψ_S.

The problem of the matrix representation of the permutations can best be studied by group theory, and this has been done by Wigner and collaborators.[1] Now the matrix representation

---

[1]E. P. Wigner, *Group Theory and Its Application to the Quantum Mechanics of Atomic Spectra,* translated by J. J. Griffin, Academic Press, New York, 1959.

of the permutation group carries over to the case of a *general* Hamiltonian H which does not separate into one-particle Hamiltonians as H_0 does, but which of course is still symmetrical in all particles. Thus for a general symmetric Hamiltonian involving three particles, and for solutions corresponding to two equal and one different orbital, we get two types of eigenfunctions: One type is totally symmetric like Ψ_S, and these eigenfunctions are nondegenerate; the other type is always doubly degenerate and transforms under permutations like the functions Ψ_4, Ψ_5 in (2-11). With a general Hamiltonian, doubly degenerate eigenvalues are in general different from the nondegenerate ones corresponding to totally symmetric eigenfunctions; a coincidence between these eigenvalues could only occur by pure accident (in contrast to the simple Hamiltonian H_0). Thus part of the n!-fold exchange degeneracy is removed when we go from H_0 to H, but some degeneracy will remain. This result holds generally, for any number of particles, and whether or not some orbitals are equal. Thus mathematically, the Schrödinger equation for n identical particles has many solutions, of different symmetry, and most of its eigenvalues are degenerate.

Now we can give the second argument why the physical solutions should be either totally symmetric or antisymmetric. Assume that the wave function of the universe has one of the complicated symmetries we discussed with respect to the electrons. This means it will be symmetric with respect to the interchange of certain electron pairs, antisymmetric with respect to others, and it will keep this symmetry for all time. Then we may consider two helium atoms, one containing two electrons with symmetric wave function, the other containing antisymmetric electrons. These two atoms will have different eigenvalues, hence different optical spectra. Thus we would find in nature two different types of helium atoms, and still more varieties of any heavier atom like carbon. But we know that all atoms of given nuclear charge are alike, both chemically and spectroscopically. Hence only one symmetry of the wave function of electrons in one atom can be allowed, and this means that the wave function of the universe must have one of the simple symmetries, either totally symmetric or antisymmetric.

To reach this conclusion we had to make use of an empirical fact. But this was a simple, qualitative, and long-established fact, namely, the identical behavior of all atoms of the same chemical element. Since atoms are composite systems containing electrons, their identical physical behavior is sufficient to establish that the symmetry of the wave function with respect to all electrons in the universe is simple. Which symmetry is to be chosen, i.e., symmetry or antisymmetry, can of course only be decided on the basis of more quantitative information. Similar arguments establish simple symmetry with respect to other fundamental particles, e.g., protons and neutrons.

SYMMETRY FOR COMPOSITE SYSTEMS

We shall presently show that particles described by a symmetric wave function obey Bose statistics and those described by an antisymmetric wave function obey Fermi statistics. They are therefore commonly called *bosons* and *fermions*, respectively. Consider now tightly bound composite objects like nuclei. It then makes sense to ask for the symmetry of the wave function of a system containing many identical objects of the same type, e.g., many He^4 nuclei. This symmetry can be deduced by imagining that the interchange of two composites is carried out particle by particle. Each interchange of fermions changes the sign of the wave function. Hence the composite will be a fermion if and only if it contains an odd number of fermions, a boson if the number of fermions in the composite is even. The number of bosons contained in the composite does not matter. If the wave function contains coordinates of several types of particles, definite symmetry exists only for interchange within each type.

CONSTRUCTION OF SYMMETRIZED WAVE FUNCTIONS

We may construct the properly symmetrized (unnormalized) wave function for n identical particles from one unsymmetric solution. For this purpose, we obtain the n! solutions by permuting the indices and form the sums

$$\sum_P \Psi[P(1, 2, \ldots, n)] \qquad (2\text{-}13)$$

$$\sum_P \epsilon_P \Psi[P(1, 2, \ldots, n)] \qquad (2\text{-}14)$$

where the summations are over all the permutations, $\epsilon_P = -1$ for odd permutations of 1, 2, . . . , n, and $\epsilon_P = +1$ for even ones. An odd permutation is one which may be obtained by carrying out an odd number of simple interchanges of pairs of particles in succession. It is seen that equation (2-13) gives a symmetric wave function and (2-14) an antisymmetric one.

If the interaction of the n identical particles in the system is weak, then the Hamiltonian H_0 in (2-5) is a good approximation to the true Hamiltonian, and the product wave function (2-7) a good approximation to the unsymmetrized eigenfunction. From this we can form a symmetric or antisymmetric wave function using (2-13) and (2-14), respectively. The normalized antisymmetric function may conveniently be written as a determinant of n rows and columns, the so-called *Slater determinant:*

$$(n!)^{-1/2} \begin{vmatrix} \phi_a(1)\,\phi_a(2)\cdots\phi_a(n) \\ \phi_b(1)\,\phi_b(2)\cdots\phi_b(n) \\ \vdots \\ \phi_k(1)\phi_k(2)\cdots\phi_k(n) \end{vmatrix} \qquad (2\text{-}15)$$

Interchange of two coordinates will interchange two columns, which changes the sign of the determinant demonstrating the required antisymmetry. In addition the interchange of two *states*, i.e., of two orbitals, interchanges two rows, which also changes the sign of the determinant. Therefore, if two particles are in the same state, i.e., if two orbitals are the same, the determinant is zero. This follows also from the well-known theorem that a determinant with two equal rows must vanish. We have thereby proved the famous *Pauli exclusion principle,* which states that no two electrons in a given system can occupy the same quantum state (orbital). This principle was postulated by Pauli to explain the periodic table of elements. In quantum mechanics, it follows automatically from the antisymmetry of the wave function. Or vice versa, knowing the periodic system, we must conclude that the electron eigenfunction must be antisymmetric rather than symmetric.

STATISTICAL MECHANICS

In a system of weakly interacting particles, we don't need to write down the wave function explicitly as we have done in (2-15), but it suffices to specify which orbitals are occupied by particles, and by how many. This is the method of *statistical mechanics*. We specify the "occupation numbers" n_a, n_b, ... of orbitals a, b, ... If the wave function is antisymmetric, we can only have $n_a = 0$ or 1; this type of statistics was first postulated by Fermi and bears his name, hence the particles obeying these statistics are called *fermions*. For a symmetric wave function, any occupation number is possible; this is known as Bose statistics *(bosons)*. There is one and only one symmetric eigenfunction for a given set of occupation numbers n_a, n_b, ...; hence the statistical weight for any such set is the same in Bose statistics (unity).

We have previously listed the particles which are empirically fermions, and it will be noted that all these particles have spin $\frac{1}{2}$. Similarly, all particles which are observed to be bosons have spin 0 (π^- and K-mesons) or 1 (photons). Pauli[2] proved that relativistic the-

[2]For a modern discussion, see R. F. Streater and A. S. Wightman, *PCT, Spin and Statistics and All That*, Benjamin/Cummings, Reading MA 1978.

ories can be consistent only if half-integer-spin particles are fermions and integer-spin particles are bosons.

EXPERIMENTAL DETERMINATION OF SYMMETRY

The symmetry of the wave function has direct, observable physical consequences. Probably the most useful is the wave function of the spatial motion of the two nuclei in a homonuclear molecule, such as C_2^{12}. It can be shown (see Chapter 9) that such a molecule will only have rotational states of even angular momentum j if the nuclei are bosons without spin, and only odd j if they are fermions without spin (if such objects existed). Observation of the rotational band spectrum shows that C_2^{12} has indeed only even rotational states; therefore C^{12} is a boson. More important, this shows that all C^{12} nuclei are truly identical. The same can be done, e.g., for molecules like H_2^3, involving two radioactive nuclei H^3. Experiment shows that all nuclei H^3 are identical, because the rotational spectrum of H_2^3 is just such as would be expected for fermions of spin $\frac{1}{2}$. If the nuclei were not identical, the wave function would be unsymmetrical in their coordinates, and *all* rotational quantum numbers would be equally likely. Now H^3 decays radioactively into He^3 plus a β^-—particle, with a half-life of 12 years. The two H^3 nuclei in a H_2^3 molecule will in general decay at different times. Yet direct experiment shows that they are identical: thus there is no "hidden variable" in the nucleus which indicates when it will decay. If there were such a variable, even if it had no influence on the energy and dynamics of the system, it would still influence the symmetry of the wave function. Wave mechanics really gives us a specific handle to determine identity of particles, not simply by the absence of observable differences between them.

There are occasional attempts to replace the probabilistic predictions of quantum mechanics by strictly causal ones.[3] It has been suggested that there exist hidden variables in terms of which these causal descriptions could be effected. These variables are "hidden" in the sense that they do not affect energy eigenvalues. The existence of identical particles and composites show that such hidden variables cannot have any observable consequences (as in the example of radioactivity) and are therefore empty. It assures us that

---

[3] A review of such attempts is given by J. S. Bell, *Rev. Mod. Phys. 38*, 447 (1966), who also proposed a critical test known as *Bell's theorem*. For a clear account see R. Peierls, *Surprises in Theoretical Physics*, Princeton Press, Princeton 1979, p. 23.

the present description must be complete to the extent that it explains present experimental data.

For electrons it is a particularly simple observation that no further quantum numbers are needed to describe possible states. If there were degrees of freedom which we have not specified, we would expect that we could find electrons in nature which are described by different values of these hidden variables. That is, the observed degeneracy of electronic states would have to be greater than that predicted by our present theory. The overwhelming success of the explanation of the periodic table in terms of the present theory indicates that this is not the case; i.e., no further degrees of freedom exist.

Another application of the symmetry of the rotational wave function of a homonuclear diatomic molecule was to determine the statistics of the nuclei involved. In particular, the fact that the N^{14} nucleus is a boson helped disprove the "nuclear electron" hypothesis, which claimed that the N^{14} nucleus was constituted of 14 protons and 7 electrons, which would make it a fermion.

Gross effects of the symmetry of the wave function are found in statistical mechanics. The Fermi statistics of the electrons in solids is fundamental to the understanding of the behavior of metals. The Bose statistics of He^4 atoms (4 nucleons, 2 electrons) is the reason for the peculiar behavior of liquid He^4 (modification II) at low temperatures, such as superfluidity. He^3 is a fermion and liquid He^3 behaves differently.

CLASSICAL LIMIT

We shall now examine the behavior of a system of identical particles in the limit of classical mechanics. The only method for distinguishing classical identical particles is by following their orbits, which are of course well defined. It is the lack of well-defined orbits for particles obeying quantum mechanics which gives rise to exchange effects. In particular if the wave functions of two particles overlap in space we can no longer distinguish the two orbits. In the classical limit the state functions become well-defined, nonoverlapping wave packets. Explicitly, if the two particles are described by

$$\Psi(1, 2) \pm \Psi(2, 1) \tag{2-16}$$

the coordinate probability density is

$$|\Psi(1, 2)|^2 + |\Psi(2, 1)|^2 \pm 2\,\mathrm{Re}[\Psi(1, 2)\Psi^*(2, 1)] \tag{2-17}$$

If $\Psi(1, 2)$ is zero unless coordinate 1 is in some region A and coordinate 2 is in some region B, and A and B do not overlap, the interference term will vanish. The coordinate probability density then becomes that of two particles which could be distinguished. Hence we do not find any exchange effects in classical descriptions.

It should be realized that even in nonclassical descriptions identical particles can be distinguished if their wave functions do not overlap; i.e., if they are in sufficiently different states. If the wave functions do overlap, there is an "exchange density" in (2-17) which leads to observable consequences. Thus, for example, when two identical particles scatter each other, their wave functions will overlap at some time, and there are then interference terms arising from the symmetry of the wave function; see Chapters 15 and 16.

PROBLEMS

1. Assume that two identical particles, interacting through a spin independent central potential, are described by a symmetric (antisymmetric) space wave function $\psi(\mathbf{r}_1, \mathbf{r}_2)$ which satisfies the Schrödinger equation

$$[-\tfrac{1}{2}\nabla_1^2 - \tfrac{1}{2}\nabla_2^2 + V(|\mathbf{r}_1 - \mathbf{r}_2|) - E]\psi(\mathbf{r}_1, \mathbf{r}_2) = 0$$

Separate the above equation into an equation for the center of mass motion, and for the relative motion of the two particles. Discuss the symmetry of the center of mass wave function, and of the relative motion wave function. If the relative motion wave function is expanded in orbital angular momentum eigenstates, which orbital angular momenta states contribute?

2. In the example discussed in the text (p. 23, equations (2-10), (2-11), and (2-12)), show that Ψ_S, Ψ_4, Ψ_5 are each normalized to unity and orthogonal to each other. Show also that any other linear combination of Ψ_1, Ψ_2, Ψ_3 can be expressed in terms of Ψ_S, Ψ_4, Ψ_5. Why is there no antisymmetric combination? Exhibit explicitly the effect of P_{ij} on Ψ_S, Ψ_4, Ψ_5; i, j = 1, 2, 3.
3. A system of two identical spin $\frac{1}{2}$ fermions is described by a space-spin function which is assumed to separate into a space part and a spin part. The spin part is described by eigenstates of the total spin, as in Table 1-1. What are the symmetry properties under interchange of the space part of the state function?
4. A system of two identical spin $\frac{1}{2}$ fermions has weak interactions, and a separated Hamiltonian of the form (2-5) serves as a good approximation to the true Hamiltonian. Therefore a good approximation to the unsymmetrized eigenfunction Ψ is a product wave function of the form (2-7); viz., $\Psi = \varphi_a(1)\,\varphi_b(2)$. The orbitals φ_a, φ_b are further assumed to separate into a space part and into a spin part. Evidently, there are four possible (unsymmetrized)

eigenfunctions corresponding to the four possible ways of choosing the spin functions. Form the four Slater determinantal state functions which arise when the unsymmetrized wave functions are antisymmetrized. Observe that two of the determinantal functions factor into a space part and a spin part, but that the remaining two do not factor in this fashion. Form linear combinations of these remaining two to obtain antisymmetric wave functions which do factor. Compare your results with the conclusions of Problem 3.

5. **Show that a Hamiltonian which describes two spin $\frac{1}{2}$ particles and is symmetric in the two spins commutes with $S^2 = (S_1 + S_2)^2$**

Chapter 3

TWO-ELECTRON ATOMS

In this chapter we study the energy levels of two-electron atoms, principally helium. A two-electron atom is a system of three interacting particles, a problem extremely difficult to analyze within classical mechanics. Since the mass of the nucleus is much larger than that of the electrons, the nuclear motion may be ignored and the problem reduces to two interacting particles moving in an external potential. This is still difficult to solve within classical mechanics. The old Bohr-Sommerfeld quantum mechanics also had little success with this problem. Thus it was a critical triumph for wave mechanics to provide correct calculations of the energy levels of two-electron atoms. Indeed it is remarkable that very little effort is required, as we shall see, to give a good estimate for the ground state energy. Further calculations for two-electron atoms provided six-figure agreement between theory and experiment, a striking proof of the validity of wave mechanics.

CLASSIFICATION OF SOLUTIONS

We shall solve (1-4) for two electrons by perturbation and variation methods. The Hamiltonian in our approximation contains no spin terms; thus we may take the solution in the form

$$\psi(1, 2) = U(\mathbf{r}_1, \mathbf{r}_2)\, \chi(1, 2) \tag{3-1}$$

where U is the space part and χ is the spin part of the total wave function ψ. U satisfies (1-4) for two electrons

$$\left[\nabla_1^2 + \nabla_2^2 + 2\left(E + \frac{Z}{r_1} + \frac{Z}{r_2} - \frac{1}{r_{12}}\right)\right] U(\mathbf{r}_1, \mathbf{r}_2) = 0 \tag{3-2}$$

Since the Hamiltonian is symmetric under interchange of space coordinates, we may take U to be entirely symmetric or antisymmetric. The former case is called *para*, the latter *ortho*.

$$U_+(\mathbf{r}_1, \mathbf{r}_2) = U_+(\mathbf{r}_2, \mathbf{r}_1) \qquad \text{(para)}$$

$$U_-(\mathbf{r}_1, \mathbf{r}_2) = -U_-(\mathbf{r}_2, \mathbf{r}_1) \qquad \text{(ortho)} \tag{3-3}$$

The over-all antisymmetry of ψ determines the symmetry properties of the spin function.

$$\chi_+(1, 2) = -\chi_+(2, 1) \qquad \text{(para)}$$

$$\chi_-(1, 2) = \chi_-(2, 1) \qquad \text{(ortho)} \tag{3-4}$$

From Table 1-1 we see that the only antisymmetric two-electron spin function is the singlet

$$\chi_+(1, 2) = 2^{-1/2}[(\alpha_1\beta_2) - (\beta_1\alpha_2)] \qquad \text{(para)} \tag{3-5a}$$

There are three linearly independent, symmetric (ortho) spin functions, and it is convenient to choose them to be eigenfunctions of S^2 and S_z,

$$\chi^{+1}(1, 2) = (\alpha_1\alpha_2)$$

$$\chi^{0}(1, 2) = 2^{-1/2}[(\alpha_1\beta_2) + (\beta_1\alpha_2)] \qquad \text{(ortho)}$$

$$\chi^{-1}(1, 2) = (\beta_1\beta_2) \tag{3-5b}$$

Since there are three spin functions, the ortho states are also called *triplet states;* while the only para state is also called a *singlet state.*

PERTURBATION CALCULATION

For purposes of perturbation calculations, the complete Hamiltonian appearing in equation (3-2) can be written as $H^0 + \lambda H^1$, where

$$H^0 = -\tfrac{1}{2}(\nabla_1^2 + \nabla_2^2) + V_1(r_1) + V_2(r_2)$$

$$\lambda H^1 = -\frac{Z}{r_1} - V_1(r_1) - \frac{Z}{r_2} - V_2(r_2) + \frac{1}{r_{12}} \tag{3-6}$$

Powers of λ will distinguish orders of the perturbation. The criteria for choosing V_1 and V_2 are that H^0 lead to a solvable equation and that the effect of λH^1 be small. In general $V_1 \neq V_2$.

Since the zero-order Hamiltonian separates into two terms, one depending on $\mathbf{r}_1$ and the other on $\mathbf{r}_2$, the zero-order solutions are properly symmetrized products of single particle solutions of the individual Hamiltonians. Evidently, three types of solution can be distinguished. (1) Each single particle solution is a ground state solution, leading to the ground state solution of the atom. (2) One solution is the ground state, while the other is an excited state. (3) Both solutions are excited states.

For helium (and other helium-like ions) only those states are of practical importance in which at least one electron is in the ground state for the following reason. One finds that the energy for any helium state where both electrons are excited is higher than the ground state energy of a He^+ ion plus a free electron. Such doubly excited states would therefore quickly disintegrate into a He^+ ion and a free electron; they are states in the continuous spectrum, not discrete states.[1] Hence we shall not consider solutions of the third category.

Let us first consider the ground state problem. Since each electron is in the same state, it is physically reasonable to take $V_1 = V_2$, viz., each electron "sees" the same potential. This has the immediate consequence that the antisymmetric solution vanishes. Thus the ground state is para.

The simplest and most obvious choice for the potential is

$$V_1(r) = V_2(r) = -\frac{Z}{r} \tag{3-7}$$

That is, only the interelectron interaction $1/r_{12}$ is treated as a perturbation. The ground state wave function is then the (symmetric) product of two hydrogenic ground state wave functions $(Z^3/\pi) \times e^{-Z(r_1+r_2)}$. The zero-order energy is $-Z^2$ while the first-order contribution is

$$\lambda E^1 = \frac{Z^6}{\pi^2} \int dr_1\, dr_2 \, \frac{e^{-2Z(r_1+r_2)}}{r_{12}} \tag{3-8}$$

To evaluate the integral, we expand $1/r_{12}$ in Legendre polynomials and use the addition theorem for spherical harmonics.

$$\frac{1}{r_{12}} = \frac{1}{r_>} \sum_{\ell=0}^{\infty} \left(\frac{r_<}{r_>}\right)^{\ell} P_\ell(\cos\Theta)$$

$$= \frac{1}{r_>} \sum_{\ell<0}^{\infty} \left(\frac{r_<}{r_>}\right)^{\ell} \frac{4\pi}{2\ell+1} \sum_{m=-\ell}^{m=\ell} Y_{\ell m}(\Omega_1)\, Y^*_{\ell m}(\Omega_2) \tag{3-9}$$

where $r_>$ $(r_<)$ is the greater (lesser) of r_1 and r_2, while Θ is the

---

[1]However, there are a few doubly excited states which disintegrate slowly and can therefore be observed spectroscopically as almost discrete states. We shall not concern ourselves with these states, but an example is discussed in Problem 3.

angle between $\mathbf{r}_1$ and $\mathbf{r}_2$. The integration then gives $\lambda E^1 = 5Z/8$. For helium, $Z = 2$, the ground state energy in this approximation is -2.75 au, which is about 5% above the experimental value -2.904 au. More accurate calculations follow by taking account of the fact that one electron screens part of the nuclear charge from the other electron. We do not pursue the perturbative treatment of the ground state of helium, since far better results are obtained by variation methods discussed below.

We next examine the excited states, that is, solutions of the second category, where one particle is in the ground state and the other is an excited state. Before proceeding with the detailed evaluation of the energy levels, let us examine the general features of two-electron atoms and ions which follow the symmetry of solutions. When two electrons coincide, $\mathbf{r}_1 = \mathbf{r}_2$, the ortho wave function vanishes identically since

$$U_-(\mathbf{r}_1, \mathbf{r}_1) = -U_-(\mathbf{r}_1, \mathbf{r}_1) = 0$$

For para states this does not happen. Therefore, the probability that the two electrons are very close to each other is much smaller for ortho than for para states. This in turn means that the energy of a para state should be larger than that of the corresponding ortho state, where by "corresponding" we mean that both states are constructed from the same two orbitals. Furthermore, we expect optical combinations between ortho and para states (or vice versa) to be forbidden for the following reason. The operator for the electric dipole moment $(x_1 + x_2)$ is spin independent. The transition matrix element, which involves a space and spin integration, vanishes by the orthogonality of the ortho and para spin states. (The spatial integration also vanishes since the dipole moment operator and the spatial para wave function are even, while the ortho wave function is odd under interchange of the two coordinates. The possible occurrence of these forbidden transitions is discussed in Chapter 11.)

To summarize: The level scheme of helium and of ions with two electrons consists of two systems of levels, one containing triplet levels (orthohelium) and the other singlet levels (parahelium) which do not combine optically with each other. The lowest energy state is is para. For all excited states, the energy eigenvalues of the para states are larger than those of the corresponding ortho states. Using standard spectroscopic notation, we shall designate these by $n^{2S+1}L$ Here L is the total orbital angular momentum specified by the letters S, P, D, etc.; the superscript $2S + 1$ is the *multiplicity* which takes on values 1 or 3; and n orders the various levels. In this notation the ground state is $1\ ^1S$, while the subsequent several states are $2\ ^{1,3}S$, $2\ ^{1,3}P$, etc.

We shall now make a specific choice for V_1 and V_2 occurring in (3-6), which is due to Heisenberg. A reasonable assumption for the effective single-particle potential would be the following. For small enough r, the electron 1 "sees" the entire nuclear charge Z. For sufficiently large r the nuclear charge is screened by electron 2 and electron 1 "sees" a charge Z − 1. Figure 3-1 exhibits the qualitative behavior of a potential describing this situation. With Heisenberg we now assume that electron 1 is in the ground state and electron 2 is in an excited state. We are thus led to the following choice of potentials:

$$V_1(r_1) = -\frac{Z}{r_1} \qquad V_2(r_2) = -\frac{Z-1}{r_2} \tag{3-10}$$

With this unsymmetric choice of potentials, the two electrons are treated unsymmetrically, and first-order perturbation theory must be modified to overcome this asymmetry.

We split the Hamiltonian H into a zero-order Hamiltonian and a small perturbing part in two ways.

$$H = H_a^0 + \lambda H_a^1 = H_b^0 + \lambda H_b^1 \tag{3-11a}$$

The first decomposition, labeled a, is identical with (3-6). The second decomposition, labeled b, is obtained by interchanging $\mathbf{r}_1$ with $\mathbf{r}_2$ in (3-6),

$$H_b^0(\mathbf{r}_1, \mathbf{r}_2) = H_a^0(\mathbf{r}_2, \mathbf{r}_1)$$

$$\lambda H_b^1(\mathbf{r}_1, \mathbf{r}_2) = \lambda H_a^1(\mathbf{r}_2, \mathbf{r}_1) \tag{3-11b}$$

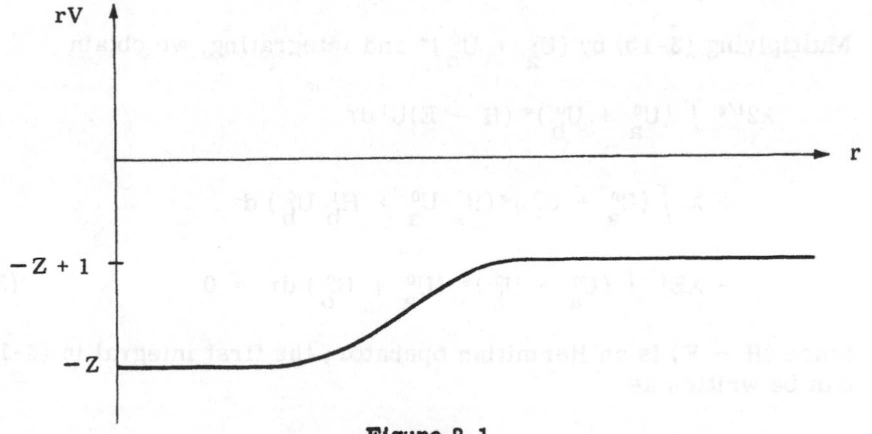

Figure 3-1

Effective single-particle potential for a two-electron atom.

Clearly H_a^0 and H_b^0 possess the same eigenvalue spectrum. All four Hamiltonians, H_a^0, H_b^0, λH_a^1, λH_b^1, are Hermitian. The two zero-order Hamiltonians differ from each other only by a term of first order in λ:

$$H_a^0 - H_b^0 = \lambda(H_b^1 - H_a^1) \tag{3-12}$$

Let U_a^0 and U_b^0 be particular normalized eigenfunctions of H_a^0 and H_b^0 respectively, belonging to the same eigenvalue E^0.

$$(H_a^0 - E^0)\,U_a^0 = (H_b^0 - E^0)\,U_b^0 = 0 \tag{3-13}$$

Since H_a^0 and H_b^0 are identical to zero order in λ, U_a^0 and U_b^0 are either identical (within phase) or can be chosen to be orthogonal, within zero order in λ. In what follows, we assume the two wave functions to be orthogonal within zero order. It will be seen below to what extent this is satisfied. The *symmetric* perturbation solution proceeds by setting

$$(H - E)\,U = 0 \qquad E = E^0 + \lambda E^1$$

$$U = U^0 + \lambda U^1 \qquad U^0 = 2^{-1/2}[\,U_a^0 + U_b^0\,] \tag{3-14}$$

By appropriate choice of U_a^0 and U_b^0, U^0 is assured to have the desired symmetry.

From (3-11a), (3-13), and (3-14) we obtain

$$\lambda 2^{1/2}(H - E)U^1 + \lambda(H_a^1\,U_a^0 + H_b^1\,U_b^0) - \lambda E^1(U_a^0 + U_b^0) = 0 \tag{3-15}$$

Multiplying (3-15) by $(U_a^0 + U_b^0)^*$ and integrating, we obtain

$$\lambda 2^{1/2} \int (U_a^0 + U_b^0)^*\,(H - E)U^1\,d\tau + \lambda \int (U_a^0 + U_b^0)^*(H_a^1\,U_a^0 + H_b^1\,U_b^0)\,d\tau - \lambda E^1 \int (U_a^0 + U_b^0)^*\,(U_a^0 + U_b^0)\,d\tau = 0 \tag{3-16}$$

Since $(H - E)$ is an Hermitian operator, the first integral in (3-16) can be written as

$$\lambda 2^{1/2} \int [(H - E)(U_a^0 + U_b^0)]^*\,U^1\,d\tau$$

The above integrand is first order in λ, since

$$(H - E)(U_a^0 + U_b^0) = (H - E)\,2^{1/2}U - (H - E)2^{1/2}\lambda U^1$$
$$= -2^{1/2}\lambda(H - E)U^1$$

Therefore the above integral is quadratic in λ. As we shall only keep terms of first order in λ, the first integral in (3-16) vanishes. The third integral in (3-16) is

$$-\lambda E^1\Big[\int |U_a^0|^2\,d\tau + \int |U_b^0|^2\,d\tau + \int U_b^{0*}U_a^0\,d\tau + \int U_a^{0*}U_b^0\,d\tau\Big] = -2\lambda E^1$$

since U_a^0, U_b^0 are orthonormal to zero order in λ. Thus (3-16) reduces to

$$\lambda E^1 = \tfrac{1}{2}\int (U_a^0 + U_b^0)^*\,(\lambda H_a^1 U_a^0 + \lambda H_b^1 U_b^0)\,d\tau \tag{3-17}$$

With Heisenberg's choice of potentials (3-10), we have

$$H_a^0 = -\tfrac{1}{2}(\nabla_1^2 + \nabla_2^2) - \frac{Z}{r_1} - \frac{Z-1}{r_2}$$

$$\lambda H_a^1 = \frac{1}{r_{12}} - \frac{1}{r_2}$$

$$H_b^0 = -\tfrac{1}{2}(\nabla_1^2 + \nabla_2^2) - \frac{Z-1}{r_1} - \frac{Z}{r_2}$$

$$\lambda H_b^1 = \frac{1}{r_{12}} - \frac{1}{r_1} \tag{3-18}$$

The zero-order Hamiltonians are sums of hydrogenic Hamiltonians; their eigenfunctions are products of hydrogenic eigenfunctions (1-5). The solutions U_a^0 and U_b^0 which lead to the proper symmetry of U^0 are

$$U_a^0 = v(r_1)\,u_{n\ell m}(r_2)$$

$$U_b^0 = \pm\, v(r_2)\,u_{n\ell m}(r_1)$$

$$U_\pm^0 = 2^{-1/2}\,[v(r_1)\,u_{n\ell m}(r_2) \pm v(r_2)\,u_{n\ell m}(r_1)]$$

$$E^0 = -\tfrac{1}{2}Z^2 - \frac{1}{2n^2}(Z-1)^2 \tag{3-19}$$

where, in accordance with the remarks on p. 35, $v(r)$ is the normalized hydrogenic ground state wave function with charge Z; and $u_{n\ell m}(\mathbf{r})$ is the normalized hydrogenic wave function with charge Z − 1 for an excited state (n, ℓ, m). As we are taking one electron in the ground state, and the other in an excited state, n is greater than 1. The plus sign refers to para states, the minus sign to ortho states.

We recall that in deriving (3-17) we took U_a^0 and U_b^0 to be orthogonal. From the explicit expressions for U_a^0 and U_b^0 in (3-19), it is seen that this orthogonality condition is equivalent to taking $u_{n\ell m}(\mathbf{r})$ and $v(\mathbf{r})$ to be orthogonal. This is exactly true when $\ell \neq 0$ due to the orthogonality of the spherical harmonics which occur in $u_{n\ell m}$ and v. When $\ell = 0$ the orthogonality condition is no longer satisfied, even though we take $n > 1$. This is so because $u_{n\ell m}$ and v are solutions corresponding to different Z. Thus we can only say that for S states u_{n00} and v are orthogonal to zero order.

Inserting the expressions for U_a^0, U_b^0 from (3-19) into (3-17) gives the first-order perturbation energy for a state with $n > 1$.

$$\lambda E^1 = 2^{-1/2} \int U_\pm^{0\,*} \left[\left(\frac{1}{r_{12}} - \frac{1}{r_2} \right) v(\mathbf{r}_1)\, u_{n\ell m}(\mathbf{r}_2) \pm \left(\frac{1}{r_{12}} - \frac{1}{r_1} \right) v(\mathbf{r}_2)\, u_{n\ell m}(\mathbf{r}_1) \right] d\mathbf{r}_1\, d\mathbf{r}_2 \tag{3-20}$$

Using the orthogonality of v and $u_{n\ell m}$, we obtain from (3-20)

$$\lambda E^1 = J \pm K$$

$$J \equiv \int d\mathbf{r}_1\, d\mathbf{r}_2 \left(\frac{1}{r_{12}} - \frac{1}{r_2} \right) v^2(\mathbf{r}_1)\, |u_{n\ell m}(\mathbf{r}_2)|^2$$

$$K \equiv \int d\mathbf{r}_1\, d\mathbf{r}_2\, \frac{1}{r_{12}}\, v(\mathbf{r}_1)\, u^*_{n\ell m}(\mathbf{r}_1)\, v(\mathbf{r}_2)\, u_{n\ell m}(\mathbf{r}_2) \tag{3-21}$$

Thus we have two energies, J + K and J − K, corresponding to the two types of spatial symmetry possible. J is called the *direct integral* and represents the Coulomb interaction between the charge distributions of the two electrons plus the interaction of the outer electron with one unit of positive charge concentrated at the nucleus. K is called the *exchange integral.* It measures the frequency with which the two electrons exchange their quantum states.

To make the physical significance of K clear, let us assume we know that at time $t = 0$ electron 1 is in the ground state and electron 2 is in an excited state. Then the wave function at $t = 0$ is

$$\Psi(0) = v(r_1)\, u_{n\ell m}(r_2) = 2^{-1/2}(U_+^0 + U_-^0)$$

At any later time

$$\Psi(t) = 2^{-1/2}(\Psi_+(t) + \Psi_-(t))$$

$$= 2^{-1/2}\left(U_+^0\, e^{-i(E'+K)t} + U_-^0\, e^{-i(E'-K)t}\right)$$

$$= e^{-iE't}\Big(v(r_1)\, u_{n\ell m}(r_2) \cos Kt$$

$$- iv(r_2)\, u_{n\ell m}(r_1) \sin Kt\Big)$$

$$E' \equiv E^0 + J$$

After the time interval $\pi/2K$ has elapsed, the two electrons have interchanged their roles: electron 1 is now excited, electron 2 is in the ground state. (This explanation assumed we could distinguish electron 1 from electron 2 at time $t = 0$, which of course violates the principle of indistinguishability. Hence the entire argument should not be taken too seriously.)

To evaluate J we write

$$J = \int_0^\infty \int_0^\infty r_1^2\, dr_1\, r_2^2\, dr_2\, R_{10}^2(r_1)\, R_{n\ell}^2(r_2)\, J(r_1, r_2)$$

$$J(r_1, r_2) \equiv \iint d\Omega_1\, d\Omega_2\, Y_{00}^2(\Omega_1)\, |Y_{\ell m}(\Omega_2)|^2 \left(\frac{1}{r_{12}} - \frac{1}{r_2}\right) \quad (3\text{-}22)$$

Here R_{10} is the radial hydrogenic ground state wave function with charge Z, and $R_{n\ell}$ is the radial hydrogenic wave function with charge $Z - 1$ for the state $n\ell$. In the evaluation of $J(r_1, r_2)$, we again expand $1/r_{12}$ in spherical harmonics, according to (3-9). The angular integrations are then trivial and yield

$$J(r_1, r_2) = \begin{cases} 1/r_1 - 1/r_2 & r_1 > r_2 \\ 0 & r_1 < r_2 \end{cases} \tag{3-23}$$

Recalling $R_{10}(r) = 2Z^{3/2} e^{-Zr}$, we obtain

$$J = \int_0^\infty r_2^2 R_{n\ell}^2(r_2)\, dr_2 \int_{r_2}^\infty 4Z^3 r_1^2 e^{-2Zr_1} \left(\frac{1}{r_1} - \frac{1}{r_2} \right) dr_1$$

$$= -\int_0^\infty \left(Z + \frac{1}{r_2} \right) e^{-2Zr_2} r_2^2 R_{n\ell}^2(r_2)\, dr_2 \tag{3-24}$$

Equation (3-24) can be evaluated for specific values of n and ℓ (Bethe and Salpeter,[2] pp. 133-134).

For large n, J takes on the limiting form

$$J \approx \frac{1}{n^3} F(Z, \ell) \tag{3-25}$$

where $F(Z, \ell)$ falls off rapidly with ℓ. To see this we make use of the WKB approximation, which is valid for large n, to obtain the dependence of $R_{n\ell}$ on n and ℓ. Setting $\mathcal{R} = rR_{n\ell}$ we have according to (1-22)

$$\frac{d^2\mathcal{R}}{dr^2} + \Phi\mathcal{R} \equiv \frac{d^2\mathcal{R}}{dr^2} + \left[-\frac{Z^2}{n^2} + \frac{2Z}{r} - \frac{\ell(\ell+1)}{r^2} \right] \mathcal{R} = 0 \tag{3-26}$$

(For simplicity we are writing Z instead of Z-1, since we are not concerned here with the explicit dependence on Z.) The coefficient Φ of $\mathcal{R}$ represents twice the kinetic energy of the electron and is positive for $a_1 < r < a_2$ where $a_{1,2}$ are the classical turning points.

$$a_{1,2} = \frac{n^2}{Z} \mp \frac{n}{Z}(n^2 - \ell(\ell+1))^{1/2} \tag{3-27}$$

In this region, the eigenfunctions are, according to (1-23),

$$\mathcal{R} = A\Phi^{-1/4} \cos\left[\int_{a_1}^{r} \Phi^{1/2}\, dr' - \frac{\pi}{4} \right] \tag{3-28}$$

---

[2] H. A. Bethe and E. E. Salpeter, *Quantum Mechanics of One- and Two-Electron Atoms*, Plenum, New York, 1977.

To obtain the normalization factor A, we need only consider the region $a_1 < r < a_2$ in the normalization integral since outside this region $\mathcal{R}$ decreases exponentially. In our case of interest, $n \gg 1$, the kinetic energy Φ is large. Thus there are many oscillations of the cosine, so that we can replace $\cos^2$ by its average value $\frac{1}{2}$. Then

$$1 = \int_0^\infty \mathcal{R}^2\, dr \approx \tfrac{1}{2} A^2 \int_{a_1}^{a_2} \frac{dr}{\Phi^{1/2}} \approx \tfrac{1}{2} A^2 \pi Z^{-2} n^3$$

by actual integration or by the observation that $\Phi^{-1/2} \sim n/Z$ and $a_2 - a_1 \sim n^2/Z$. Therefore,

$$A \sim n^{-3/2} \tag{3-29}$$

Now for small r (of order 1/Z) and large n, the term Z^2/n^2 is negligible compared with the other terms in (3-26). Then it is easily seen that both (3-28) and the exact solution of (3-26) will have the form

$$rR_{n\ell} \equiv \mathcal{R} = Af(r, Z, \ell) \tag{3-30}$$

where f is independent of n and only A depends on n as above. Since only values of r_2 of order 1/Z contribute to (3-24), this integral becomes

$$J \approx F(Z, \ell)/n^3 \tag{3-31}$$

$F(Z, \ell)$ falls off rapidly for large ℓ, since according to (3-23) only the region $r_2 < r_1$ contributes to J. Electrons with small orbital quantum number are more likely to penetrate the 1s shell than those with large ℓ. (The probability of finding an electron in the neighborhood of the nucleus is proportional to $r^{2\ell}$.) Defining $\delta_C = F(Z, \ell)/(Z-1)^2$, we can sum up the unperturbed energy and the perturbation J of the outer electron in the field of nuclear charge $Z - 1$:

$$-\frac{(Z-1)^2}{2n^2} + J = -\frac{(Z-1)^2}{2n^2}\left[1 - \frac{2\delta_C}{n}\right]$$

$$\approx -\frac{(Z-1)^2}{2n^2}\left[1 + \frac{\delta_C}{n}\right]^{-2}$$

$$= -\frac{(Z-1)^2}{2(n+\delta_C)^2} \tag{3-32}$$

This is Rydberg's form for an energy level, customary in spectroscopy. δ_C is closely related to the phase shift in scattering theory, see p. 79.

To evaluate K we write

$$K = \int_0^\infty \int_0^\infty r_1^2\, dr_1\, r_2^2\, dr_2\, R_{10}(r_1) R_{n\ell}(r_1) \times R_{10}(r_2) R_{n\ell}(r_2) K(r_1, r_2)$$

$$K(r_1, r_2) \equiv \iint d\Omega_1\, d\Omega_2\, Y_{00}(\Omega_1) Y_{00}(\Omega_2) Y_{\ell m}^*(\Omega_1) Y_{\ell m}(\Omega_2) \frac{1}{r_{12}} \tag{3-33}$$

Expanding $1/r_{12}$ in spherical harmonics according to (3-9), we see that only the term $\ell' = \ell$, $m' = m$ contributes; all others vanish by orthogonality of the spherical harmonics upon integration over $d\Omega_1$ (or $d\Omega_2$). Thus we obtain

$$K(r_1, r_2) = \frac{1}{2\ell + 1} \frac{r_<^\ell}{r_>^{\ell+1}} \tag{3-34}$$

and

$$K = \frac{2}{2\ell + 1} \int_0^\infty r_2^{\ell+2}\, dr_2\, R_{10}(r_2) R_{n\ell}(r_2) \times \int_{r_2}^\infty r_1^{-\ell+1}\, dr_1\, R_{10}(r_1) R_{n\ell}(r_1) \tag{3-35}$$

where the symmetry of the integrand in r_1 and r_2 has been used.

The same qualitative considerations apply to K as to J. Evidently, we can write

$$K \approx \frac{G(Z, \ell)}{n^3} \tag{3-36}$$

where $G(Z, \ell)$ falls off quickly for large ℓ. Defining

$$\delta_A = \frac{G(Z, \ell)}{(Z - 1)^2}$$

we obtain by an argument similar to (3-32)

$$E + \frac{1}{2} Z^2 = -\frac{1}{2} \frac{(Z - 1)^2}{(n + \delta_C \pm \delta_A)^2} \tag{3-37}$$

The positive sign belongs to parahelium, the negative to orthohelium. The term $\delta_C \pm \delta_A$ is called the *Rydberg correction;* our calculation shows that this attains a limit for large n.

Table 3-1

Rydberg Corrections for Helium States with Large Principal Quantum Number

	$-(\delta_C + \delta_\pi)$	$-\frac{1}{2}(\delta_o + \delta_p)$	δ_A	$\frac{1}{2}(\delta_p - \delta_o)$
S	0.216	0.218	0.376	0.078
P	0.0248	0.0279	0.0351	0.0398
D	0.00262	0.00252	0.00066	0.00035
F	8×10^{-5}	13×10^{-5}	$< 10^{-5}$	8×10^{-5}

The correct helium wave functions cannot be of the simple form of a product of two independent single-particle wave functions. The presence of one electron at a particular position affects the wave function of the other. The Coulomb repulsion due to one electron polarizes the charge distribution of the other such as to increase their mutual separation. One calculates the effect of polarization on the Coulomb correction δ_C, and on the exchange term δ_A. The correction to δ_A is very small (Bethe and Salpeter,[2] p. 140), the correction δ_π to the Coulomb term is sizeable. It is found that

$$E + \tfrac{1}{2}Z^2 = -\frac{1}{2}\frac{(Z-1)^2}{(n + \delta_C + \delta_\pi \pm \delta_A)^2} \tag{3-39}$$

δ_π is constant for large n and decreases with increasing ℓ, but not as rapidly as δ_C (Bethe and Salpeter,[2] pp. 137-139).

Table 3-1 shows the comparison between observed and calculated values of the Rydberg correction $\delta_C + \delta_\pi \pm \delta_A$ for helium. δ_p is the observed Rydberg correction for parahelium; δ_o, for orthohelium. The columns $\delta_C + \delta_\pi$ and $\frac{1}{2}(\delta_o + \delta_p)$ represent, respectively, the calculated and observed *average* Rydberg correction and should agree; likewise, the last two columns of the table should agree.

The agreement is fair for P and D states. For the exchange part δ_A of the S states it is seen that the Heisenberg method fails. (The good agreement for the Coulomb part is accidental.) If the lack of orthogonality of 1s and ns states is taken into account, the result for the exchange term even has the wrong sign. Better wave functions are therefore necessary. Such functions are provided by Fock's method, discussed in detail in Chapter 4. The calculated Rydberg corrections are then -0.289 and -0.160 for orthohelium and parahelium, respectively. The experimental values are -0.296 and -0.140.

VARIATION CALCULATION

By far the most successful method for obtaining an accurate value for the ground state of two-electron atoms is the Ritz variation method.

The simplest trial function for a variation calculation is the product of two hydrogenic wave functions,

$$\psi = \frac{\alpha^3}{\pi} e^{-\alpha(r_1+r_2)} \tag{3-40}$$

with α as the variation parameter. Then some algebra yields (Schiff,[3] pp. 257 - 258)

$$\langle H \rangle = \alpha^2 - 2Z\alpha + \tfrac{5}{8}\alpha \tag{3-41}$$

The first term on the right-hand side represents the expectation value of the kinetic energy; the second, the expectation value of the nuclear potential energy. The third term is the expectation value of the interaction energy between the electrons. The relevant integral for this term was evaluated by expanding $1/r_{12}$ in the usual fashion in spherical harmonics.

Minimizing (3-41) as a function of α gives

$$\alpha = Z - \tfrac{5}{16} \tag{3-42}$$

Thus the hydrogenic wave functions give the best energy when Z is replaced by $Z - \frac{5}{16}$, indicating that each electron screens the nucleus from the other electron. We find for the ground state energy

$$E = -\left(Z - \tfrac{5}{16}\right)^2 \quad \text{au} \tag{3-43}$$

The most naive perturbation calculation, which considers $1/r_{12}$ as the perturbation, obtains (3-40) for the zero-order (unnormalized) wave function, and (3-41) for the ground state energy correct to first order, both with $\alpha = Z$(see p. 33). This gives an energy of $-Z^2 + \frac{5}{8}Z$. Our variation calculation has decreased this value by $(5/16)^2 = 0.098$.

The experimentally measured quantity is not the ground state energy E of a helium-like atom, but its ionization potential I. I equals $E_0 - E$, where E_0 is the ground state energy of the singly ionized (hydrogen-like) atom $E_0 = -Z^2/2$.

$$\begin{aligned} I &= \frac{Z^2}{2} - \frac{5Z}{8} + \frac{25}{256} \text{ au} \\ &= Z^2 - \frac{5Z}{4} + \frac{25}{128} \text{ Ry} \end{aligned} \tag{3-44}$$

---

[3] L. I. Schiff, *Quantum Mechanics*, 3rd ed., McGraw-Hill, New York, 1968.

Table 3-2
Ionization Potential for Helium-Like Atoms in Rydbergs

	H^-	He	Li^+	Be^{++}
Theoretical	−0.055	1.695	5.445	11.195
Experimental	0.055	1.807	5.560	11.312
Difference	0.110	0.112	0.115	0.117

[We recall that 1 Ry unit of energy is one-half an atomic unit (au) of energy.]

For Z = 2, the calculated ionization potential is 1.695 Ry; the observed is 1.807 Ry. The naive perturbation calculation discussed above would give 1.500 Ry. Table 3-2 lists the calculated (variational) and the measured values for the ionization potential in terms of the Rydberg. It is interesting to note that the difference between this simple variation result and the experimental value is almost independent of Z.

Higher approximations in the variation treatment begin with the assumption that

$$\psi(\mathbf{r}_1, \mathbf{r}_2) = e^{-\alpha(r_1 + r_2)}\, P(\mathbf{r}_1, \mathbf{r}_2) \tag{3-45}$$

The first approximation (3-40) corresponds to a constant $P = \alpha^3/\pi$. Higher approximations result in expanding $P(\mathbf{r}_1, \mathbf{r}_2)$. One could try writing

$$P(\mathbf{r}_1, \mathbf{r}_2) = \sum_{\ell} f_{\ell}(r_1, r_2)\, P_{\ell}(\cos \Theta_{12}) \tag{3-46}$$

This is not successful, for reasons which will become clear below.

It was Hylleraas who suggested that the trial function ought to depend on r_{12}. He introduced symmetric coordinates

$$s = \alpha(r_1 + r_2) \qquad t = \alpha(r_1 - r_2) \qquad u = \alpha r_{12} \tag{3-47}$$

and

$$\psi = e^{-s}\, P(s, t, u) \tag{3-48}$$

The "effective charge" α is fixed by the condition that $\langle H \rangle$ be minimum. It is also now clear why (3-46) was an unsatisfactory expansion.

The difficulty lay in the fact that the expansion of r_{12} in terms of $P_\ell(\cos\Theta_{12})$ converges very slowly.

The variation calculation proceeds by choosing a specific dependence of P on the coordinates s, t, u and on several parameters c_i. Proper symmetry must be maintained. For the ground state we showed above that the wave function must be symmetric under interchange of r_1 and r_2. In the present context this means that only even powers of t may occur in P. $\langle H \rangle$ is obtained as a function of the c_i and of α. This then is minimized, and the parameters c_i and α are obtained from the minimizing conditions

$$\frac{\partial\langle H\rangle}{\partial c_i} = 0 \qquad \frac{\partial\langle H\rangle}{\partial \alpha} = 0 \tag{3-49}$$

Such variation calculations have been performed by Kinoshita and Pekeris. Kinoshita uses an 80-parameter wave function. Pekeris uses 210 terms in an expansion of P in powers of $r_1 + r_2 - r_{12}$, $r_{12} + r_1 - r_2$, and $r_{12} - r_1 + r_2$. This permits him to determine the coefficients in the power series for P by recursion formulas. The energy is then obtained as the eigenvalue of a determinant. Pekeris shows that his method is equivalent to the variational principle but facilitates accurate evaluation on an electronic computer. His result for I in the case of helium is[4]

$$I = 198317.374 \pm 0.022\ \text{cm}^{-1} \tag{3-50}$$

In judging the accuracy of Pekeris' result (3-50), it should be pointed out that 0.012 cm^{-1} of the error is due to the inaccuracy of the Rydberg constant for helium, only 0.010 cm^{-1}, or 1 part in 20 million, is due to Pekeris' approximations.

The experimental result is

$$I = 198310.82 \pm 0.15\ \text{cm}^{-1} \tag{3-51}$$

It is seen to be lower than the variation calculation result (3-50), and far outside the two stated limits of error. At first sight this is not very satisfactory, since the variation method ought to give a lower bound for I (recall $I = -E + E_0$).

The discrepancy is in part due to the motion of the nucleus which we have been ignoring so far. To account for this we write the total Hamiltonian for the system in ordinary units

---

[4] C. L. Pekeris, *Phys. Rev.* **112**, 1649 (1958).

$$H = \frac{p_N^2}{2M} + \frac{p_1^2}{2m} + \frac{p_2^2}{2m} + V(r_1 - R, r_2 - R, r_1 - r_2) \tag{3-52}$$

where p_N is the momentum, M the mass, and R the position of the nucleus. By a canonical transformation, we introduce a center-of-mass coordinate ρ and relative coordinates R_i.

$$\rho = \frac{1}{M + 2m}(MR + mr_1 + mr_2)$$

$$R_i = r_i - R \qquad i = 1, 2 \tag{3-53}$$

Conjugate to these coordinates are the total momentum P and the relative momenta P_i.

$$P = p_N + p_1 + p_2$$

$$P_1 = \frac{(M + m)\, p_1 - mp_2 - mp_N}{M + 2m}$$

$$P_2 = \frac{(M + m)p_2 - mp_1 - mp_N}{M + 2m} \tag{3-54a}$$

The expression for the old momenta in terms of the new is

$$p_N = \frac{M}{M + 2m} P - P_1 - P_2$$

$$p_i = \frac{m}{M + 2m} P + P_i \tag{3-54b}$$

The Hamiltonian (3-52) in terms of new variables becomes

$$H = \frac{P^2}{M + 2m} + \frac{1}{2\mu}(P_1^2 + P_2^2) + \frac{1}{M} P_1 \cdot P_2$$
$$+ V(R_1, R_2, R_1 - R_2) \tag{3-55}$$

where the reduced mass μ has been introduced

$$\mu = \frac{Mm}{M + m} \tag{3-56}$$

The potential does not depend on the center-of-mass coordinate, consequently the total momentum P is a constant of motion. The center-of-mass motion is free, and may be separated from the remaining equation. The time-independent Schrödinger equation for relative motion is

$$\left[-\frac{\hbar^2}{2\mu}\left(\nabla^2_{R_1} + \nabla^2_{R_2}\right) - \frac{\hbar^2}{M}\nabla_{R_1} \cdot \nabla_{R_2} + V(R_1, R_2, R_1 - R_2)\right]\psi = E\psi \tag{3-57}$$

Thus the motion of the nucleus modifies the Schrödinger equation in two ways. In the first place, the effective mass of the electron μ replaces the actual mass m. This is taken into account when we express the energy in terms of the reduced Rydberg unit,

$$R_M = \frac{M}{M + m} R_\infty \approx R_\infty\left(1 - \frac{m}{M}\right) \tag{3-58}$$

The second effect is the addition of a perturbation term $-(\hbar^2/M) \times \nabla_{R_1} \cdot \nabla_{R_2}$ to the energy. This can be evaluated by perturbation theory (Bethe and Salpeter,[2] pp. 166-170).

Further corrections to (3-50) involve relativistic corrections and other corrections due to the interaction of the electron with its own field (Lamb shift).

Pekeris gives for the nuclear motion plus relativistic correction the result $-5.348 \pm 0.0005\ \text{cm}^{-1}$. The Lamb shift contribution was calculated by Salpeter and Zaidi[8] to be $-1.360 \pm 0.02\ \text{cm}^{-1}$. The corrected ionization potential is then

$$I_{\text{theory}}^{\text{total}} = 198310.665 \pm 0.04\ \text{cm}^{-1} \tag{3-59}$$

The phenomenal agreement of (3-59) with (3-51) is one of the most striking proofs of the validity of wave mechanics, in a definitely non-trivial problem.

Kinoshita gives an estimate of the accuracy of his trial wave function in the following way. He writes the trial function ψ as

$$\psi = \sqrt{1 - \eta^2}\ \psi_0 + \eta f \tag{3-60}$$

where ψ_0 is the exact wave function and f satisfies[5]

$$\int \psi_0 f\, d\tau = 0 \qquad \int |f|^2\, d\tau = 1 \tag{3-61}$$

In this case it can be verified that $\langle H \rangle$ differs from E by terms of order η^2. Kinoshita finds $\eta = 1.1 \times 10^{-3}$. Hence η^2 is about 10^{-6}

We can now see why the Heisenberg perturbation theory fails to give the correct ground state energy. A symmetrized wave function of the sort we were using for the perturbation calculations could never depend on $u = r_{12}$. However, we see from the variation calculations that ψ must depend on u. Specifically for $s = 1, t = 0$, the Kinoshita wave function behaves as

$$\psi \sim 1 + 0.498u + \cdots \tag{3-62}$$

---

[5]See Chapter 1, Problem 5.

It can also be shown that in the approximation r_1, r_2 large, r_{12} small Schrödinger's equation leads to a solution of the form $\psi \sim 1 + 0.5u$, which agrees very well with (3-62) (Slater,[6] Vol. II, p. 38).

IONIZATION POTENTIAL FOR HIGHER Z ATOMS

The variation method can be used to obtain the ground state energies of helium-like ions. For large Z we use the following procedure. In our Schrödinger equation (3-2) we change variables:

$$\rho_{1,2} = 2Zr_{1,2} \qquad \epsilon = \frac{E}{2Z^2} \tag{3-63}$$

Then we write Schrödinger's equation in the form

$$(H^0 + \lambda H^1 - \epsilon)U = 0$$

$$H^0 = -\left(\nabla_1^2 + \nabla_2^2 + \frac{1}{\rho_1} + \frac{1}{\rho_2}\right)$$

$$H^1 = \frac{1}{\rho_{12}} \qquad \lambda = \frac{1}{Z}$$

$$\epsilon = \epsilon^0 + \frac{\epsilon^1}{Z} + \frac{\epsilon^2}{Z^2} + \cdots$$

$$U = U^0 + \frac{U^1}{Z} + \frac{U^2}{Z^2} + \cdots \tag{3-64}$$

Then, defining $\sigma = 2Zs$, $\tau = 2Zt$, $\nu = 2Zu$, where s, t, u are defined in (3-47) (with $\alpha = Z$), we have

$$U^0 = \tfrac{1}{2} e^{-1/2\,\sigma}$$

$$\epsilon^0 = -\tfrac{1}{2}$$

$$\epsilon^1 = \tfrac{5}{16} \tag{3-65}$$

To obtain ϵ^2, we write

$$U^1 = U^0\Phi$$

or

$$U \approx \tfrac{1}{2}e^{-Zs}(1 + \Phi/Z) \tag{3-66}$$

---

[6] J. C. Slater, *Quantum Theory of Atomic Structure*, McGraw-Hill, New York, 1960.

Φ is obtained by variational methods, and $\varepsilon^2 = -0.0788278$ (Bethe and Salpeter,[2] pp. 151-153).

In principle one can calculate higher-order corrections also. Unfortunately the variational procedures become very cumbersome. Nevertheless, we now have exact values for ϵ^0 and ϵ^1 and a very accurate (variational) estimate for ϵ^2. One can then get an excellent semiempirical expansion for ϵ by using these values of ϵ^0 to ϵ^2 and by fitting ϵ^3 to ϵ^6 to the values determined directly for Z = 1, 2, 3, and 8 by Hylleraas. Then the ionization potential in Rydberg is, according to Hylleraas,

$$I = Z^2 - \tfrac{5}{4} Z + 0.315311 - 0.01707 \frac{1}{Z} + 0.00068 \frac{1}{Z^2} + 0.00164 \frac{1}{Z^3} + 0.00489 \frac{1}{Z^4} \quad (3\text{-}67)$$

The coefficients of 1/Z and the following terms are remarkably small.

EXCITED STATES

The energies for the excited states of helium can be calculated by the variation method provided the trial functions of the excited state are orthogonal to the eigenfunctions of all lower states. In general, this subsidiary condition makes the calculation quite difficult. However, cases do exist in which the subsidiary condition is satisfied automatically if the form of the wave function is prescribed by the character of the term to be calculated. A case in point is the $2\,^3S$ term. Every trial function must be chosen to be antisymmetric in the two-electron space coordinates. This in itself is sufficient to assure orthogonality to the symmetric ground state eigenfunction. In general, the eigenfunctions belonging to two states of an atom are automatically orthogonal if either the total orbital angular momentum **L** or the total spin **S** (or both) have different values for the two states. Hence the $2\,^3S$, $2\,^1P$, $2\,^3P$, etc., states of helium can be treated by the Ritz procedure without additional conditions. However, for the $2\,^1S$ term, we must specifically provide for the orthogonality of the eigenfunction to that of the ground state $1\,^1S$.

Table 3-3 lists the theoretical and experimental values for the energy of various helium states.[7] The degree of agreement is directly related to the amount of calculational labor expended on each of these states.

---

[7]The calculation for 2P states is reproduced in K. Gottfried, *Quantum Mechanics*, Vol. I, Benjamin/Cummings, Reading MA, 1966, p. 376.

Table 3-3
Ionization Potential for Various Levels of Helium in Rydbergs

State	Theoretical	Experimental
$2\,^3S$	0.35044	0.35047
$2\,^1S$	0.2898	0.2920
$2\,^3P$	0.262	0.266
$2\,^1P$	0.245	0.247

PROBLEMS

1. **In deriving the energy levels of excited states of two-electron atoms by perturbation methods, equation (3-21), we assumed that the two zero-order solutions U_a^0 and U_b^0, or equivalently v and $u_{n\ell m}$, are orthogonal. For S states this is not so. Repeat the derivation of equation (3-21) taking into account this lack of orthogonality. Show that λE^1 may be written as**

$$\lambda E^1 = (1 + \delta)^{-1}[J \pm (K - L)]$$

where J and K are defined as before, while δ and L arise from the lack of orthogonality of v and u_{n00}. For large n, show that $\delta \ll 1$, but that L is comparable to K. Evaluate L and show that the result for the exchange term when put in the form (3-38) gives the wrong sign when compared to experiment given in Table 3-1. The relevant integrals over hydrogenic wave functions may be evaluated with the help of formulas found in Landau and Lifshitz, *Quantum Mechanics,* 3rd ed., Pergamon, Oxford, 1977, pp. 662-665.
2. **Perform a variation calculation for the ground state of helium by using an arbitrary trial function depending only on $r_1 + r_2$ and varying the functional form in an arbitrary fashion. Compare this result to the simple variation calculation on p. 44.**
3. **Consider a helium atom with both electrons described by hydrogen orbitals corresponding to excited states $n_1 = n_2 = 2$; $\ell_1 = \ell_2 = 1$; $m_{\ell 1} = 1$, $m_{\ell 2} = 0$; $m_{s1} = m_{s2} = \frac{1}{2}$. Evidently, the total wave function for this state is given by $\psi(\mathbf{r}_1, \mathbf{r}_2)\chi$, where ψ is the space part, and χ is the spin part. Using the above quantum numbers write out explicitly ψ and χ in terms of hydrogenic wave functions and one-electron spin functions. Prove that this is a ^3P state. Letting Z, the total electron charge, be a variational**

parameter, perform a variation calculation for the ground state and obtain the effective charge. Discuss the orthogonality of this trial function to the wave functions of other states.

4. Using (3-9), derive (3-23) from (3-22), and (3-34) from (3-33).
5. Show that when a helium wave function is approximated by a simple (unsymmetrized) product of hydrogen orbitals, the effect of nuclear motion vanishes within first-order perturbation theory. Obtain an expression for the first-order energy shift when the helium wave function is approximated by a symmetrized product of hydrogenic orbitals as in (3-19).
6. Verify the entries in Table 3-2. What is the meaning of the negative value for the theoretical ionization potential for H^-? Give an argument to explain the fact that the difference between the experimental and theoretical values of the ionization potential is almost independent of Z.
7. (a) Verify equations (3-64) and (3-65).
   (b) Evaluate the semiempirical formula for I, (3-67), when Z = 1, 2, 3, 4 and compare with the experiment in Table 3-2.

Chapter 4

SELF-CONSISTENT FIELD

In the previous chapter we saw that the Ritz method can give very accurate results for the ground state energy of two-electron atoms, and with sufficient labor can be extended to yield good results for some excited states. The variational wave function is also determined, although it is not expected to be as accurate as the energy eigenvalue. For complex atoms, it is evident that this approach would become prohibitively cumbersome if carried out to a satisfactory degree of accuracy. For these complex atoms there exists the self-consistent field method, which we shall study in the present chapter. This method is useful for obtaining energy levels and wave functions for atoms.

INTUITIVE PRELIMINARIES

We assume with Hartree that each electron in a multielectron system is described by its own wave function. This implies that each electron is subject to an equivalent potential due to the other electrons and to the nucleus. This equivalent potential is obtained by postulating that there is a charge density associated with each electron which is $-e$ times its position probability density. The equivalent potential for the ith electron is then

$$V_i(\mathbf{r}_1) = \sum_{k \neq i} \int d\tau_2 \frac{1}{r_{12}} |u_k(\mathbf{r}_2)|^2 - \frac{Z}{r_1} \tag{4-1}$$

Here the subscript k indicates a set of quantum numbers describing the state of the kth electron. The summation extends over all the electrons except the ith.

If there are N electrons, this leads to N simultaneous nonlinear integrodifferential equations of the form

$$[-\tfrac{1}{2}\nabla^2 + V_i(\mathbf{r})]u_i(\mathbf{r}) = \epsilon_i u_i(\mathbf{r}) \tag{4-2}$$

The next approximation is made by replacing $V(\mathbf{r})$ by its average over the angles of $\mathbf{r}$, thus making it spherically symmetric:

$$V_i(r) = \frac{1}{4\pi}\int V_i(\mathbf{r})\, d\Omega \tag{4-3}$$

This is the so-called *central field approximation*. (We shall see below that this is in fact a very mild approximation.) The solutions of (4-2) can then be expressed as products of radial functions and spherical harmonics; see p. 5.

$$u_i(\mathbf{r}) = u_{n\ell m}(\mathbf{r}) = \frac{\mathcal{R}_{n\ell}(r)}{r} Y_{\ell m}(\Omega) \tag{4-4}$$

$\mathcal{R}_{n\ell}(r)$ satisfies the differential equation

$$\frac{1}{2}\frac{d^2\mathcal{R}_{n\ell}}{dr^2} + \left[\epsilon_{n\ell} - V_{n\ell}(r) - \frac{\ell(\ell+1)}{2r^2}\right]\mathcal{R}_{n\ell} = 0 \tag{4-5}$$

It is clear that even with all these assumptions we cannot solve the N equations of the form (4-5) exactly. Hartree's procedure is to solve this system by successive approximations, subject to the requirement of self-consistency. That is, an initial potential is constructed by an intelligent guess for the wave functions (or by Thomas-Fermi techniques to be discussed in the next chapter). This potential is then used to obtain wave functions from (4-5), which in turn determine a new potential from (4-1) and (4-3). The procedure is continued until the final wave functions determine a potential which is self-consistent to a high order of accuracy.

It will be shown below, that the Hartree method is equivalent to a variation calculation, where the trial function is taken to be a simple product of single particle orbitals, and the variation is performed by varying each orbital in an arbitrary fashion. The use of single particle wave functions has the consequence that correlations between electrons are neglected. The use of a simple product for the total wave function ignores symmetry considerations. Nevertheless the Pauli principle can be satisfied by choosing the quantum numbers of the one-electron orbitals appropriately.

The two shortcomings of the Hartree method mentioned above, viz., ignoring symmetry and correlation effects, can be overcome in the following way. Proper symmetry of the wave function is assured by taking the trial function to be a Slater determinant of single particle orbitals. This is the Hartree-Fock theory and is discussed below. The correlation effects, which arise from the $1/r_{ij}$ terms in the Hamiltonian, can be evaluated by perturbation theory. This is done in Chapter 7.

VARIATION DERIVATION

Hartree arrived at (4-2) and (4-5) by physically reasonable, intuitive arguments. We shall now show how one can obtain similar results from a variation principle. We shall generalize Hartree's equations by including the symmetry requirements. This generalization is known as the *Hartree-Fock theory* and is due to Fock and Slater.

For the variational trial function we choose a determinantal wave function of the form

$$\Psi = (N!)^{-1/2} \begin{vmatrix} u_1(1) & u_1(2) & \cdots & u_1(N) \\ u_2(1) & u_2(2) & \cdots & u_2(N) \\ & & \cdots & \\ u_N(1) & u_N(2) & \cdots & u_N(N) \end{vmatrix} \tag{4-6}$$

Each orbital occurring in (4-6) is a product of a space part $u(\mathbf{r})$, and a spin part $\chi(\sigma)$ which is either α or β. Here σ is the argument of χ in spin space, and takes on two values 1, 2. Explicitly if $\chi = \alpha = \binom{1}{0}$, then $\chi(1) = 1$, $\chi(2) = 0$, and similarly if $\chi = \beta = \binom{0}{1}$, then $\chi(1) = 0$, $\chi(2) = 1$. We require that all orbitals be orthonormal:

$$u_i(j) = u_i(\mathbf{r}_j)\chi_i(\sigma_j)$$

$$\int u_i^*(1)u_j(1)\, d\tau_1 = \delta_{ij} \tag{4-7}$$

The integration is over space and spin coordinates. Since orbitals of different spins are automatically orthogonal, (4-7) reduces to the condition that space orbitals corresponding to the same spin function be orthonormal. This assures that Ψ is normalized and the variation condition becomes

$$\delta \int \Psi^* H \Psi\, d\tau = 0 \tag{4-8}$$

MATRIX ELEMENTS BETWEEN DETERMINANTAL WAVE FUNCTIONS

In order to evaluate $\int\Psi^* H\Psi d\tau$, and for other applications, we calculate the matrix elements of an arbitrary operator F, involving all the electrons, between determinantal wave functions. We recall that (4-6) can be written as

$$\Psi = (N!)^{-1/2} \sum_P \epsilon_P \prod_{i=1}^{N} u_{Pi}(i)$$

$$= (N!)^{-1/2} \sum_P \epsilon_P \prod_{i=1}^{N} u_i(Pi) \tag{4-9}$$

The sum extends over all permutations, and ϵ_P is + or − depending on whether P1, P2, ..., PN is an even or odd permutation of 1, 2, ..., N. In order to calculate $\int \Psi_b^* F \Psi_a \, d\tau$, we assume that the orbitals corresponding to Ψ_b are $u_i(j)$, and Ψ_a corresponds to $v_i(j)$. Then

$$\langle F\rangle = \frac{1}{N!} \int \sum_Q \epsilon_Q \prod_{i=1}^{N} u_i^*(Qi)\, F \sum_P \epsilon_P \prod_{j=1}^{N} [v_{Pj}(j)\, d\tau_j] \tag{4-10a}$$

Note that for the final wave function we permute the electrons, and for the initial wave function we permute the states. Observe that F must be symmetric in the coordinates of all the electrons, since these electrons are identical. To simplify (4-10a) it is convenient to group the terms referring to the same electron coordinates together. We therefore set j = Qi in the product over j; this does not change anything, since Qi will run over all values 1 to N when i does. Therefore,

$$\langle F\rangle = \frac{1}{N!} \int \sum_Q \sum_P \epsilon_Q \epsilon_P \prod_{i=1}^{N} u_i^*(Qi)\, F v_{PQi}(Qi)\, d\tau_{Qi} \tag{4-10b}$$

We note that $\epsilon_Q \epsilon_P = \epsilon_{PQ}$. Now Qi is only a dummy variable of integration; therefore the integral (for each given P and Q) is not changed if we change the label Qi into i. (Here the symmetry of F in all electron coordinates is important.) Also we can sum over all permutations PQ, for given Q, and thus cover all permutations P. Then

$$\langle F \rangle = \frac{1}{N!} \sum_{Q} \sum_{PQ} \epsilon_{PQ} \int \prod_{i=1}^{N} u_i^*(i) F v_{PQi}(i)\, d\tau_i \qquad (4\text{-}10c)$$

Now the integral (and the coefficient ϵ_{PQ}) is entirely independent of Q, for each PQ; therefore each Q gives the same contribution, and the total is N! times the contribution of the simplest permutation, the identity. This eliminates the normalization factor $N!^{-1}$. Replacing now again the label PQ by P we obtain

$$\langle F \rangle = \sum_{P} \epsilon_P \int \prod_{i=1}^{N} u_i^*(i) F v_{Pi}(i)\, d\tau_i \qquad (4\text{-}11)$$

We now consider particular forms of F.

1. F = 1. Owing to the orthogonality of the one-electron wave functions, we obtain $\langle F \rangle = 0$, unless for some one P, $v_{Pi} = u_i$ for *all* i. There can be at most one such P, since (4-7) holds. We shall assume here, and in everything that follows, that the ordering of the determinant Ψ_a is such that the v's that are identical with the u's are arranged in the same order. Then P is the identity permutation, and

$$\langle F \rangle = 1 \qquad (4\text{-}12)$$

2. $F = \Sigma_{j=1}^{N} f_j$, where f_j is a one-electron operator, operating on electron j. If $u_i \neq v_i$ for more than one i, we get $\langle F \rangle = 0$. If $u_i \neq v_i$ for some i, but $u_j = v_j$ for all j except j = i, we get

$$\langle F \rangle = \langle i | f | i \rangle = \int u_i^*(1) f_1 v_i(1)\, d\tau_1 \qquad (4\text{-}13)$$

If $u_i = v_i$ for all i,

$$\langle F \rangle = \sum_i \langle i | f | i \rangle \qquad (4\text{-}14)$$

Only the identity permutation P = I contributes to (4-13) and (4-14).

3. $F = \Sigma_{i<j} g_{ij}$. The summation extends over all different pairs $i \neq j$ and g_{ij} is an operator operating on electrons i and j. If $u_i \neq v_i$ for more than two i's we get $\langle F \rangle = 0$. Suppose now that $u_i = v_i$ for all i. Then

$$\langle F \rangle = \sum_{i<j} [\langle ij | g | ij \rangle - \langle ij | g | ji \rangle] \qquad (4\text{-}15)$$

where the sum extends over all pairs, and

$$\langle ij|g|k\ell\rangle \equiv \int u_i^*(1)u_j^*(2)g_{12}v_k(1)v_\ell(2)\,d\tau_1\,d\tau_2 \tag{4-16}$$

The first term in (4-15) comes from $P = I$, the second from $P = P_{ij}$, i.e., the interchange of electrons i and j; this is, of course, an odd permutation, $\epsilon_P = -1$. In (4-16), the integration is over space and spin coordinates, and 1 and 2 are dummy variables. If for some i, $u_i \neq v_i$, but for all j except $j = i$, we have $u_j = v_j$, then

$$\langle F\rangle = \sum_{j\neq i}[\langle ij|g|ij\rangle - \langle ij|g|ji\rangle] \tag{4-17}$$

If for some i and j, $u_i \neq v_i$ and $u_j \neq v_j$, but for all k except $k = i$, $k = j$, we have $u_k = v_k$, then

$$\langle F\rangle = \langle ij|g|ij\rangle - \langle ij|g|ji\rangle \tag{4-18}$$

DERIVATION OF THE HARTREE-FOCK EQUATIONS

We return now to the specific problem of evaluating $\int\Psi^* H\Psi\,d\tau$. Here $\Psi_a = \Psi_b = \Psi$. Hence $u_i = v_i$ for all i. We write $H = F_1 + F_2$, where

$$F_1 = \sum_i f_i \qquad f_i = -\tfrac{1}{2}\nabla_i^2 - \frac{Z}{r_i}$$

$$F_2 = \sum_{i<j} g_{ij} \qquad g_{ij} = \frac{1}{r_{ij}} \tag{4-19}$$

Then from (4-14) and (4-15)

$$\langle F_1\rangle = \sum_i \langle i|f|i\rangle \tag{4-20a}$$

$$\langle F_2\rangle = \sum_{i<j}[\langle ij|g|ij\rangle - \langle ij|g|ji\rangle] \tag{4-20b}$$

We shall write (4-20b), indicating explicitly the space integrations and spin summations.

$$\langle F_2\rangle = \sum_{i<j}\Bigg[\sum_{\sigma_1,\sigma_2}\int d\tau_1\, d\tau_2\, u_i^*(\mathbf{r}_1)u_j^*(\mathbf{r}_2)g_{12}u_i(\mathbf{r}_1)u_j(\mathbf{r}_2)$$

$$\times\, |\chi_i(\sigma_1)|^2\,|\chi_j(\sigma_2)|^2$$

$$-\sum_{\sigma_1,\sigma_2}\int d\tau_1\, d\tau_2\, u_i^*(\mathbf{r}_1)u_j^*(\mathbf{r}_2)g_{12}u_i(\mathbf{r}_2)u_j(\mathbf{r}_1)$$

$$\times\, \chi_i^*(\sigma_1)\chi_j^*(\sigma_2)\chi_i(\sigma_2)\chi_j(\sigma_1)\Bigg] \qquad (4\text{-}21)$$

[We have taken $u_i(j) = u_i(\mathbf{r}_j)\chi_i(\sigma_j)$.] Recalling that

$$\sum_\sigma \chi_i^*(\sigma)\chi_j(\sigma) = \delta(m_{si}, m_{sj})$$

we obtain

$$\langle H\rangle = \sum_i \int d\tau\, u_i^*(\mathbf{r})\left(-\tfrac{1}{2}\nabla^2 - \frac{Z}{r}\right)u_i(\mathbf{r})$$

$$+\sum_{i<j}\Bigg[\int d\tau_1\, d\tau_2\, \frac{1}{r_{12}}|u_i(\mathbf{r}_1)|^2\,|u_j(\mathbf{r}_2)|^2$$

$$-\,\delta(m_{si}, m_{sj})\int d\tau_1\, d\tau_2\, \frac{1}{r_{12}}u_i^*(\mathbf{r}_1)u_j^*(\mathbf{r}_2)u_j(\mathbf{r}_1)u_i(\mathbf{r}_2)\Bigg] \qquad (4\text{-}22)$$

The first term in the second sum is called the direct term, the second the exchange term. We notice that the exchange term is 0, unless the spins are the same for the two states of each pair. (This is another example of the fact that exchange effects do not exist for identical particles if their state functions are nonoverlapping; see p. 27).

According to (4-8) and (4-7) we require

$$\delta\langle H\rangle = 0 \qquad (4\text{-}7)$$

with the subsidiary orthonormality conditions for orbitals with the same spin

$$\int u_i^*(\mathbf{r})u_j(\mathbf{r})\, d\mathbf{r} = \delta_{ij} \qquad (4\text{-}8)$$

The subsidiary conditions normally are implemented by the method of Lagrange multipliers; i.e. one demands

$$\delta(\langle H\rangle + \sum_{i<j} \delta(m_{s_i}, m_{s_j})[\lambda_{ij}\int u_i^* u_j d\tau + \lambda_{ij}^* \int u_j^* u_i d\tau]) = 0 \tag{4-23}$$

(The Lagrange multiplier matrix $\{\lambda_{ij}\}$ may be chosen to be Hermitian.) However, one can in fact proceed in a simpler way. Only the normalization condition is imposed,

$$\int u_i^* u_i d\tau = 1 \tag{4-24}$$

i.e., a diagonal Lagrange multiplier matrix is used in the variational principle

$$\delta(\langle H\rangle - \sum_i \varepsilon_i \int u_i^* u_i d\tau) = 0 \tag{4-25}$$

That the resulting equations imply orthogonality of the orbitals is verified a posteriori. (This is analogous to the variational derivation of the static Schrödinger equation: one needs to impose only the normalization condition, as in (1-19); orthogonality of wave functions then follows. The possibility of using (4-25) instead of (4-23) is related to the Hermiticity of the Lagrange multiplier matrix $\{\lambda_{ij}\}$, which may be diagonalized by a unitary transformation; see Problem 2.)

We now vary in (4-25) u_i and u_i^* arbitrarily. Making use of the Hermiticity and symmetry of the various operators, we arrive at

$$\int \delta u_i^*(\mathbf{r}_1)\{-\varepsilon_i u_i(\mathbf{r}_1) + f_1 u_i(\mathbf{r}_1) + \sum_j \int u_j^*(\mathbf{r}_2) g_{12} [u_i(\mathbf{r}_1) u_j(\mathbf{r}_2)$$

$$- \delta(m_{si}, m_{sj}) u_i(\mathbf{r}_2) u_j(\mathbf{r}_1)] d\tau_2\} d\tau_1 \tag{4-26}$$

$$+ \text{complex conjugate} = 0$$

Since the variations δu_i^* and δu_i are arbitrary (4-26) requires

$$-\tfrac{1}{2}\nabla_1^2 u_i(\mathbf{r}_1) - \frac{Z}{r_1} u_i(\mathbf{r}_1) + \left[\sum_j \int d\tau_2 \frac{1}{r_{12}} |u_j(\mathbf{r}_2)|^2\right] u_i(\mathbf{r}_1)$$

$$-\sum_j \delta(m_{si}, m_{sj}) \left[\int d\tau_2 \frac{1}{r_{12}} u_j^*(\mathbf{r}_2) u_i(\mathbf{r}_2)\right] u_j(\mathbf{r}_1)$$

$$= \epsilon_i u_i(\mathbf{r}_1) \tag{4-27}$$

This is called the *Hartree-Fock equation*. Orthogonality of different orbitals will be established shortly.

The Hartree-Fock equation differs from the Hartree equation, (4-1) and (4-2), by

$$\int d\tau_2 \frac{1}{r_{12}} |u_i(\mathbf{r}_2)|^2 u_i(\mathbf{r}_1)$$

$$- \sum_j \delta(m_{si}, m_{sj}) \left[\int d\tau_2 \frac{1}{r_{12}} u_j^*(\mathbf{r}_2) u_i(\mathbf{r}_2) \right] u_j(\mathbf{r}_1)$$

$$= -\sum_{j \neq i} \delta(m_{si}, m_{sj}) \left[\int d\tau_2 \frac{1}{r_{12}} u_j^*(\mathbf{r}_2) u_i(\mathbf{r}_2) \right] u_j(\mathbf{r}_1) \qquad (4\text{-}28)$$

The right-hand side is the *exchange integral*. The expression (4-28) arises from the fact that we used determinantal trial functions [see (4-15)]. Had we used a product trial wave function this term would be missing and we would have obtained Hartree's equation.[1] We shall discuss the physical significance of the exchange term below.

DISCUSSION OF THE EXCHANGE TERM

That part of the Hartree-Fock equation which coincides with the Hartree equation has the same physical significance; i.e., it represents an electron moving in an equivalent potential due to the other electrons and to the nucleus.

The exchange term, which we write as

$$-\int \sum_j u_j^*(\mathbf{r}_2) \frac{1}{r_{12}} u_j(\mathbf{r}_1) u_i(\mathbf{r}_2)\, d\tau_2 \qquad (4\text{-}29)$$

is an example of a *nonlocal potential*. [We have suppressed the factor $\delta(m_{si}, m_{sj})$ for convenience. This means that all subsequent sums run over orbitals corresponding to the same spin.]

The time-independent Schrödinger equation for a particle moving in a local potential $V(\mathbf{r})$ and a nonlocal potential $U(\mathbf{r}, \mathbf{r}')$ has, by definition, the form (in atomic units)

---

[1] In order to obtain Hartree's equation variationally, it is also necessary to relax the orthogonality condition on the orbitals, but unlike the Hartree-Fock equation, the Hartree equation (4-2) does not imply orthogonality of the orbitals; see Problem 3.

$$-\tfrac{1}{2}\nabla^2\psi(\mathbf{r}) + V(\mathbf{r})\psi(\mathbf{r}) + \int U(\mathbf{r}, \mathbf{r}')\psi(\mathbf{r}')\, d\tau' = E\psi(\mathbf{r}) \tag{4-30}$$

Any local potential V(r) can be considered as a special case of a nonlocal potential by setting $U(r, r') = V(r)\delta(r-r')$. The general properties of solutions to (4-30) are as follows.

The Laplacian is no longer necessarily zero when ψ vanishes. Hence the inflection points of ψ need not coincide with its zeroes.

To assure reality of the eigenvalue, U must satisfy the Hermiticity requirement

$$U(\mathbf{r}_1, \mathbf{r}_2) = U^*(\mathbf{r}_2, \mathbf{r}_1) \tag{4-31}$$

Solutions of (4-30) belonging to different E_i are orthogonal. To see this we proceed with the usual proof; i.e., we multiply the equation for ψ_i by ψ_k^*, the equation for ψ_k^* by ψ_i, and subtract, obtaining

$$(E_i - E_k^*) \int \psi_k^*\psi_i\, d\tau = \int [\psi_k^*(\mathbf{r}_1) U(\mathbf{r}_1, \mathbf{r}_2)\psi_i(\mathbf{r}_2) - \psi_i(\mathbf{r}_1) U^*(\mathbf{r}_1, \mathbf{r}_2)\psi_k^*(\mathbf{r}_2)]\, d\tau_1\, d\tau_2 \tag{4-32}$$

Using the Hermiticity of U and relabeling dummy variables in the second term of the right-hand integral we obtain the desired result.

The Hartree-Fock equation (4-27) has the form

$$-\tfrac{1}{2}\nabla^2 u_i(r) + V(r)u_i(r) + \int U(r,r')u_i(r')\, d\tau' = \varepsilon_i U_i(r) \tag{4-33}$$

with the local potential

$$V(\mathbf{r}_1) = -\frac{Z}{\mathbf{r}_1} + \sum_{j=1}^{N} \int |u_j(\mathbf{r}_2)|^2 \frac{1}{\mathbf{r}_{12}}\, d\tau_2 \tag{4-34a}$$

and the nonlocal potential

$$U(\mathbf{r}_1, \mathbf{r}_2) = -\frac{1}{\mathbf{r}_{12}} \sum_{j=i}^{N} u_j^*(\mathbf{r}_2) u_j(\mathbf{r}_1) = U^*(\mathbf{r}_2, \mathbf{r}_1) \tag{4-34b}$$

We have indicated explicitly that the sum extends over the occupied orbitals (of a given spin). Thus (4-33) has the form of a Schrödinger equation with a nonlocal potential. We may immediately conclude from our previous discussion that the ϵ_i are real and solutions of (4-33) belonging to different ϵ_i are orthogonal. This indicates that our truncated method of Lagrange multipliers was successful.

For further discussion, it is useful to define a quantity called the *density matrix* $\rho(\mathbf{r}_1,\mathbf{r}_2)$,

$$\rho(\mathbf{r}_1,\mathbf{r}_2) = \sum_{i=1}^{N} u_i^*(\mathbf{r}_2)u_i(\mathbf{r}_1) \tag{4-35}$$

If all orbitals were occupied, we would have by closure

$$\rho(\mathbf{r}_1,\mathbf{r}_2) = \sum_{i=1}^{\infty} u_i^*(\mathbf{r}_2)u_i(\mathbf{r}_1) = \delta(\mathbf{r}_2-\mathbf{r}_1)$$

and

$$\int\rho(\mathbf{r}_1,\mathbf{r}_2)d\tau_2 = 1 \tag{4-36}$$

Even with a physical atom, where not all states are occupied, (4-36) is approximately true, as we shall show in the next chapter. In terms of the density matrix, the nonlocal potential is

$$U(\mathbf{r}_1,\mathbf{r}_2) = -\frac{1}{r_{12}}\rho(\mathbf{r}_1,\mathbf{r}_2) \tag{4-37}$$

It is interesting to consider an *average exchange potential*, which is defined by

$$\overline{V}(\mathbf{r}_1) = \int U(\mathbf{r}_1,\mathbf{r}_2)d\tau_2 \tag{4-38}$$

This may be interpreted as the potential at $\mathbf{r}_1$ due to a charge density $-\rho(\mathbf{r}_1,\mathbf{r}_2)$. Since $\int\rho(\mathbf{r}_1,\mathbf{r}_2)d\tau_2 = 1$, $\overline{V}$ is the potential at $\mathbf{r}_1$ arising from the absence of one electron. Thus the total potential energy in the Hartree-Fock theory arises from the nuclear interaction, from the interaction with all electrons of spin opposite to that of the electron considered, and from an interaction with a charge distribution of electrons of the same spin as the electron considered. This charge distribution of electrons of the same spin adds up to one less than the total number of electrons in this spin state. It is as if the electron under consideration carried a hole with it. This so-called *Fermi hole* is a result of the Pauli principle (antisymmetry of the wave function), which keeps electrons of the same spin separated. The potential energy is lower in the Hartree-Fock model than in the Hartree model, since in the former the other electrons stay further away from the electron in question than in the latter.

Finally we define an *effective potential*, which will be useful below in an approximate treatment of the exchange term.

$$V_{eff}(r_1)u_i(r_1) = \int U(r_1,r_2)u_i(r_2)d\tau_2 \qquad (4\text{-}39)$$

The effective potential depends on i, viz., on the orbital considered. Both in the expectation value of the Hamiltonian (4-22) and in the Hartree-Fock equation (4-27), the contribution of the non-local exchange term may be written in terms of V_{eff}.

PHYSICAL SIGNIFICANCE OF THE EIGENVALUE

We now inquire into the physical significance of the ϵ_i occurring in the Hartree-Fock equations. We multiply (4-27) by $u_i^*(\mathbf{r})$ and integrate obtaining

$$\epsilon_i = \langle i|f|i\rangle + \sum_j [\langle ij|g|ij\rangle - \langle ij|g|ji\rangle] \qquad (4\text{-}40)$$

This is the expectation value of that part of the energy which involves electron i. Furthermore, from (4-22) we see that

$$E = \sum_i \epsilon_i - \sum_{i<j} [\langle ij|g|ij\rangle - \langle ij|g|ji\rangle] \qquad (4\text{-}41)$$

Let us consider the removal energy of electron i, which equals the energy of the ion minus the energy of the atom. If we assume that the orbitals for the ion are the same as for the atom, this energy difference is just the expectation value of those terms in the Hamiltonian which involve the removed electron i. Then, by (4-40) we obtain $-\epsilon_i$ for the removal energy. This result is known as *Koopmans' theorem*.

If the ion wave function when built up out of atomic orbitals is in error by δ, the ion energy is in error by an amount of order δ^2, since the variation method gives a stationary value for the energy. For many-electron atoms δ will be quite small. Even for helium the error committed in the energy is less than 0.1 Ry.

One should realize that the removal energy so calculated is no longer an upper bound for the exact removal energy. This is because we have taken the difference of two upper bounds. This difference is of course still a good approximation.

SPHERICAL SYMMETRY AND THE HARTREE-FOCK EQUATIONS

We shall prove that atoms with closed shells lead to a spherically symmetric theory. By a closed shell is meant that all the $4\ell + 2$ states corresponding to a given $n\ell$ are occupied. We prove this by assuming a solution of the form $[\mathcal{R}_{n\ell}(r)/r]Y_{\ell m}(\Omega)$. This will determine the equivalent potential, which we shall find to be spherically symmetric. This shows the self-consistency of the claim that for closed shells the effective potential is spherically symmetric. We write

$$u_i(\mathbf{r}) = \frac{\mathcal{R}_{n\ell}(r)}{r} Y_{\ell m}(\Omega) \qquad u_j(\mathbf{r}) = \frac{\mathcal{R}_{n'\ell'}(r)}{r} Y_{\ell' m'}(\Omega) \tag{4-42}$$

Then, using the addition theorem of the spherical harmonics,

$$\sum_{m=-\ell}^{\ell} |Y_{\ell m}(\theta, \varphi)|^2 = \frac{2\ell + 1}{4\pi}$$

we get

$$\sum_{m_s} \sum_{m'=-\ell'}^{\ell'} |u_j(\mathbf{r})|^2 = \frac{\mathcal{R}^2_{n'\ell'}(r)}{r^2} \frac{2(2\ell' + 1)}{4\pi} \tag{4-43}$$

$$\begin{aligned} V_{\text{Coulomb}} &= \sum_j \int |u_j(\mathbf{r}_2)|^2 \frac{1}{r_{12}} d\tau_2 \\ &= \sum_{n'\ell'} \int \frac{\mathcal{R}^2_{n'\ell'}(r_2)}{r_2^2} \frac{2(2\ell' + 1)}{4\pi} \frac{1}{r_{12}} d\tau_2 \\ &= \sum_{n'\ell'} \int_0^\infty \mathcal{R}^2_{n'\ell'}(r_2)\, 2(2\ell' + 1) \frac{1}{r_>} dr_2 \end{aligned} \tag{4-44}$$

where the factor 2 comes from the fact that we have two spin orientations. This shows that the Coulomb potential is spherically symmetric.

For the exchange term

$$\sum_{m'=-\ell'}^{\ell'} u_j^*(\mathbf{r}_2)\, u_j(\mathbf{r}_1) = \frac{\mathcal{R}_{n'\ell'}(r_2)\,\mathcal{R}_{n'\ell'}(r_1)}{r_2 r_1}$$

$$\times \sum_{m'=-\ell'}^{\ell'} Y^*_{\ell' m'}(\Omega_2)\, Y_{\ell' m'}(\Omega_1)$$

$$= \frac{\mathcal{R}_{n'\ell'}(r_2)\,\mathcal{R}_{n'\ell'}(r_1)}{r_2 r_1}\,\frac{2\ell'+1}{4\pi}\,P_{\ell'}(\cos\theta_{12}) \tag{4-45}$$

Here there is no factor 2 because the exchange term is summed only over the electrons with one spin. Then

$$\int \sum_{m'} \frac{u_j^*(\mathbf{r}_2)\, u_j(\mathbf{r}_1)\, u_i(\mathbf{r}_2)}{r_{12}}\, d\tau_2$$

$$= \int_0^\infty \frac{\mathcal{R}_{n'\ell'}(r_2)\,\mathcal{R}_{n'\ell'}(r_1)\,\mathcal{R}_{n\ell}(r_2)}{r_1}\, dr_2$$

$$\times \int \frac{d\Omega_2}{r_{12}}\,\frac{2\ell'+1}{4\pi}\,P_{\ell'}(\cos\theta_{12})\, Y_{\ell m}(\Omega_2) \tag{4-46}$$

We first consider the integral over Ω_2. We expand $1/r_{12}$ in terms of $P_k(\cos\theta_{12})$. The terms $P_k(\cos\theta_{12})\,P_{\ell'}(\cos\theta_{12})$ are polynomials of $\cos\theta_{12}$ and therefore can be expanded in $P_\lambda(\cos\theta_{12})$

$$P_k(\cos\theta_{12})\,P_{\ell'}(\cos\theta_{12})$$

$$= \sum_{\lambda=|\ell'-k|}^{\lambda=\ell'+k} \sqrt{\frac{(2\lambda+1)}{(2\ell'+1)}}\; c^k(\ell' 0, \lambda 0)\, P_\lambda(\cos\theta_{12}) \tag{4-47a}$$

$$c^k(\ell' 0, \lambda 0)$$

$$= \tfrac{1}{2}\sqrt{(2\ell'+1)(2\lambda+1)} \int_{-1}^{1} P_\lambda(w)\, P_k(w)\, P_{\ell'}(w)\, dw \tag{4-47b}$$

The expansion coefficients c^k in (4-47) are special cases of[2]

$$c^k(\ell m, \ell' m') = \sqrt{\frac{4\pi}{2k+1}} \int d\Omega$$

$$\times\, Y^*_{\ell m}(\Omega) Y_{\ell' m'}(\Omega) Y_{k,\, m-m'}(\Omega) \qquad (4\text{-}48)$$

It is seen that they are nonzero only when

$$|\ell - \ell'| \le k \le \ell + \ell'$$

$$\ell + \ell' + k = \text{even integer} \qquad (4\text{-}49)$$

Interesting particular cases are

$$c^0(\ell m, \ell' m') = \sqrt{4\pi} \int Y^*_{\ell m}(\Omega) Y_{\ell' m'}(\Omega) Y_{00}(\Omega)\, d\Omega$$

$$= \delta_{\ell\ell'}\, \delta_{mm'} \qquad (4\text{-}50a)$$

$$\sum_{m=-\ell}^{\ell} c^k(\ell m, \ell m) = \sum_m \sqrt{\frac{4\pi}{2k+1}} \int |Y_{\ell m}(\Omega)|^2 Y_{k0}(\Omega)\, d\Omega$$

$$= \sqrt{\frac{4\pi}{2k+1}}\, \frac{2\ell+1}{4\pi} \int Y_{k0}(\Omega)\, d\Omega$$

$$= (2\ell + 1)\, \delta_{k0} \qquad (4\text{-}50b)$$

---

[2] The c^k coefficients can be expressed in terms of Clebsch-Gordan coefficients C which occur in the addition of angular momentum. (See Chapter 6 for definition of the Clebsch-Gordan coefficients.) The connection is given by

$$c^k(\ell m, \ell' m') = (-1)^{m'} \sqrt{\frac{(2\ell+1)(2\ell'+1)}{(2k+1)^2}}$$

$$\times\, C(\ell\, -m\ell'\, m', k\, m'-m)\, C(\ell 0 \ell' 0, k0)$$

Therefore

$$\int_{-1}^{1} P_\ell(w) P_{\ell'}(w) P_k(w)\, dw = \frac{2}{2k+1} C^2(\ell 0 \ell' 0, k0)$$

$$\int d\Omega\, Y_{\ell m}(\Omega)\, Y_{\ell' m'}(\Omega)\, Y_{k,\, -m-m'}(\Omega)$$

$$= \frac{(2\ell+1)(2\ell'+1)}{4\pi(2k+1)}\, C(\ell m \ell' m', k\, m+m')\, C(\ell 0 \ell' 0, k0)$$

Returning to (4-47a), we now expand

$$P_\lambda(\cos\theta_{12}) = \frac{4\pi}{2\lambda+1} \sum_{\mu=-\lambda}^{\lambda} Y^*_{\lambda\mu}(\Omega_2) Y_{\lambda\mu}(\Omega_1) \tag{4-51}$$

When we integrate (4-46) over Ω_2, only $\ell = \lambda$, $m = \mu$ will give a nonvanishing result. Collecting terms, the angular integral in (4-46) becomes

$$\sqrt{\frac{2\ell'+1}{2\ell+1}} \sum_k \frac{r_<^k}{r_>^{k+1}} c^k(\ell'0, \ell 0) Y_{\ell m}(\Omega_1) \tag{4-52}$$

We see that the angular dependence of the exchange term is such that it becomes equivalent to a central potential. Substituting these results into the Hartree-Fock equation, we obtain for the radial wave equation,

$$\begin{aligned}
\mathcal{R}''_{n\ell}(r_1) &- \frac{\ell(\ell+1)}{r_1^2}\mathcal{R}_{n\ell}(r_1) + (2\epsilon_{n\ell} - 2V_C)\mathcal{R}_{n\ell}(r_1) \\
&= -2\sum_{n'\ell'}\sum_k \sqrt{\frac{2\ell'+1}{2\ell+1}}\, c^k(\ell'0, \ell 0) \\
&\quad \times \int_0^\infty \mathcal{R}_{n'\ell'}(r_2)\mathcal{R}_{n\ell}(r_2) \frac{r_<^k}{r_>^{k+1}}\, dr_2\, \mathcal{R}_{n'\ell'}(r_1)
\end{aligned}$$

$$V_C = -\frac{Z}{r_1} + \sum_{n'\ell'} 2(2\ell'+1) \int_0^\infty \mathcal{R}^2_{n'\ell'}(r_2)\frac{dr_2}{r_>}$$

$$|\ell - \ell'| \le k \le |\ell + \ell'| \tag{4-53}$$

For closed shells the *central field approximation* is not an approximation but is exact within the framework of the Hartree-Fock scheme. For noncomplete shells it is justified by noticing that mostly there is only one incomplete shell. Also, in the lowest energy state of the atom, electrons tend to go into a shell with their spins oriented in one direction as long as this is permitted by the Pauli principle (see Chapter 7). Thus even a half-complete shell leads to a spherically symmetric equivalent potential. The most unfavorable case is then when we have ℓ electrons in a shell. The shells occurring in actual atoms are $\ell = 1, 2, 3$. For $\ell = 1$, and one electron in the shell, the potential acting on this electron is obviously spherically symmetric, because the electron does not act on itself (Coulomb and exchange term cancel). For $\ell = 2$ and two electrons, only the one interaction between these two needs to be considered, so the assumption of spherical symmetry should still be very good; and even for $\ell = 3$ it should be acceptable, especially because $\ell = 3$ occurs only for very large Z. Thus the central field assumption is only a very mild approximation for atoms. (It is different for nuclei where the values of ℓ are much larger, and nuclear spins tend to be opposite.)

The *central field approximation* for an atom with one incomplete shell $n''\ell''$ with $N_{n''\ell''}$ electrons in this incomplete shell proceeds as follows. (We assume only one incomplete shell for simplicity.) The orbitals are taken to be of the spherical form (4-42). The sum over orbitals j in the Coulomb and exchange terms of the Hartree-Fock equation is split into a sum over the complete shells $n'\ell'$ and into a sum over the incomplete shell $n''\ell''$. The complete shells sum up into a spherically symmetric result. The incomplete shell gives the contribution

$$\sum_{m''m''_s}{}' \left(\int d\tau_2 \frac{1}{r_{12}} |u_{n''\ell''m''}(\mathbf{r}_2)|^2 u_{n\ell m}(\mathbf{r}_1) - \delta(m''_s, m_s) \right.$$
$$\left. \times \int d\tau_2 \frac{1}{r_{12}} u^*_{n''\ell''m''}(\mathbf{r}_2) u_{n\ell m}(\mathbf{r}_2) u_{n''\ell''m''}(\mathbf{r}_1) \right) \qquad (4\text{-}54)$$

The prime on the summation sign indicates that the sum extends only over those values of m″ and m''_s which are present in the incomplete shell under consideration.

To obtain a central field approximation the above expression must be averaged. We distinguish two cases. First, let us assume that $n\ell$ refers to one of the closed shells; i.e., that the particular Hartree-Fock equation we are considering determines an orbital in one of the closed shells. Then (4-54) is replaced by an average which is obtained by extending the sum over the entire shell $n''\ell''$ and multiplying by $N_{n''\ell''}(4\ell'' + 2)^{-1}$. For $n\ell \neq n'\ell'$, this gives

$$N_{n''\ell''}\left[\int_0^\infty \mathcal{R}^2_{n''\ell''}(r_2)\frac{dr_2}{r_>}\mathcal{R}_{n\ell}(r_1)\frac{1}{r_1}Y_{\ell m}(\Omega_1)\right.$$
$$-\frac{1}{2\sqrt{(2\ell+1)(2\ell''+1)}}\sum_k c^k(\ell 0, \ell'' 0)$$
$$\left.\times \int_0^\infty \mathcal{R}_{n''\ell''}(r_2)\mathcal{R}_{n\ell}(r_2)\frac{r_<^k}{r_>^{k+1}}\, dr_2\, \mathcal{R}_{n''\ell''}(r_1)\frac{1}{r_1}Y_{\ell m}(\Omega_1)\right] \qquad (4\text{-}55)$$

Clearly, this averaging over m'' and m''_s has accomplished the desired result of making the interaction central.

For the second case, when $n\ell$ refers to the incomplete shell, (4-54) is replaced by an average which is obtained by extending the sum over the entire shell and multiplying by $(N_{n''\ell''} - 1)(4\ell'' + 1)^{-1}$. This gives ($n\ell = n''\ell''$)

$$\frac{N_{n''\ell''} - 1}{4\ell'' + 1}\left[(4\ell'' + 2)\int_0^\infty \mathcal{R}^2_{n''\ell''}(r_2)\frac{dr_2}{r_>} - \sum_{k=0}^{2\ell''} c^k(\ell''0, \ell''0)\right.$$
$$\left.\times \int_0^\infty \mathcal{R}^2_{n''\ell''}(r_2)\frac{r_<^k}{r_>^{k+1}}\,dr_2\right]\mathcal{R}_{n''\ell''}(r_1)\frac{1}{r_1}Y_{\ell''m}(\Omega_1)$$
$$= (N_{n''\ell''} - 1)\left[\int_0^\infty \mathcal{R}^2_{n''\ell''}(r_2)\frac{dr_2}{r_>} - \frac{1}{4\ell'' + 1}\sum_{k=2}^{2\ell''} c^k(\ell''0, \ell''0)\right.$$
$$\left.\times \int_0^\infty \mathcal{R}^2_{n''\ell''}(r_2)\frac{r_<^k}{r_>^{k+1}}\,dr_2\right]\mathcal{R}_{n''\ell''}(r_1)\frac{1}{r_1}Y_{\ell''m}(\Omega_1) \qquad (4\text{-}56)$$

In arriving at the second expression in (4-56) we used the fact that $c^0(\ell''0, \ell''0) = 1$.

The reason for the difference in the weighting factors between (4-55) and (4-56) is that in the latter case, there are only $N_{n''\ell''} - 1$ nonzero terms in the sum over the incomplete shell, and $4\ell'' + 1$ nonzero terms in the sum over the complete shell. Physically, this reflects the fact that the particle cannot interact with itself. This point will be clarified in a late section of Chapter 7, entitled "Average Energy."

In summary, the central field approximation for the Hartree-Fock[3] equations is then for $n\ell \neq n''\ell''$

---

[3]The central field approximation may be arrived at also in another fashion. Instead of approximating the Hartree-Fock equations (4-27), we may take the orbitals to be of the spherical form (4-42) and perform the averaging over spin directions and angles already in the expectation value of the Hamiltonian (4-22). We then proceed with the variation of the Hamiltonian together with the subsidiary normalization condition as in (4-33). The quantity which is varied is now only the radial part of the wave function. This results in

$$\mathcal{R}''_{n\ell}(r_1) - \frac{\ell(\ell+1)}{r_1^2}\mathcal{R}_{n\ell}(r_1) + (2\epsilon_{n\ell} - 2V_C)\mathcal{R}_{n\ell}(r_1)$$

$$= -\sum_{n'\ell'} \frac{N_{n'\ell'}}{\sqrt{(2\ell'+1)(2\ell+1)}} \sum_{k=|\ell-\ell'|}^{\ell+\ell'} c^k(\ell'0, \ell 0)$$

$$\times \int_0^\infty \mathcal{R}_{n'\ell'}(r_2)\mathcal{R}_{n\ell}(r_2)\frac{r_<^k}{r_>^{k+1}}\,dr_2\,\mathcal{R}_{n'\ell'}(r_1) \qquad \text{(4-57a)}$$

$$V_C = -\frac{Z}{r_1} + \sum_{n'\ell'} N_{n'\ell'}\int_0^\infty \mathcal{R}^2_{n'\ell'}(r_2)\frac{dr_2}{r_>} \qquad \text{(4-57b)}$$

The sum in $n'\ell'$ extends over all shells; $N_{n'\ell'}$ is the number of electrons in the shell $n'\ell'$ and equals $4\ell' + 2$ for a closed shell and $N_{n''\ell''}$ for the incomplete shell. For $n\ell = n''\ell''$ the Hartree-Fock equation in the central field approximation is

$$\mathcal{R}''_{n''\ell''}(r_1) - \frac{\ell''(\ell''+1)}{r_1^2}\mathcal{R}_{n''\ell''}(r_1) + (2\epsilon_{n''\ell''} - 2V_C)\mathcal{R}_{n''\ell''}(r_1)$$

$$= -\sum_{\substack{n'\ell' \\ \text{complete} \\ \text{shells}}} 2\sqrt{\frac{2\ell'+1}{2\ell''+1}} \sum_{k=|\ell''-\ell'|}^{\ell''+\ell'} c^k(\ell'0, \ell''0)$$

$$\times \int_0^\infty \mathcal{R}_{n'\ell'}(r_2)\mathcal{R}_{n''\ell''}(r_2)\frac{r_<^k}{r_>^{k+1}}\,dr_2\,\mathcal{R}_{n'\ell'}(r_1)$$

$$-\frac{2(N_{n''\ell''}-1)}{4\ell''+1}\sum_{k=2}^{2\ell''} c^k(\ell''0, \ell''0)\int_0^\infty \mathcal{R}^2_{n''\ell''}(r_2)$$

(4-57) and (4-58), which do not imply orthogonality of wave functions. On the other hand if one begins with the Lagrange multipliers which insure orthonormality, as in (4-23), it is impossible to diagonalize the matrix $\{\lambda_{ij}\}$ when the central field approximation is made. Thus the procedure described in Problem 2 cannot be carried out, and additional terms are present in the equations. For details, see J. C. Slater, *Quantum Theory of Atomic Structure*, McGraw-Hill, New York, 1960; Vol. II pp. 23-30.

$$\times \frac{r_<^k}{r_>^{k+1}}\, dr_2\, \mathcal{R}_{n''\ell''}(r_1) \quad (4\text{-}58a)$$

$$V_C = -\frac{Z}{r_1} + \sum_{\substack{n'\ell' \\ \text{complete} \\ \text{shells}}} 2(2\ell' + 1) \int_0^\infty \mathcal{R}^2_{n'\ell'}(r_2)\, \frac{dr_2}{r_>}$$

$$+ (N_{n''\ell''} - 1) \int_0^\infty \mathcal{R}^2_{n''\ell''}(r_2)\, \frac{dr_2}{r_>} \quad (4\text{-}58b)$$

It should be noted that orthogonality of wave functions is lost in the central field approximation.[3]

APPROXIMATE TREATMENT OF EXCHANGE TERM

We can obtain an estimate for the exchange term in the Hartree-Fock theory by using electron wave functions given by the Thomas-Fermi model of the atom. The principal results of this model, which we discuss in detail in Chapter 5, are the following. We assume the electrons move in a constant potential and hence the wave functions are plane waves. All momenta $k \leq k_F$ are present where $\hbar k_F$ is the Fermi momentum,

$$k_F = (3\pi^2 \rho)^{1/3} \quad (4\text{-}59)$$

in which ρ is the density of electrons at the point considered. We then find that[4]

$$\rho(\mathbf{r}_1, \mathbf{r}_2) = \frac{1}{2\pi^2 r^3} (\sin k_F r - k_F r \cos k_F r)$$

$$r = r_{12} = |\mathbf{r}_1 - \mathbf{r}_2| \quad (4\text{-}60)$$

$$\int \rho(\mathbf{r}_1, \mathbf{r}_2)\, d\tau_2 = \frac{4\pi}{2\pi^2} \int_0^\infty \frac{dx}{x} (\sin x - x \cos x) \quad (4\text{-}61)$$

To evaluate the integral, we must introduce a convergence factor such as $e^{-\alpha x}$ and take the limit $\alpha \to 0$. Then $\int \rho(\mathbf{r}_1,\mathbf{r}_2)d\tau_2 = 1$, which verifies (4-36).

---

[4] The dependence of k_F on position in the expression for $\rho(\mathbf{r}_1,\mathbf{r}_2)$ in (4-60) is as follows. For $\mathbf{r}_1$ near $\mathbf{r}_2$, k_F may be evaluated at some common point, e.g., $\frac{1}{2}(\mathbf{r}_1+\mathbf{r}_2)$. When $|\mathbf{r}_1-\mathbf{r}_2|$ is large, the dependence of k_F on position is unimportant

The above expressions may be used to simplify the Hartree-Fock equations themselves, viz., to replace the nonlocal potential by an approximate local potential. To arrive at this approximation, we return to the expectation value of the Hamiltonian (4-22). The interelectron interaction contributes the amount

$$\sum_{i<j} \Big[\int d\tau_1\, d\tau_2 \frac{1}{r_{12}} |u_i(\mathbf{r}_1)|^2 |u_j(\mathbf{r}_2)|^2$$

$$- \delta(m_{si}, m_{sj}) \int d\tau_1 d\tau_2 \frac{1}{r_{12}} u_i^*(\mathbf{r}_1) u_j^*(\mathbf{r}_2) u_j(\mathbf{r}_1) u_i(\mathbf{r}_2)\Big]$$

$$= \tfrac{1}{2} \sum_{i,j} \int d\tau_1\, d\tau_2 \frac{1}{r_{12}} |u_i(\mathbf{r}_1)|^2 |u_j(\mathbf{r}_2)|^2$$

$$- \tfrac{1}{2} \sum_{i,j} \delta(m_{si}, m_{sj}) \int d\tau_1\, d\tau_2 \frac{1}{r_{12}} u_i^*(\mathbf{r}_1) u_j^*(\mathbf{r}_2) u_j(\mathbf{r}_1) u_i(\mathbf{r}_2) \tag{4-62}$$

The exchange energy, the last line of (4-62), may be written in terms of the mixed density matrix, (4-35).

$$E_{ex} = - \tfrac{1}{2}\, 2 \int d\tau_1\, d\tau_2\, (1/r_{12})\, \rho^2\, (r_1, r_2) \tag{4-63}$$

The factor 2 arises because of the two possible values of $m_{si} = m_{sj}$. We now use the approximate expression (4.60) for $\rho(\mathbf{r}_1, \mathbf{r}_2)$ and change the integral to one over the relative coordinate $\mathbf{r} = \mathbf{r}_1 - \mathbf{r}_2$ and the center-of-mass coordinate $\mathbf{R} = \frac{1}{2}(\mathbf{r}_1 + \mathbf{r}_2)$. Then we have

$$E_{ex} = - \int dR\, dr \rho^2 (k_F(R),\, r)/r$$

with k_F given by (4-59), and now taken to depend[3] on R, through $\rho(R)$, the ordinary density at R.

---

since $\rho(r_1, r_2)$ goes rapidly to zero. Hence, for purposes of evaluating integrals over r_2 which contain $\rho(r_1, r_2)$, like (4-61), we may take k_F to be a function of r_1 only, hence a constant in the integration; see also Chapter 5.

The integration over r is elementary and gives

$$\int \frac{dr}{r}\,\rho^2(k_F,r) = \frac{1}{4\pi^3}\,k_F{}^4 \tag{4-65}$$

Thus the exchange energy may be expressed in terms of the local density by use of (4-59)

$$E_{ex} = -\frac{3}{4}\left(\frac{3}{\pi}\right)^{1/3} \int \rho^{4/3}(R)\,dR \tag{4-66}$$

Remembering that

$$\rho(\mathbf{r}) = \sum_i u_i{}^*(\mathbf{r})\,u_i(\mathbf{r})$$

the total interelectron interaction energy (4-62) is now written as

$$\frac{1}{2}\sum_{i,j} \int d\tau_1 d\tau_2\, \frac{1}{r_{12}}\,|u_i(\mathbf{r}_1)|^2\,|u_j(\mathbf{r}_2)|^2$$

$$-\frac{3}{4}\left(\frac{3}{\pi}\right)^{1/3} \int d\tau \left(\sum_i u_i^*(\mathbf{r})u_i(\mathbf{r})\right)^{4/3} \tag{4-67}$$

Variation of the orbitals produces the new, approximate Hartree-Fock equation

$$\left[-\tfrac{1}{2}\nabla_1^2 - \frac{Z}{r_1} + \int d\tau_2\,\frac{\rho(\mathbf{r}_2)}{r_{12}} - \left(\frac{3}{\pi}\,\rho(\mathbf{r}_1)\right)^{1/3} - \epsilon_i\right] u_i(\mathbf{r}_1) = 0 \tag{4-68}$$

in which the non-local interaction term $-\int d\tau_2\,\rho(\mathbf{r}_1,\mathbf{r}_2)u_i(\mathbf{r}_2)/r_{12}$ has been replaced by the local interaction

$$-\left(\frac{3}{\pi}\,\rho(\mathbf{r}_1)\right)^{1/3} u(\mathbf{r}_1).$$

The idea of this approximation is due to Slater. However, he performed the Thomas-Fermi approximation in the Hartree-Fock *equations* themselves rather than in the exchange *energy*, and got a result with an additional factor of 3/2. Eq. (4-68) was first found by Kohn and Sham. Apart from the better logic of its derivation, it has also been shown to be a much better approximation to the true Hartree-Fock theory by direct numerical solution of the equations. The Slater modification is known as the HFS theory, that of Kohn and Sham as HFKS.

RESULTS OF CALCULATIONS

As we mentioned before, the exchange term lowers the value of ϵ_i and modifies the curvature of the Hartree-Fock wave function as compared to the Hartree wave function. We can represent this graphically as follows. If $\Delta\epsilon_i$ is the increase of the Hartree-Fock relative to the Hartree energy, then $\Delta\epsilon_i - V_{eff}$ is the term which tends to increase the curvature of the Hartree-Fock relative to the Hartree wave functions. Now $-V_{eff} \sim \rho^{1/3}$; ρ falls off with increasing r. Hence if we plot $\Delta\epsilon_i - V_{eff}$ versus r (assuming V_{eff} has been averaged over angles) we obtain Figure 4-1. Therefore for $r < r_0$ the curvature of the Hartree-Fock wave function is increased over the Hartree function, and for $r > r_0$ it is decreased. This means that for $r \to 0$ the wavelength is shorter, for $r \to \infty$, longer. The first loop of the Hartree-Fock wave function is compressed, the last loop extended.

In the Hartree model the total charge inside a spherical shell of radius r is

$$Z(r) = Z - 4\pi \int_0^r r'^2 \rho(r')\, dr' \tag{4-69}$$

where $\rho(r)$ is the total electronic charge density, so that Z(r) is a decreasing function of r. For small r, $\mathcal{R}_{n\ell} \sim r^{\ell+1}$. Electrons with small ℓ can get closer to the nucleus than those with large ℓ; they "see" a larger effective charge. Therefore the binding energy, which is largely determined by the average of Z(r)/r, is larger for small ℓ than for large ℓ.

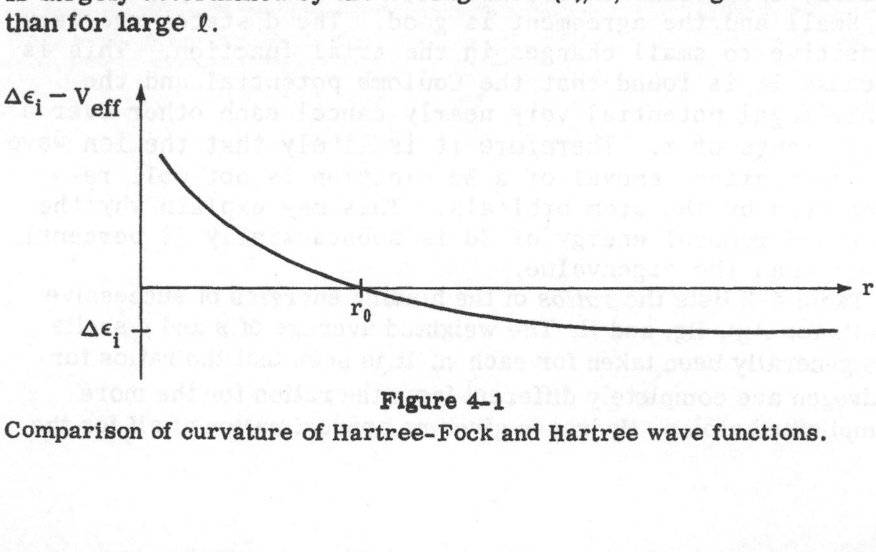

Figure 4-1
Comparison of curvature of Hartree-Fock and Hartree wave functions.

Table 4-1
Removal Energies of Ag^+ (in Ry)

	Calculated	Observed
1s	1828	1879.7
2s	270	282.0
2p	251	260.1, 247.2
3s	52.2	53.4
3p	44.3	46.0, 43.6
3d	29.8	27.8, 27.4
4s	8.46	7.3
4p	5.82	5.8, 4.9
4d	1.69	1.57

Table 4-1 gives a comparison of the Hartree-Fock results for Ag^+ with observation. The observed energies are the X-ray absorption limits. Two values are generally given, for $j = \ell - \frac{1}{2}$ and $j = \ell + \frac{1}{2}$. The observed energy for 4d is the ionization potential of Ag^+.

The observed removal energies are generally larger than the calculated ones. This is mainly due to the relativistic corrections which increase the removal energies. This increase is greatest for $j = \frac{1}{2}$; therefore the s states and the (first-listed) $p_{1/2}$ states show appreciably higher observed than calculated energies. For the $p_{3/2}$ states, the second-listed number, the relativity correction is small and the agreement is good. The d states are very sensitive to small changes in the trial function. This is because it is found that the Coulomb potential and the centrifugal potential very nearly cancel each other over a large range of r. Therefore it is likely that the ion wave function after removal of a 3d electron is not well represented by the atom orbitals. This may explain why the observed removal energy of 3d is substantially (8 percent) lower than the eigenvalue.

Table 4-2 lists the *ratios* of the binding energies of successive shells for Ag^+, Hg, and H. The weighted average of s and p shells has generally been taken for each n. It is seen that the ratios for hydrogen are completely different from the ratios for the more complicated atoms. Hydrogen gives no approximation at all for the

Table 4-2
Ratio of the Binding Energies of Successive Shells for Various Atoms

Ratio of principal quantum numbers	Ag^+	Hg	H
1:2	7	6.2	4
2:3	$5\frac{1}{2}$	4.4	2.25
3:4	7	5.1	1.78
4:5		7.8	1.56

more complicated atoms. For these the outer electrons are strongly screened by the inner ones so that they "see" a much smaller effective nuclear charge. It is a useful approximation to say that the binding energy for successive n decreases on the average by about a factor of 6 for complex atoms.

It is instructive to calculate the quantities Zr_1, r_2/r_1, and r_3/r_2, where r_i is the ith node of the radial wave function for a 4s electron. Table 4-3 lists these for several Z. It is seen that they vary rather slowly with Z. But we see again the tendency, repeatedly pointed out, that the innermost loop gets shorter and the outer loop longer in going from H to the other atoms. The effect is most pronounced for Ca because here the 4s electron is least strongly bound.

Hartree has defined a screening constant $\sigma(n, \ell)$ by

$$\langle r \rangle = \int \mathcal{R}^2_{n\ell} r \, dr = \frac{\langle r_H \rangle}{Z - \sigma} \tag{4-70}$$

where r_H is the radius of the corresponding hydrogen orbital and $\mathcal{R}_{n\ell}$ is normalized. Table 4-4 lists $\sigma(n\ell)$ for various Z. The $\sigma(2p)$ is just slightly higher than $\sigma(2s)$. Hartree has shown that $\mathcal{R}_{n\ell}$, plotted against $r/\langle r \rangle$, is a slowly variable function of Z which can easily be interpolated.

Table 4-3

	Zr_1	r_2/r_1	r_3/r_2
H	1.87	3.7	2.25
Hg	1.93	3.8	2.6
Ag^+	1.95	3.9	3.75
Ca	2.04	4.5	3.4

Table 4-4
Screening Constant for Various Z

Z =	11	14	17	19	26	29
σ(2s)	3.29	3.30	3.42	3.48	3.30	3.74
σ(2p)	4.72	4.57	4.67	4.68	4.43	4.89

Further self-consistent field calculations are described in various monography.[5]

HIGH n

Calculations for high n, i.e., for optically excited states, are facilitated by two simplifications. First of all, the position probability density for a highly excited state is small near the nucleus. Therefore to a very good approximation the wave functions for the nonexcited electrons are the ion wave functions. Second, the optical electron moves in an equivalent potential which is $-1/r$ for r greater than some radius R. Hence for $r > R$, we have a hydrogenic wave function which is shifted in phase. The phase shift $\pi\delta(E_n)$ is a slowly varying function of the energy. The energy dependence is weak because the E_n for various n are not much different from each other, and because each E_n is much less than the potential energy.

If we use the WKB approximation for the wave function of the optical electron, we get $\Phi^{-1/4} \cos \alpha(r)$ where $\Phi^{-1/4}$ is the usual square root of the momentum and the phase $\alpha(r)$ is some function of r [see (1-24)]. As this wave function is a phase shifted hydrogenic wave function, we may write $\alpha(r)$ as $\alpha_H(r) + \pi\delta$, where $\alpha_H(r)$ is the phase of a hydrogen wave function at the same energy. At the outer turning point r_2,

[5]C. Froese Fisher, *The Hartree-Fock Method for Atoms*, Wiley, New York, 1977; D. R. Hartree, *The Calculation of Atomic Structures*, Wiley, New York, 1957. F. Herman and S. Skillman, *Atomic Structure Calculations*, Prentice-Hall, New York, 1963, have published numerical tables of HFS functions. Useful calculations using HFKS and further modifications of it are by D.A. Liberman, *Phys. Rev.* **171**, 1(1968); R.D. Cowan, *Phys. Rev.* **163**, 54(1967); and D.C. Griffin, R.D. Cowan, and K.L. Andrews, *Phys. Rev. A* **3**, 1233(1971). With care, it is possible to get the total binding energy of all electrons in an atom correct to about 0.1 Ry, that of the last electron to 0.01 - 0.04 Ry.

the WKB conditions require that $\alpha(r_2) = (n' + \frac{1}{4})\pi$, where $n' = n - \ell - 1$ is the number of nodes in $\mathcal{R}_{n\ell}$. Hence $\alpha_H(r_2) = (n' - \delta + \frac{1}{4})\pi$. An effective quantum number n^* for a wave function, with just the hydrogen phase α_H, can be defined by setting the energy eigenvalue $E_n = -Ry/n^{*2}$. Again the WKB condition at the outer turning point gives $\alpha_H(r_2) = (n^* - \ell - 1 + \frac{1}{4})\pi$. Hence we get

$$n - \delta = n^* \tag{4-71}$$

with δ independent of n. This is the original *Rydberg formula*.

PERIODIC SYSTEM

Combining the Pauli principle with the Hartree-Fock calculation, we can understand the periodic system. Electron shells get filled in order of increasing energy, with the Pauli principle limiting the number of electrons in a given shell.

Table 4-5 lists the ionization potentials for the first electron (1st and 2nd columns) and for the second electron (3rd column) of various atoms. We see that the alkalis (Li, Na) and the corresponding ions like Ca^+, having one electron outside closed shells, have the lowest

Table 4-5
Ionization Potentials (in eV) for the First Electron (1st and 2nd Columns) and for the Second Electron (3rd Column) of Various Atoms

Li	5.40	Na	5.14	Ca^+	11.9
Be	9.32	Mg	7.64	Sc^+	12.8
B	8.28	Al	5.97	Ti^+	13.6
C	11.27	Si	8.15	V^+	14.1
N	14.55	P	10.9	Cr^+	16.7
O	13.62	S	10.36	Mn^+	15.6
F	17.47	Cl	12.90	Fe^+	16.5
Ne	21.56	Ar	15.76	Co^+	17.4
				Ni^+	18.2
				Cu^+	20.2
				Zn^+	18.0
				Ga^+	20.5
				Ge^+	16.0

Table 4-6
Order of Filling of Levels According to Madelung's Rule

$n + \ell$	Level	Shell
1	1s	1
2	2s	2
3	2p	
	3s	3
4	3p	
	4s	4
5	3d	
	4p	
	5s	5
6	4d	
	5p	
	6s	6
7	4f	
	5d	
	6p	
	7s	7

binding energy. The second s electron (Be, Mg) is bound more strongly than the first. The next electron must go into a p shell and has again lesser binding energy (B, Al). As we build up the p shell, the binding increases, as might be expected. But we note that there is a break in the ionization potential after three electrons have been put into the p shell (i.e., just before O, S). This is due to the fact that the first three electrons can be put in with the same spin, whereas the fourth electron must go in with opposite spin. The system with three electrons of the same spin is a symmetric spin state, thus an antisymmetric space state. The antisymmetry keeps the electrons apart, hence decreases their electrostatic interaction and yields a greater binding energy (see pp. 34, 139).

In the third column we list the ionization potentials of ions which have an outer 3d shell. We do not choose neutral atoms, because for these there is competition between 3d and 4s. The binding energy increases until it breaks after five electrons have been put in (just before Mn) for the same reason as before. The shell is completed at Cu^+. In Zn^+, the 4s shell is started, which shows in a smaller binding energy. The first 4p electron (Ge^+) again has less binding than the second 4s (Ga^+).

The order of filling of all the shells is determined as above, by the Pauli principle and by energy considerations. Madelung formulated the following heuristic rule for remembering the order of filling of levels in neutral atoms. Fill up in order of increasing $n + \ell$. For each $n + \ell$ fill up in order of increasing n. A shell (in the chemical sense of the word) is closed after the p electrons have been filled up, except for the first shell, which is closed at $1s^2$. Table 4-6 lists the order of filling of levels according to this rule.

The d shells are always very sensitive, and compete with the s shell of the next higher n. Schiff's table, p. 428, [6] shows this competition. The 4s shell is first filled by two electrons (Ca); then three electrons are put into the 3d shell (V). The next atom, Cr, has one 4s electron replaced by an extra 3d, giving the *configuration* [7] $4s3d^5$, and showing again the special stability of a half-filled shell. Mn has again $4s^2$, together with the stable $3d^5$, and $4s^2$ persists through Ni. In Cu, the extra stability of the complete shell $3d^{10}$ again leads to a single 4s electron, which is responsible for the fact that Cu is often monovalent, often divalent. In the next transition period, the equilibrium is generally shifted in favor of the 4d shell, a single 5s electron being the rule from $5s4d^4$ (Nb) to $5s4d^8$ (Rh). In the last similar period, the shift is in the opposite direction; we have $6s^2\,5d^x$ up to $x = 6$ (Os). The change of pattern from 3d to 4d is probably due to the increased nuclear charge, from 4d to 5d to the intervening 4f shell, which effectively screens the 5d.

It should be remembered that the antisymmetrization of the wave function makes it impossible to assign a particular set of quantum numbers to a definite electron. Such statements as "the $n\ell$ electron" refer merely to the occurrence in the Slater determinant of an orbital with quantum numbers n and ℓ.

---

[6] L. I. Schiff, *Quantum Mechanics*, 3rd ed., McGraw-Hill, New York, 1968.

[7] The term configuration is used to describe an assignment of electrons to specific levels $n\ell$. Thus we say phosphorus has the configuration $1s^2\,2s^2\,2p^6\,3s^2\,3p^3$.

PROBLEMS

1. Carry out in detail the derivation of (4-27).
2. Begin with (4-23) and derive a set of equations involving the "eigenmatrix" $\{\lambda_{ij}\}$. Redefine the orbitals with the unitary transformation which diagonalizes $\{\lambda_{ij}\}$, thus derive (4-27).
3. Show explicitly that the solutions of the Hartree equation (4-2) corresponding to different ϵ_i are not in general orthogonal.
4. A nonlocal potential which satisfies $U(\mathbf{r}, \mathbf{r}') = u(\mathbf{r})v(\mathbf{r}')$ is called a *separable, nonlocal potential*. Show that the Hermiticity condition (4-31) requires that a separable nonlocal potential be of the form $\lambda u(\mathbf{r})u^*(\mathbf{r}')$, where λ is real. Solve the Schrödinger equation for negative energy where the potential is nonlocal and separable.

$$-\tfrac{1}{2}\nabla^2\psi(\mathbf{r}) + \lambda u(\mathbf{r})\int u^*(\mathbf{r}')\psi(\mathbf{r}')\,d\tau' = -|E|\psi(\mathbf{r}) \tag{1}$$

Assume the $u(\mathbf{r})$ possesses a well-behaved Fourier transform. Hint: Observe that the term $\int u^*(\mathbf{r}')\psi(\mathbf{r}')\,d\tau'$ is a constant, hence equation (1) is of the form

$$(\tfrac{1}{2}\nabla^2 - |E|)\psi(\mathbf{r}) = \lambda a u(\mathbf{r}) \tag{2}$$

$$a = \int u^*(\mathbf{r})\psi(\mathbf{r})\,d\tau \tag{3}$$

This is just the free particle equation with an inhomogeneous term. Construct the Green's function for the operator $\frac{1}{2}\nabla^2 - |E|$ and solve equation (2); alternatively, take Fourier transforms of equation (2). Show that the energy eigenvalue is determined by equation (3) and a is determined by the normalization condition on $\psi(\mathbf{r})$.

5. Evaluate the integral in (4-61) to show that $\int \rho(\mathbf{r}_1, \mathbf{r}_2)\,d\tau_2 = 1$.
6. What is the electron configuration in the ground state of Mn (Z = 25)? (There are two electrons in the 4s state.)
7. A solvable two-body problem: Consider two spin 1/2 particles interacting by harmonic forces, governed by the Hamiltonian

$$H = \frac{1}{2}\left[\frac{p_1^2}{m} + \alpha r_1^2\right] + \frac{1}{2}\left[\frac{p_2^2}{m} + \alpha r_2^2\right] + \frac{1}{2}\kappa\,[\mathbf{r}_1 - \mathbf{r}_2]^2$$

(a) Solve for the ground state energy E and wave function $\psi(\mathbf{r}_1, \mathbf{r}_2)$

(Hint: use variables $R = \frac{1}{\sqrt{2}}(r_1+r_2)$; $r = \frac{1}{\sqrt{2}}(r_1-r_2)$.)

(b) Derive the Hartree-Fock equation for this problem.

(c) Assume that the ground state of the Hartree-Fock problem is a singlet, where the *spatial* wave function ψ_{HF} is given by $\psi_{HF}(r_1,r_2) = \phi(r_1)\phi(r_2)$, where $\phi(r)$ is spherically symmetric. Solve for $\phi(r)$; determine the Hartree-Fock eigenvalue ε and the ground state energy E_{HF}.

(d) Plot E_{HF}/E as well as $\int\psi^*(r_1,r_2)\psi_{HF}(r_1,r_2)d^3r_1d^3r_2$ as a function of κ. What are the limiting forms as $\kappa\to 0$ and $\kappa\to\infty$? Which Hartree-Fock quantity is in closer agreement with the exact result, the energy or the wave function?

Chapter 5

STATISTICAL MODELS

In the previous chapter we discussed the self-consistent field method for finding energy levels and wave functions for atoms. We saw that any numerical calculation is very cumbersome, especially for atoms with many electrons. For these, there exists a simpler method to obtain at least a fair approximation. Developed by Thomas and Fermi, it is based on Fermi-Dirac statistics. The results admittedly are less accurate than those of the Hartree-Fock calculations. The Thomas-Fermi method nevertheless is very useful for calculating form factors and for obtaining effective potentials which can be used as initial trial potentials in the self-consistent field method. It is also applicable to the study of nucleons in nuclei and electrons in a metal. Exchange is not treated in the Thomas-Fermi model. However, a statistical model for the exchange effects can be given also. We discuss this at the end of this chapter, after the presentation of the model without exchange.

THOMAS-FERMI MODEL

The goal of the Thomas-Fermi statistical method is to obtain the effective potential energy which is experienced by an infinitesimal test charge, and to find the electron density $\rho(r)$ around the nucleus of an atom.

Consider a number of electrons moving in a volume Ω_0, subject to a spherically symmetric potential energy V(r) which varies sufficiently slowly with r so that the system can be treated by Fermi-Dirac free particle statistics. The electrons are supposed to interact with each other sufficiently to establish statistical equilibrium, but still so little that we can speak of the kinetic and potential energy of each individual electron. We assume $\lim_{r \to \infty} V(r) = 0$. The distribution function f is

$$f = \frac{1}{e^{(E-\zeta)/kT} + 1} \tag{5-1}$$

Here ζ is the chemical potential, k is Boltzmann's constant, and T is the absolute temperature. If we assume that $T = 0$,

$$f = \begin{cases} 1 & E < \zeta \\ 0 & E > \zeta \end{cases} \tag{5-2}$$

In the zero-temperature limit, therefore, ζ is the energy of the most energetic electrons; the Pauli principle forces the electrons to occupy all states from the ground state to the state of energy ζ. ζ is not a function of r; if it were, electrons would migrate to that region of space where ζ is smallest, because this would make the total energy of the system decrease. By this process, ζ would tend to equalize. Clearly,

$$\zeta = V(r) + \frac{p_F^2(r)}{2m} \tag{5-3}$$

(In this and subsequent chapters we use ordinary units.) Here $p_F(r)$ is the maximum momentum of the electrons, the so-called Fermi momentum, which must depend on r to make ζ constant.

We can obtain an expression connecting p_F with ρ by considering the number of quantum states of translational motion of a completely free electron with a momentum whose absolute value lies between p and p + dp. For this purpose we consider the electron as moving in a box of volume Ω without any forces. Then the number of quantum states is equal to

$$2 \frac{\Omega}{(2\pi)^3} 4\pi k^2 \, dk \tag{1-18}$$

where $hk = p$ and the factor 2 is due to the two spin orientations that an electron can have. We integrate the above from 0 to k_F and this must equal N, the total number of electrons within the box.

$$2\frac{\Omega}{(2\pi)^3} \frac{4\pi}{3} k_F^3 = N \qquad k_F^3 = 3\pi^2 \rho \qquad \rho = \frac{N}{\Omega} \tag{5-4}$$

It is then assumed that we can construct such a box with volume Ω, within the big volume Ω_0 originally considered, which is large enough to make (5-4) valid, and yet small enough that the potential energy does not vary too much within the box. Then we may consider (5-4) and (5-3) simultaneously valid. We now perform a gauge transformation on the potential energy

$$V - \zeta \rightarrow V_1 \tag{5-5}$$

From (5-3) and (5-4) it then follows that

$$\rho = \frac{1}{3\pi^2}\,\frac{(2m)^{3/2}}{\hbar^3}\,(-V_1)^{3/2} \tag{5-6}$$

The Poisson equation connects the electrostatic potential $-(1/e)V$ with the charge density $-e\rho$. With suitable rearrangement this is

$$\nabla^2 V_1 = -4\pi e^2 \rho \tag{5-7}$$

Combining (5-6) with (5-7) we get

$$\frac{1}{r}\,\frac{d^2}{dr^2}(rV_1) = -\frac{4e^2}{3\pi\hbar^3}(2m)^{3/2}(-V_1)^{3/2} \tag{5-8}$$

For $r \rightarrow 0$ the leading term of V must be $-Ze^2/r$. Hence (5-8) is supplemented by the boundary condition

$$\lim_{r \rightarrow 0}\,(rV_1) = -Ze^2 \tag{5-9}$$

We make the change of variable

$$r = xb \qquad rV_1 = -Ze^2\Phi$$

$$b = \frac{(3\pi)^{2/3}}{2^{7/3}}\,\frac{\hbar^2}{me^2}\,Z^{-1/3} = 0.885a_0 Z^{-1/3} \tag{5-10}$$

with $a_0 = \hbar^2/me^2$, the Bohr radius. Therefore the equation we must solve is

$$\frac{d^2\Phi}{dx^2} = \frac{\Phi^{3/2}}{\sqrt{x}} \tag{5-11}$$

$$\Phi(0) = 1 \tag{5-12}$$

This is the *Thomas-Fermi* equation.

Physically (5-11) holds only for positive Φ; for negative Φ the electron density vanishes because there is no state with $E > \zeta$, c.f. (5-2) and (5-6). The correct differential equation for negative Φ is therefore $d^2\Phi/dx^2 = 0$ ($\Phi < 0$). If x_0 is the point where Φ crosses the x axis, we have $\Phi = A(x - x_0)$, where A is a negative constant which by continuity is equal to $\Phi'(x_0)$. Thus the solution is completely determined if we know it for $\Phi > 0$, and we therefore consider only this portion of it. At x_0, $\Phi(x_0) = 0$, and $\Phi'(x_0)$ cannot vanish; for if it did, equation (5-11) indicates that Φ'' and all higher derivatives would vanish, giving the unacceptable trivial solution $\Phi = 0$.

SOLUTIONS OF THE THOMAS-FERMI EQUATION

Equation (5-11) is a second-order, nonlinear, differential equation. It is important to notice that it is independent of Z. Evidently, we shall obtain a whole family of solutions, since we have specified only one boundary condition. These solutions can be classified according to the initial slope, which is clearly arbitrary in the differential equation (5-11).

Certain properties of the solutions can be obtained by examining (5-11). All solutions are initially concave upward. Hence if a particular solution doesn't become zero, it will remain concave upward, and will either diverge for large x or approach the x axis asymptotically. If a solution becomes zero for finite $x = x_0$, the differential equation (5-11) stops being valid as explained above. Finally, we note that Φ does not have a Taylor expansion in x about 0 since $\Phi''(0)$ diverges. The behavior of the various possible solutions is illustrated in Figure 5-1.

Numerical integration of equation (5-11) for various initial slopes indicates that the solutions can be expressed in a semiconvergent series

$$\Phi = 1 - a_2 x + a_3 x^{3/2} + a_4 x^2 + \ldots$$

with

$$a_3 = \tfrac{4}{3} \qquad a_4 = 0 \qquad \Phi'(0) = -a_2 \tag{5-13}$$

Three distinct types of solutions are found as predicted above. One class consists of solutions which vanish for finite $x = x_0$. These have a_2 larger than a critical value. With a_2 smaller than the critical value, another class of solutions is found, which do not vanish anywhere and diverge for large x. Finally when a_2 is precisely at the critical value, a unique solution is obtained which is asymptotic to the x axis. Numerical determination of the critical initial slope gives $a_2 = -\Phi'(0) = 1.5880710\ldots \approx \pi/2$, and this critical solution

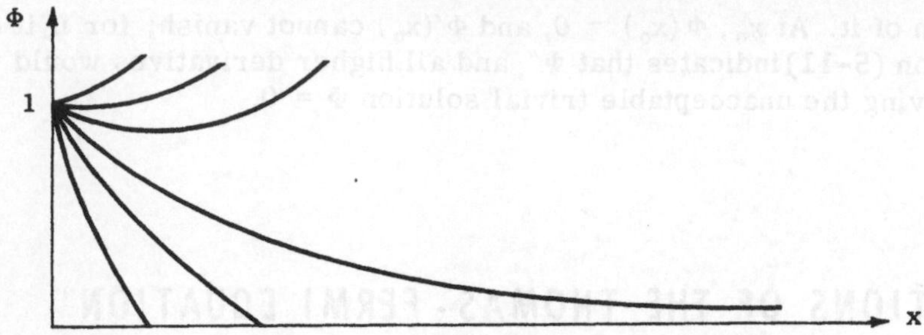

Figure 5-1
Behavior of the solutions to (5-12)

has been tabulated (Landau and Lifshitz[1], p. 261). These three different types of solutions correspond to different physical circumstances for the atom, which we now discuss.

DETERMINATION OF ADDITIONAL BOUNDARY CONDITIONS

Further analysis of specific physical situations leads to a determination of additional boundary conditions on Φ, which in turn determine unique solutions. We examine several such situations.

Let the electronic configuration of a neutral atom or charged ion under consideration be confined to a sphere of radius r_0. The total number of electrons N in the electronic configuration is given by

$$N = 4\pi \int_0^{r_0} \rho r^2\, dr = -\frac{1}{e^2}\int_0^{r_0} r \frac{d^2}{dr^2}(rV_1)\, dr$$

$$= Z \int_0^{x_0} x\, \Phi''\, dx$$

$$\frac{N}{Z} = (\Phi' x - \Phi)\Big|_0^{x_0} \qquad (5\text{-}14a)$$

Using the boundary condition at $x = 0$ this reduces to

$$\Phi(x_0) - x_0 \Phi'(x_0) = \frac{Z - N}{Z} = \frac{z}{Z} \qquad (5\text{-}14b)$$

where z is the net charge of the neutral atom or charged ion. (For a neutral atom $z = 0$.)

[1]L.D. Landau and E.M. Lifshitz, *Quantum Mechanics*, 3rd ed., Pergamon, Oxford, 1977.

Let us first consider free atoms or ions, viz., systems not subject to external pressure. In this case $\rho = 0$ at $r = r_0$ which implies

$$\Phi(x_0) = 0 \qquad (5\text{-}15a)$$

$$x_0 \Phi'(x_0) = -\frac{z}{Z} \qquad (5\text{-}15b)$$

For neutral atoms, $z = 0$ and (5-15) has the consequence that

$$\Phi(x_0) = 0 \qquad (5\text{-}16a)$$

$$x_0 \Phi'(x_0) = 0 \qquad (5\text{-}16b)$$

If x_0 is a finite number, (5-16) gives $\Phi(x_0) = \Phi'(x_0) = 0$. Since no nontrivial solutions with this property exist, a solution for a free neutral atom with finite radius cannot be obtained. To achieve a solution for a free neutral atom, we must assume that the surface of the atom lies at infinity and interpret (5-16) as

$$\lim_{x \to \infty} \Phi = 0 \qquad (5\text{-}17a)$$

$$\lim_{x \to \infty} x\Phi' = 0 \qquad (5\text{-}17b)$$

(5-17a) by itself determines a unique solution, the only solution which is asymptotic to the x axis. This is the critical solution mentioned earlier. Since the function Φ vanishes only at infinity, the neutral free atom has no boundaries in the Thomas-Fermi model.

It can be verified that $144\, x^{-3}$ satisfies the differential equation, as well as the bounding conditions (5-17). The boundary condition at the origin is not satisfied. Sommerfeld showed that this solution is the asymptotic form of the correct solution for a free neutral atom.

For free ions ($z \neq 0$), the boundary condition (5-15a) indicates that the solutions of (5-11) which vanish at finite $x = x_0$ correspond to ions of radius r_0. Boundary condition (5-15b) gives the net charge of the ion. Since the slope of Φ must be negative at x_0 (see Figure 5-1), (5-15) implies that the theory cannot handle negative free ions.

Next, we consider neutral atoms under pressure. We do not concern ourselves with ions under pressure, since an assembly of many ions under pressure would lead to physical difficulties, because of the large, cumulative Coulomb forces. When the atom is subject to an external pressure, $\rho(x_0)$ no longer equals zero, and we must remain with (5-14b). Evidently the solutions that do not vanish for any x correspond to this case. Equation (5-14b) determines x_0 and therefore the radius of such systems. Since atoms are neutral, (5-14b) becomes

$$\frac{\Phi(x_0)}{x_0} = \Phi'(x_0). \tag{5-18}$$

This defines the point x_0 at which the tangent to Φ passes through the origin. For $x > x_0$ the differential equation (5-12) is no longer a description of the physical situation.

APPLICATIONS

All atoms in the Thomas-Fermi model have the same electron distribution, except for a different scale of length and total number of electrons. Equations (5-10) show that the length scale for any atom is proportional to $Z^{-1/3}$. Thus the radius of the entire atom decreases as $Z^{-1/3}$. However it can be shown that the radius of the sphere which contains all but one electron is roughly proportional to $Z^{1/6}$.

An interesting application of the Thomas-Fermi method is to calculate for which Z bound atomic states with a given angular momentum first appear. We consider the reduced radial equation

$$\frac{d^2\mathcal{R}}{dr^2} + \frac{2m}{\hbar^2}(E - V_r)\mathcal{R} = 0$$

$$V_r = V(r) + \frac{\hbar^2}{2m}\frac{(\ell + \frac{1}{2})^2}{r^2} \tag{5-19}$$

[We have made the usual WKB substitution $\ell(\ell + 1) \to (\ell + \frac{1}{2})^2$.] Bound states exist only if $E - V_r > 0$ for some range of r. Since $E < 0$, this means we must have

$$-\frac{2m}{\hbar^2} V(r) r^2 > (\ell + \tfrac{1}{2})^2$$

$$\frac{2me^2}{\hbar^2} Zr\Phi = 0.885\, Z^{2/3}\, 2x\Phi > (\ell + \tfrac{1}{2})^2 \tag{5-20}$$

for some range of r.

The following broad maximum is found in $2x\Phi$.

$\sqrt{2x}$	1	1.96	2.04	2.12	2.20	3.0
$2x\Phi$	0.607	0.972	0.973	0.968	0.968	0.829

A necessary condition for (5-20) to hold is that

$$0.885Z^{2/3}\,\mathrm{Max}(2x\Phi) > (\ell + \tfrac{1}{2})^2$$

$$0.861Z^{2/3} > (\ell + \tfrac{1}{2})^2$$

$$Z > 0.157(2\ell + 1)^3 \qquad (5\text{-}21a)$$

This formula determines the value of Z for which an electron with a given ℓ is first bound. We should expect that we can change the "greater" sign in (5-21a) to an "equal" sign if we increase the coefficient somewhat. If we take 0.17 instead of 0.157:

$$Z = 0.17(2\ell + 1)^3 \qquad (5\text{-}21b)$$

we get

for ℓ = 1		2	3	4
	Z = 4.6	21.25	58.3	123.9

Rounding to the nearest integer we obtain 5, 21, 58, 124. Comparing with experiment, we find the first three results to be correct and the last one predicting that g electrons can appear only in the 124th element. This is six places beyond the predicted noble gas Z = 118, with the heaviest element so far discovered having Z = 103.

The maximum of Vr^2 was found above to be quite flat. We expect therefore that there will be close cancellation between V(r) and the centrifugal potential term for the largest ℓ which can be bound by a given atom. In this situation a small change in Z would effect a large change in the wave function; see pp. 81.

VALIDITY OF THE THOMAS-FERMI METHOD

The Fermi model is useful for calculating properties that depend on the average electron such as form factor (see Chapter 13), total energy of all electrons, electrostatic potential produced by all electrons at the nucleus, and average excitation potential. The latter occurs in the theory of atomic stopping power and is defined by

$$\log E_{AV} = \frac{1}{Z}\sum_j \log E_j$$

where E_j is the average excitation potential of the jth orbital. The Thomas-Fermi method, even when corrected for exchange effects (as discussed below), is very poor for calculating properties that depend on the outer electrons, such as the ionization potential, or the mean square radius of the atom, which is important for diamagnetism.

Because of the statistical nature of this model, best results are obtained for large Z. Experience shows that calculations for $Z < 10$ are unreliable.

For fixed Z, the Thomas-Fermi results are inaccurate for large and small r. For large r the electron density is overestimated. Indeed, we saw that the Thomas-Fermi atom has no boundaries, with the electron density $\rho \sim (\Phi/r)^{3/2}$ decreasing as $1/r^6$ at large distance. On the other hand, we expect the electron density of a physical atom should decrease exponentially. For small r the Thomas-Fermi density diverges as $1/r^{3/2}$ rather than remaining finite.

CORRECTION FOR EXCHANGE; THE THOMAS-FERMI-DIRAC EQUATION

The Thomas-Fermi equation (5-11) does not take into account the exchange interaction. This was done by Dirac. We shall here give a simple derivation of this correction.

We recall the exchange term in the Hartree-Fock theory,

$$\int U(\mathbf{r}_1, \mathbf{r}_2)\, u_1(\mathbf{r}_2)\, d\tau_2$$

$$U(\mathbf{r}_1, \mathbf{r}_2) = -\frac{e^2}{r_{12}}\,\rho(\mathbf{r}_1,\mathbf{r}_2)$$

$$\rho(\mathbf{r}_1, \mathbf{r}_2) = \sum_j u_j^*(\mathbf{r}_2)\, u_j(\mathbf{r}_1)$$

In the spirit of the Thomas-Fermi method we consider the electrons to be free (subject to a constant potential). Thus we set

$$u_j(\mathbf{r}_1) = \Omega^{-1/2} e^{i\mathbf{k}_j \cdot \mathbf{r}_1} \tag{5-22}$$

$$\rho(\mathbf{r}_1, \mathbf{r}_2) = \Omega^{-1} \sum_j e^{i\mathbf{k}_j \cdot (\mathbf{r}_1 - \mathbf{r}_2)}$$

$$\approx \frac{1}{(2\pi)^3} \int e^{i\mathbf{k}\cdot\mathbf{r}_{12}}\, d^3k \qquad (\mathbf{r}_{12} = \mathbf{r}_1 - \mathbf{r}_2)$$

$$= \frac{4\pi}{(2\pi)^3} \int_0^{k_F} \frac{\sin kr_{12}}{kr_{12}} k^2 dk$$

$$= \frac{1}{2\pi^2} \frac{1}{r_{12}^3} (\sin k_F r_{12} - k_F r_{12} \cos k_F r_{12}) \qquad (5\text{-}23)$$

which verifies (4-60).

In the expression (5-23) for $\rho(\mathbf{r}_1,\mathbf{r}_2) k_F$ depends on position in the following fashion; see also footnote 4 in Chapter 4. When $\mathbf{r}_1$ and $\mathbf{r}_2$ are sufficiently close together so that they lie in the same cell Ω in which we take the potential to be constant, k_F is evaluated at the radius vector of that cell, which we may take to be $\frac{1}{2}(\mathbf{r}_1+\mathbf{r}_2)$. When $\mathbf{r}_1$ and $\mathbf{r}_2$ lie in different cells, we may assume $|\mathbf{r}_1 - \mathbf{r}_2|$ to be large. In that case the dependence of k_F on position may be ignored as $\rho(\mathbf{r}_1,\mathbf{r}_2)$ goes rapidly to zero. In evaluating integrals over $\mathbf{r}_2$, which involve $\rho(\mathbf{r}_1,\mathbf{r}_2)$, we may take k_F to be a function of $\mathbf{r}_1$ only, hence constant in the integration.

It is seen that

$$\rho(\mathbf{r},\mathbf{r}) = \frac{k^3{}_F}{6\pi^2} = \frac{1}{2}\rho \qquad (5\text{-}24)$$

The factor $\frac{1}{2}$ is present due to the fact that the density matrix $\rho(\mathbf{r},\mathbf{r})$ is defined as a sum of $|u_i(\mathbf{r})|^2$ taken over all orbitals with the same value of the spin projection, while the density ρ is the sum of $|u_i(\mathbf{r})|^2$ over all orbitals with both spin projections. In the absence of interactions half the electrons will be in one spin state and half in the other.

We recall that for the ith electron

$$V_{\text{eff}}(\mathbf{r}_1) = -\int e^2 \frac{\rho(\mathbf{r}_1, \mathbf{r}_2)}{r_{12}} \frac{u_i(\mathbf{r}_2)}{u_i(\mathbf{r}_1)} d\tau_2 \qquad (4\text{-}39)$$

Hence in the Thomas-Fermi spirit

$$V_{\text{eff}}(\mathbf{r}_1) = -e^2 \int \frac{\rho(\mathbf{r}_1, \mathbf{r}_2)}{r_{12}} e^{i\mathbf{k}_i \cdot \mathbf{r}_{21}} d\tau_2 \qquad (5\text{-}25a)$$

This integral is evaluated to be

$$V_{\text{eff}}(\mathbf{r}_1) = -2\left(\frac{3}{\pi}\rho\right)^{1/3} e^2 F(\eta)$$

$$= -\frac{2}{\pi} e^2 k_F F(\eta) \qquad (5\text{-}25b)$$

where

$$\eta = \frac{k_i}{k_F}$$

$$F(\eta) = \frac{1}{2} + \frac{1-\eta^2}{4\eta} \log \frac{1+\eta}{1-\eta} \tag{5-26}$$

$F(0) = 1$, $F(1) = \frac{1}{2}$, and F decreases from 1 to $\frac{1}{2}$ monotonically as η goes from 0 to 1.

We now must determine a relation between V, the electrostatic potential energy, and ρ. The simplest argument is that the total energy of the most energetic electron is now, instead of (5-3),

$$\zeta = V(r) - \frac{2}{\pi} e^2 k_F F(1) + \frac{p_F^2}{2m} = V(r) - \frac{e^2}{\pi}(3\pi^2\rho)^{1/3} + \frac{\hbar^2}{2m}(3\pi^2\rho)^{2/3} \tag{5-27}$$

Here the effective exchange potential energy has been added to V(r); V_{eff} depends on the momentum according to (5-25). In our case of the most energetic electron, $\eta = 1$ and $F(1) = \frac{1}{2}$, according to (5-26). The result (5-27) is identical with (5-30) to be derived below.

For variety, we shall now derive (5-27) in another fashion. We shall consider the total energy E of the system of electrons, vary this quantity as a function of ρ, and obtain the desired relations from the requirement that E be stationary. This variational approach could have been used in deriving the Thomas-Fermi model.

The total energy is the sum of the kinetic energy E_k and potential energy E_p. The total kinetic energy of the electrons can be obtained by multiplying the number of states by $\hbar^2 k^2/2m$, integrating over all momenta from 0 to k_F, and then integrating over all volume. The result is easily found to be

$$E_k = \int d\tau \left[\frac{3}{5}\frac{\hbar^2\pi^2}{2m}\left(\frac{3}{\pi}\rho\right)^{2/3}\rho\right] \tag{5-28}$$

The potential energy is

$$E_p = \int \left[\left(-\frac{Ze^2}{r}\right)\rho + \left(\frac{1}{2}\int d\tau_2 \frac{e^2}{r_{12}}\rho(r_2)\right)\rho - \frac{3}{4}e^2\left(\frac{3\rho}{\pi}\right)^{1/3}\rho\right] d\tau \tag{5-29}$$

The first term is due to the nuclear charge; the second term is due to the interelectron interaction, the factor ½ being inserted to avoid counting the electron pairs twice. The third term is the exchange energy, already computed when the Thomas-Fermi approximation was applied to the Hartree-Fock theory; see (4-66).

We now set the arbitrary variation of $E = E_k + E_p$ with respect to ρ equal to 0 and obtain

$$a_0 e^2 \frac{\pi^2}{2}\left(\frac{3}{\pi}\rho\right)^{2/3} + V - e^2\left(\frac{3\rho}{\pi}\right)^{1/3} = 0$$

$$V(r_1) = \frac{Ze^2}{r_1} + \int d\tau_2 \frac{e^2}{r_{12}}\rho(r_2) \qquad a_0 = \frac{\hbar^2}{me^2} \tag{5-30}$$

[In point of fact, the variation of E is not totally unrestricted, since we must also have $\int\rho\, d\tau = N$, the total number of electrons. We can introduce this subsidiary condition by the Lagrange multiplier method. This adds a term λ to (5-30), where λ is the multiplier. We now can make a gauge transform $V + \lambda \rightarrow V$ and obtain (5-30).]

We now solve (5-30) for the density. Setting $y = a_0(3\rho/\pi)^{1/3} = (a_0/\pi)k_F$, we get

$$y = \frac{1}{\pi^2}\left(1 + \sqrt{1 - 2\pi^2\frac{Va_0}{e^2}}\right) \tag{5-31}$$

The plus sign is chosen in front of the radical to assure agreement with the Thomas-Fermi theory and to avoid negative density. With

$$\Psi = \frac{1}{2\pi^2} - \frac{Va_0}{e^2} \tag{5-32}$$

we obtain

$$y = \frac{\sqrt{2}}{\pi}\left(\sqrt{\Psi} + \frac{1}{\pi\sqrt{2}}\right) \tag{5-33}$$

Poisson's equation now gives

$$\frac{d^2}{dr^2}(r\Psi) = 4\pi a_0 \rho r = \frac{4\pi^2}{3a_0^2} y^3 r$$

$$\frac{d^2}{dr^2}(r\Psi) = \frac{2^{7/2}}{3a_0^2\pi} r\left(\sqrt{\Psi} + \frac{1}{\pi\sqrt{2}}\right)^3 \tag{5-34}$$

Finally, changing variables as in (5-10),

$$r = xb \qquad r\Psi = a_0 Z \Phi \qquad b = 0.885 a_0 Z^{-1/3}$$

we obtain

$$\Phi'' = x\left(\sqrt{\frac{\Phi}{x}} + \beta\right)^3$$

$$\beta = \sqrt{\frac{b}{a_0 Z}}\,\frac{1}{\pi\sqrt{2}} = 0.2118 Z^{-2/3} \tag{5-35}$$

This is the *Thomas-Fermi-Dirac* equation. Unlike (5-12), it depends on Z through β. We see that for $Z \to \infty$, (5-35) becomes (5-11) Indeed it can be shown quite rigorously that the Thomas-Fermi model becomes exact as $Z \to \infty$. Moreover, there are non-asymptotic corrections, present at finite Z, which fall off more slowly with Z than the exchange term.[2]

The boundary conditions for the Thomas-Fermi-Dirac equation are the following:

$$\Phi(0) = 1$$

$$\Phi(x_0) - x_0\Phi'(x_0) = \frac{Z - N}{Z} = \frac{z}{Z} \tag{5-36}$$

For free atoms and ions we can no longer define the surface by $\rho(x_0) = 0$, since it is seen from (5-33) that ρ never vanishes. However, we can define x_0 by requiring that the pressure vanish there. To do this we write the specific energy ϵ (energy per particle) from (5-28) and (5-29). [There is actually some problem about the electrostatic interaction of the electrons, but (5-37) gives the correct result.]

$$\epsilon = \frac{3}{5}\,\frac{\pi^2}{2}\,\frac{e^2}{a_0}\,y^2 + V(r) - \frac{3}{4}\,\frac{e^2}{a_0}\,y \tag{5-37}$$

The pressure $P = -(\partial\epsilon/\partial v)_S$, where v is the specific volume and S is the entropy. Since (5-37) has been derived for T = 0, the entropy is equal to zero, hence already constant, so we merely need to differentiate with respect to v. Recalling $y = [(3/\pi)\rho]^{1/3} a_0$, $v = 1/\rho$, we obtain

$$P = \rho\,\frac{e^2}{a_0}\left[\frac{\pi^2 y^2}{5} - \frac{y}{4}\right] \tag{5-38}$$

---

[2]A systematic study of corrections to the Thomas-Fermi theory, as well as an assessment of its validity may be found in E.H. Lieb, Rev. Mod. Phys. *53*, 603(1981).

Table 5-1
Comparison of Energy Levels of Ag as Calculated by Hartree-Fock Method and Thomas-Fermi-Dirac Method (Values are given in Ry)

	Hartree-Fock	Thomas-Fermi-Dirac
1s	1828	1805
2s	270	263
2p	251	245
3d	29.8	29.2
4s	8.46	7.95

This vanishes at

$$y = \frac{5}{4\pi^2}$$

or

$$\rho(x_0) = 2.13 \times 10^{-3} a_0^{-3} \tag{5-39}$$

Had we not included the exchange effect, P would vanish at $y = 0$, implying $\rho(x_0) = 0$ as before. A lower density than (5-39) is unphysical in the Thomas-Fermi-Dirac model since this would correspond to negative pressure. Substituting (5-39) in (5-33) we find with some algebra that

$$\frac{\Phi(x_0)}{x_0} = \frac{\beta^2}{16} \tag{5-40}$$

These results imply that in the Thomas-Fermi-Dirac theory atoms as well as ions have a finite radius. Equation (5-40) does not apply of course to systems under external pressure, since then the density can be larger than that given in (5-39). [No solution which goes to zero at $x = \infty$ exists, as can be seen from (5-35). This causes no difficulty since atomic solutions no longer satisfy (5-17).] As before, the differential equation applies only to $x \leq x_0$. Moreover, from numerical calculations it is found that in the Thomas-Fermi-Dirac model negative free ions are not treated.

One can solve the radial, one-electron Schrödinger equation for all normally occupied orbitals n, ℓ for many atoms, using as the potential that given by the Thomas-Fermi-Dirac method. This gives both energy levels and wave functions for these orbitals. From these wave functions, one could then construct a potential using the Hartree-Fock prescription, and this should be very good starting data for a Hartree-Fock calculation. For atoms for which no Hartree-Fock solution is available, the one-electron Thomas-Fermi-Dirac wave functions are the best available. That they are indeed very good is shown by comparing their eigenvalues with the Hartree-Fock results. We list these for a few orbitals in silver for which both types of calculation exist in Table 5-1.

PROBLEMS

1. Derive the Thomas-Fermi equation, using the variation principle as in the derivation of the Thomas-Fermi-Dirac theory.
2. A solvable Hartree-Fock problem: Consider a N electron "gas" moving in a large box of volume Ω in which there is a background positive charge density of eN/Ω. The electrons interact with this background through a potential of the form

$$-e^2 \int \frac{\rho(\mathbf{r}')}{|\mathbf{r} - \mathbf{r}'|} \, d\tau' \tag{1}$$

where ρ is the density of the electron gas.

$$\rho = \frac{N}{\Omega} \tag{2}$$

They also interact with each other through the usual Coulomb interaction. Assume the number of spin up (down) electrons is $N_+ (N_-)$.

$$N_+ + N_- = N \tag{3}$$

Let the spin up (down) electrons be described by individual wave functions $u_i^+ \alpha (u_i^- \beta)$, where $u_i^+ (u_i^-)$ is the space part, and $\alpha(\beta)$ the spin part of the wave function. Write down the Hartree-Fock equations for $u_i^\pm$. Show that they are solved by plane waves

$$u_i^\pm(\mathbf{r}) = \frac{1}{\sqrt{\Omega}} \, e^{i\mathbf{k}_i \cdot \mathbf{r}} \tag{4}$$

Procedure: Assume the solution is of the form (4). Evaluate the direct and exchange interactions

$$V(\mathbf{r}) = e^2 \int \sum_i \frac{|u_i^+(\mathbf{r}')|^2}{|\mathbf{r} - \mathbf{r}'|} d\tau' + e^2 \int \sum_i \frac{|u_i^-(\mathbf{r}')|^2}{|\mathbf{r} - \mathbf{r}'|} d\tau'$$

$$U_+(\mathbf{r}, \mathbf{r}') = -e^2 \sum_i \frac{u_i^{+*}(\mathbf{r}')\, u_i^+(\mathbf{r})}{|\mathbf{r} - \mathbf{r}'|} \qquad \text{for spin up electrons}$$

$$U_-(\mathbf{r}, \mathbf{r}') = -e^2 \sum_i \frac{u_i^{-*}(\mathbf{r}')\, u_i^-(\mathbf{r})}{|\mathbf{r} - \mathbf{r}'|} \qquad \text{for spin down electrons} \tag{5}$$

The sums over i are performed with the usual assumption that the spin up (down) electrons fill up the available spaces up to the Fermi momentum k_F^+ (k_F^-). [Recall equations (5-23) to (5-26).] Thus, show the following:

$$V(\mathbf{r}) = e^2 \int \frac{\rho_+}{|\mathbf{r} - \mathbf{r}'|} d\tau' + e^2 \int \frac{\rho_-}{|\mathbf{r} - \mathbf{r}'|} d\tau'$$

$$\rho_\pm = \frac{N_\pm}{\Omega} = \frac{k_F^{\pm 3}}{6\pi^2}$$

$$U_\pm(\mathbf{r}_1, \mathbf{r}_2) = \frac{-e^2 \rho_\pm(\mathbf{r}_1, \mathbf{r}_2)}{r_{12}}$$

$$= \frac{-e^2}{2\pi^2} \frac{1}{r_{12}^4} [\sin k_F^\pm r_{12} - k_F^\pm r_{12} \cos k_F^\pm r_{12}]$$

$$\int U_\pm(\mathbf{r}_1, \mathbf{r}_2)\, u_i(\mathbf{r}_2)\, d\tau_2$$

$$= -\frac{2}{\pi} e^2 k_F^\pm F(\eta^\pm)\, u_i^\pm(\mathbf{r}_1)$$

$$F(\eta) = \tfrac{1}{2} + \frac{1 - \eta^2}{4\eta} \log \frac{1 + \eta}{1 - \eta}$$

$$\eta^\pm = \frac{k}{k_F^\pm} \tag{6}$$

(Many of the integrals that arise in this problem are evaluated in Chapter 5.) Noting that the direct interaction potential V(r) cancels the interaction with the background (1), show that the Hartree-Fock eigenvalue is

$$\epsilon_k^{\pm} = \frac{\hbar k^2}{2m} - \frac{2e^2 k_F^{\pm}}{\pi} F(\eta^{\pm})$$

and the expectation value of the Hamiltonian is

$$\langle H \rangle = E = \frac{\Omega}{(2\pi)^3} \left\{ \frac{2\pi\hbar^2}{5m} (k_F^{+5} + k_F^{-5}) - e^2 (k_F^{+4} + k_F^{-4}) \right\}$$

(This provides a plausible, but not very accurate, picture of valence electrons in a conducting metal.)

3. Show directly from the Thomas-Fermi equation (5-12), that a solution which remains finite at infinity must go to zero, rather than to a constant.
4. Verify the manipulations leading to (5-11).
5. Assuming a solution to (5-11) of the form (5-13), verify that a_2 is undetermined, $a_3 = 4/3$, $a_4 = 0$.

Chapter 6

ADDITION OF ANGULAR MOMENTA

A consequence of the central field approximation, which we used in Chapters 4 and 5, is that an atom with an incomplete shell is a highly degenerate system with respect to energy. For the shell $n\ell$ there are $g = 2(2\ell + 1)$ orbitals (states) with the same energy. If the shell is occupied by $N \leq g$ electrons, then

$$\binom{g}{N} = \frac{g!}{N!(g - N)!} \tag{6-1}$$

is the number of different ways the electrons can be distributed in this shell. Hence the degeneracy is $\binom{g}{N}$. For example, phosphorus with 15 electrons has the electron configuration $1s^2\,2s^2\,2p^6\,3s^2\,3p^3$. The degeneracy therefore is 20. Only for complete shells is there no degeneracy: $\binom{g}{g} = 1$.

To break this degeneracy, we must go beyond the Hartree-Fock approximation in the calculation of energy levels. Although the Hartree-Fock method is variational, we may consider it as the starting point of a perturbation expansion by adopting the following point of view. The exact Hamiltonian may be written in the form

$$H = \sum_i f_i + \sum_i h_i + \sum_{i<j} g_{ij} - \sum_i h_i \tag{6-2}$$

Here the f_i are the one electron operators, kinetic energy and electron-nucleus interaction; the g_{ij} are the two electron operators, inter-electron interaction. If the h_i are chosen to be the Hartree-Fock potentials averaged over angles, the first two terms of (6-2) correspond to the Hartree-Fock central field Hamiltonian. These two terms can

also be considered to be an unperturbed Hamiltonian, with unperturbed eigenfunctions which are just the determinantal Hartree-Fock solutions. The perturbing Hamiltonian consists of the remaining two terms $\Sigma_{i<j} g_{ij} - \Sigma_i h_i$. Evidently, this perturbation is just the electrostatic interaction of the electrons, $\Sigma_{i<j} e^2/r_{ij}$ minus a suitable average of it.

To evaluate the first-order perturbation contribution to the degenerate energy levels of the unperturbed Hamiltonian, we must find proper linear combinations of the zero-order degenerate eigenfunctions, such that the perturbing Hamiltonian is diagonal, or equivalently, such that the entire Hamiltonian (6-2) is diagonal. To facilitate the determination of the proper linear combinations, we shall consider the vectors of the total orbital and spin angular momentum of all electrons in the atom. Each Cartesian component of these two vectors we shall show to be a constant of motion. By choosing combinations of the zero-order wave functions, such that a commuting set of these constants of motion is diagonal, we shall have simplified the secular problem of diagonalizing H.

ANGULAR MOMENTUM

We consider [H, **L**], where H is given by (1-3) and **L** by (1-30). For simplicity let us consider a two-electron system. Clearly **L** commutes with p_i^2 and $1/r_i$ since $\mathbf{L}_i$ does. But since $\mathbf{L}_i$ is proportional to the rotations generator of $\mathbf{r}_i$, it cannot commute with $1/r_{12}$. However since the total angular momentum operator generates rotations in all the position coordinates $\mathbf{r}_i$, we expect it to commute with $1/r_{12}$. To see this explicitly consider

$$[\mathbf{L}, f(r_{12}^2)] = [\mathbf{L}_1, f(r_{12}^2)] + [\mathbf{L}_2, f(r_{12}^2)] \tag{6-3}$$

where f is an arbitrary, differentiable function.

$$[L_{1x}, f(r_{12}^2)] = -i\hbar\left(y_1\frac{\partial}{\partial z_1} - z_1\frac{\partial}{\partial y_1}\right) f(r_{12}^2)$$

$$= i\hbar\, 2(y_1 z_2 - z_1 y_2) f'(r_{12}^2) \tag{6-4a}$$

By symmetry,

$$[L_{2x}, f(r_{12}^2)] = -i\hbar\, 2(y_1 z_2 - z_1 y_2) f'(r_{12}^2) \tag{6-4b}$$

Thus L_x, and all other components of **L** commute with an arbitrary function of r_{12}. In particular

$$\left[\mathbf{L}, \frac{1}{r_{12}}\right] = 0 \tag{6-4c}$$

This obviously generalizes to more than two electrons, and we obtain

$$[H, \mathbf{L}] = 0 \tag{6-5}$$

Before concluding from (6-5) that $\mathbf{L}$ can be diagonalized together with the Hamiltonian, a word of caution is in order. By imposing a symmetry condition on the admissible eigenfunctions, the Hilbert space of states becomes restricted. The only operators that can be diagonalized are those that do not take states out of the restricted Hilbert space, viz., operators that do not change the symmetry of the state. Therefore, only operators that depend symmetrically on single particle operators are admissible; a condition consistent with the supposition that the particles are identical.

We see that $\mathbf{L}$ is a symmetric operator and this together with (6-5) implies that $\mathbf{L}$ can be diagonalized with H.

In the case of spin, the Hamiltonian commutes with the spin operator of each electron $\mathbf{S}_i$, since H does not depend on $\mathbf{S}_i$. However we may not conclude from this that $\mathbf{S}_i$ can be diagonalized, because the use of antisymmetric wave functions couples the spins to the symmetry of the spatial wave function and thus to the electrostatic energy (see pp. 34 and 63). Indeed $S_{ix}\psi$ does not lie in the Hilbert space of allowed antisymmetric states. Hence it cannot be expanded in terms of antisymmetric wave functions, which would be required if $\mathbf{S}_i$ were diagonal. Total spin $\mathbf{S} = \Sigma_i \mathbf{S}_i$ is an admissible, symmetric operator, which also commutes with the Hamiltonian.

$$[H, \mathbf{S}] = 0 \tag{6-6}$$

We conclude that $\mathbf{S}$ can be diagonalized with H.

Inasmuch as L^2, L_z, S^2, and S_z all commute with each other, we can find a representation in which they are diagonal. In this representation, the matrix elements of H, labeled by L, M_L, S, and M_S are, in view of (1-26)

$$\langle L'M'_L, S'M'_S | H | LM_L, SM_S \rangle$$

$$= \delta_{LL'}\delta_{M_L M'_L}\delta_{M_S M'_S}\delta_{SS'} h(L, M_L; S, M_S) \tag{6-7a}$$

with

$$h(L, M_L; S, M_S) = \langle LM_L, SM_S | H | LM_L, SM_S \rangle \tag{6-7b}$$

For later use, we now prove that h is independent of the magnetic quantum numbers. We assume $L \neq 0$ and consider the dependence on M_L. First take $M_L \neq L$. Then

$$h(L, M_L + 1) = \langle L, M_L + 1 | H | L, M_L + 1\rangle \tag{6-8}$$

(We suppress the S, M_S dependence.) Using the raising operator L_+

$$|L, M_L + 1\rangle = [(L - M_L)(L + M_L + 1)]^{-1/2} L_+ | L, M_L\rangle \tag{6-9}$$

Therefore, since H and L_+ commute

$$h(L, M_L + 1) = [(L - M_L)(L + M_L + 1)]^{-1/2} \times \langle L, M_L + 1 | L_+ H | L, M_L\rangle \tag{6-10}$$

But

$$\langle L, M_L + 1 | L_+ = \langle L, M_L | [(L + M_L + 1)(L - M_L)]^{1/2} \tag{6-11}$$

This gives for $M_L \neq L$

$$h(L, M_L + 1) = \langle L, M_L | H | L, M_L\rangle = h(L, M_L) \tag{6-12}$$

For $M_L = L$ a similar proof can be given using the lowering operator. Therefore in general $h(L, M_L)$ is independent of M_L. Repeating the same argument for the spin quantum numbers, we conclude that $h(L, M_L; S, M_S)$ is independent of M_L and M_S. The same is true of course for the matrix elements between angular momentum states of any operator which commutes with all the components of the angular momentum.

ADDITION OF ANGULAR MOMENTA

We now must construct the eigenfunctions of L^2, L_z, S^2, and S_z. First, we examine the problem of constructing proper linear combinations of products of eigenfunctions of L_i^2, L_{iz}, S_i^2, S_{iz}. For example, we suppose we have two different spinless particles. The system is described by a product of single particle orbitals with angular momenta ℓ_1 and ℓ_2 (we suppress $\hbar$); z-component m_1 and m_2, respectively. The product $|\ell_1 m_1\rangle|\ell_2 m_2\rangle = |\ell_1 m_1, \ell_2 m_2\rangle$ is an eigenfunction of the operator $L_Z = L_{1Z} + L_{2Z}$ with eigenvalue $M_L = m_1 + m_2$. However the operator $L^2 = (L_1 + L_2)^2$ is not in general diagonal. But out of all the possible product wave functions which are eigenfunctions belonging to M_L, we can form linear combinations such that L^2 is diagonalized. To illustrate this we list in the first three columns of Table 6-1 m_1, m_2, and M_L for a few values of M_L.

Examining the table, we conclude the following: Since there is only one possible combination for $M_L = \ell_1 + \ell_2$, L^2 must be diagonal for this wave function, its eigenvalue being $L(L + 1)$, with $L = \ell_1 + \ell_2$. Next, the two states with $M_L = \ell_1 + \ell_2 - 1$ lead to a secular equation to diagonalize L^2, the eigenvalues being of course of the form $L(L + 1)$, with $L = \ell_1 + \ell_2$ and $\ell_1 + \ell_2 - 1$. Similarly for $M_L = \ell_1 + \ell_2 - 2$ one linear combination of the product wave functions will lead to $L = \ell_1 + \ell_2$, another to $L = \ell_1 + \ell_2 - 1$, and the third to $L = \ell_1 + \ell_2 - 2$. In the fourth column of Table 6-1 we have listed the values of L that can result from the linear combinations. If we assume $\ell_1 \geq \ell_2$, Table 6-1 continues until $M_L = \ell_1 - \ell_2$. For this M_L, m_2 will take all values from $-\ell_2$ to ℓ_2, with $m_1 = M_L - m_2$. This leads to a $(2\ell_2 + 1) \times (2\ell_2 + 1)$ secular problem, giving all values of L from $\ell_1 + \ell_2$ to $\ell_1 - \ell_2$. After this the next case would be

$$M_L = \ell_1 - \ell_2 - 1$$

Table 6-1
Addition of Angular Momenta

m_1	m_2	M_L	L
ℓ_1	ℓ_2	$\ell_1 + \ell_2$	$\ell_1 + \ell_2$
$\ell_1 - 1$	ℓ_2	$\ell_1 + \ell_2 - 1$	$\ell_1 + \ell_2$
ℓ_1	$\ell_2 - 1$	$\ell_1 + \ell_2 - 1$	$\ell_1 + \ell_2 - 1$
$\ell_1 - 2$	ℓ_2	$\ell_1 + \ell_2 - 2$	$\ell_1 + \ell_2$
$\ell_1 - 1$	$\ell_2 - 1$	$\ell_1 + \ell_2 - 2$	$\ell_1 + \ell_2 - 1$
ℓ_1	$\ell_2 - 2$	$\ell_1 + \ell_2 - 2$	$\ell_1 + \ell_2 - 2$

with

$$m_1 = \ell_1 - 1, \ldots, \ell_1 - 2\ell_2 - 1$$

$$m_2 = -\ell_2, \ldots, \ell_2 \tag{6-13}$$

Since $|m_2| \leq \ell_2$ we can go no further with m_2, and obtain again a $(2\ell_2 + 1) \times (2\ell_2 + 1)$ secular problem. Thus we get no new values for L. All the wave functions obtainable by linear combination of the functions listed in (6-13) are needed for the values of L already obtained. As M_L decreases further to $-\ell_1 - \ell_2$ we never get any further values for L. Hence the possible values of L are

$$L = \ell_1 + \ell_2, \ell_1 + \ell_2 - 1, \ldots, |\ell_1 - \ell_2| \tag{6-14}$$

where the absolute-value sign takes care of the eventuality $\ell_2 > \ell_1$. Equation (6-14) is a result of the vector addition of angular momenta known from the old quantum theory.

To add three angular momenta, we first add two of them, and then add the third one to the sum. Spin behaves the same way and we can add spin angular momenta and orbital angular momenta indiscriminately.

CLEBSCH-GORDAN COEFFICIENTS

We shall now develop the formal theory of addition of angular momenta and derive a method for obtaining the eigenvectors of L^2 and L_z. We write

$$|L, M\rangle = \sum_{m_1 m_2} C(\ell_1 m_1 \ell_2 m_2, LM) \times |\ell_1 m_1, \ell_2 m_2\rangle \tag{6-15}$$

The above is a transformation equation connecting the eigenvectors of L^2 and L_z with those of L_1^2, L_{1z}, L_2^2, L_{2z}. The summation is carried out only over m_1 and m_2, for ℓ_1 and ℓ_2 are assumed fixed. The problem of adding angular momenta reduces to determining the elements of the transformation matrix, called the Clebsch-Gordan coefficients (also known as vector addition or Wigner coefficients). An immediate simplification occurs when we operate with $L_z = L_{1z} + L_{2z}$ on (6-15) and recall

$$L_z |L, M\rangle = M |L, M\rangle$$

$$(L_{1z} + L_{2z}) |\ell_1 m_1, \ell_2 m_2\rangle = (m_1 + m_2) |\ell_1 m_1, \ell_2 m_2\rangle \tag{6-16}$$

Therefore $C(\ell_1 m_1, \ell_2 m_2, LM)$ is 0 unless $M = m_1 + m_2$, and the summation in (6-15) is actually over only m_1 or m_2.

Aside from looking them up in tables, there are two convenient methods of obtaining Clebsch-Gordan coefficients. The first is the *step-down* method. We start with $L = \ell_1 + \ell_2$, $M = L$. We know that there is only one term in the sum (6-15)

$$|L, M\rangle = |\ell_1 + \ell_2, \ell_1 + \ell_2\rangle$$
$$= |\ell_1 m_1, \ell_2 m_2\rangle = |\ell_1 \ell_1, \ell_2 \ell_2\rangle \tag{6-17}$$

If we operate with $L_- = L_{1-} + L_{2-}$ on (6-17) we obtain

$$L_-|L, M\rangle = \sqrt{(L + M)(L - M + 1)}\,|L, M - 1\rangle$$
$$= \sqrt{(\ell_1 + m_1)(\ell_1 - m_1 + 1)}\,|\ell_1\ m_1{-}1, \ell_2 m_2\rangle$$
$$+ \sqrt{(\ell_2 + m_2)(\ell_2 - m_2 + 1)}\,|\ell_1 m_1, \ell_2\ m_2{-}1\rangle$$

or

$$\sqrt{\ell_1 + \ell_2}\,|L, M - 1\rangle$$
$$= \sqrt{\ell_1}\,|\ell_1\ \ell_1{-}1, \ell_2 \ell_2\rangle + \sqrt{\ell_2}\,|\ell_1 \ell_1, \ell_2\ \ell_2{-}1\rangle$$

Therefore

$$|L, M - 1\rangle = \sqrt{\frac{\ell_1}{\ell_1 + \ell_2}}\,|\ell_1\ \ell_1{-}1, \ell_2 \ell_2\rangle$$
$$+ \sqrt{\frac{\ell_2}{\ell_1 + \ell_2}}\,|\ell_1\ \ell_1, \ell_2\ \ell_2{-}1\rangle$$
$$\equiv |\ell_1 + \ell_2, \ell_1 + \ell_2 - 1\rangle \tag{6-18}$$

There is another eigenfunction of $M = \ell_1 + \ell_2 - 1$, namely, the one corresponding to $L = \ell_1 + \ell_2 - 1$. It must be orthonormal to (6-18).

$$|\ell_1 + \ell_2 - 1, \ell_1 + \ell_2 - 1\rangle$$
$$= \sqrt{\frac{\ell_2}{\ell_1 + \ell_2}}\,|\ell_1\ \ell_1{-}1, \ell_2 \ell_2\rangle$$
$$- \sqrt{\frac{\ell_1}{\ell_1 + \ell_2}}\,|\ell_1 \ell_1, \ell_2\ \ell_2{-}1\rangle \tag{6-19}$$

We can now use lowering operators on equations (6-18) and (6-19) and obtain new linear combinations for $M = \ell_1 + \ell_2 - 2$ and so on.

A second method is due to Racah, who obtained a closed formula.

$$C(LM_L\,SM_S,\,JM) = \sqrt{2J+1}\,V(L,\,S,\,J;\,M_L,\,M_S,\,M) \tag{6-20}$$

$$V(a,\,b,\,c;\,\alpha,\,\beta,\,\gamma) = \sqrt{\frac{(a+b-c)!\,(a-b+c)!\,(-a+b+c)!}{(a+b+c+1)!}}$$
$$\times\sqrt{(a+\alpha)!\,(a-\alpha)!\,(b+\beta)!\,(b-\beta)!\,(c+\gamma)!\,(c-\gamma)!}$$
$$\times\sum_z (-1)^z/[z!(a+b-c-z)!(a-\alpha-z)!(b+\beta-z)!(-b+c+\alpha+z)!(-a+c-\beta+z)!] \tag{6-21}$$

where z runs through zero and the integers for as long as the arguments of all factorials are nonnegative. We have made the following relabeling:

$$\ell_1 \rightarrow L \qquad m_1 \rightarrow M_L$$
$$\ell_2 \rightarrow S \qquad m_2 \rightarrow M_S$$
$$L \rightarrow J \qquad M \rightarrow M$$

A useful recursion for the Clebsch-Gordan coefficients is obtained if we operate with J_- or J_+ on (6-15)

$$\sqrt{(J\pm M)(J\mp M+1)}\,C(LM_L\,SM_S,\,J\,M\mp 1)$$
$$= \sqrt{(L\mp M_L)(L\pm M_L+1)}\,C(L\,M_L\pm 1\,SM_S,\,JM)$$
$$+ \sqrt{(S\mp M_S)(S\pm M_S+1)}\,C(LM_L\,S\,M_S\pm 1,\,JM) \tag{6-22}$$

It is clear that the Clebsch-Gordan coefficients have an arbitrary phase factor common to all of them. It is common to choose C(LLS J − L, JJ) real and positive. Then all the coefficients will be real.

Since (6-15) represents a unitary transformation, the usual orthogonality relations hold:

$$\sum_{M_L} C(LM_L\,SM_S,\,JM)C^*(LM_L\,SM_S,\,J'M') = \delta_{MM'}\delta_{JJ'}$$
$$\sum_J C(LM_L\,SM_S,\,JM)C^*(LM'_L\,SM'_S,\,JM) = \delta_{M_LM'_L}\,\delta_{M_SM'_S} \tag{6-23}$$

We observe that the number of eigenvectors in the JM representation must be equal to the number in the $LM_L SM_S$ representation. That this is indeed so follows from the fact that in the $LM_L SM_S$ representation there are (2L + 1) possible M_L, (2S + 1) possible M_S, a total of (2L + 1)(2S + 1) vectors. In the JM representation there are 2J + 1 states for each value of J which runs from L + S to |L − S| and

$$\sum_{|L-S|}^{L+S} (2J+1) = (2L+1)(2S+1)$$

Further properties of Clebsch-Gordan coefficients and a more group-theoretical presentation of angular momentum addition is available in monographs.[1]

SPECIAL CASES

We shall now discuss several useful special cases.

For the case J = L + S, (6-20) and (6-21) reduce to

$$C(LM_L SM_S, L+S\ M_L+M_S) = \sqrt{\frac{(2L)!\,(2S)!}{(2J)!}}\ \sqrt{\frac{(J+M)!}{(L+M_L)!\,(S+M_S)!}} \times \sqrt{\frac{(J-M)!}{(L-M_L)!\,(S-M_S)!}} \tag{6-24}$$

For fermions we wish to obtain the total angular momentum **J** = **L** + **S**, where S = ½. J can equal either L + ½ or L − ½. The following formulas are then true.

$$|L+\tfrac{1}{2}, M\rangle = \sqrt{\frac{L+\frac{1}{2}+M}{2L+1}}\ |L\ M-\tfrac{1}{2}, \tfrac{1}{2}\tfrac{1}{2}\rangle + \sqrt{\frac{L+\frac{1}{2}-M}{2L+1}}\ |L\ M+\tfrac{1}{2}, \tfrac{1}{2}-\tfrac{1}{2}\rangle \tag{6-25}$$

[1]M.E. Rose, *Elementary Theory of Angular Momentum*, Wiley, New York, 1957; A.R. Edmonds, *Angular Momentum in Quantum Mechanics*, Princeton, Princeton, NJ, 1960.

$$|L - \tfrac{1}{2}, M\rangle = -\sqrt{\frac{L + \frac{1}{2} - M}{2L + 1}}\,|L\,M-\tfrac{1}{2}, \tfrac{1}{2}\,\tfrac{1}{2}\rangle + \sqrt{\frac{L + \frac{1}{2} + M}{2L + 1}}\,|L\,M+\tfrac{1}{2}, \tfrac{1}{2}\,-\tfrac{1}{2}\rangle \qquad (6\text{-}26)$$

We prove (6-25) by induction. We observe that it holds for $M = L + \frac{1}{2}$. We then operate on (6-25) with J_- and the result, when divided by $\sqrt{(L + \frac{1}{2} + M)(L + \frac{3}{2} - M)}$ can be shown to be of the form (6-25), with M replaced by $M - 1$. To verify (6-26) we merely need to observe that it is orthonormal to (6-25). Since the vector space is two-dimensional (6-26) must be correct.

We next consider the addition to two angular momenta equal in magnitude, $L = S$. Then J can equal $2L, 2L - 1, \ldots, 0$. The wave function for the highest $M = 2L$, corresponding to $J = 2L$ and $M_L = M_S = L$, must be given by

$$|2L, 2L\rangle = |L, L\rangle \qquad (6\text{-}27)$$

(We abbreviate $|LM_L, SM_S\rangle$ as $|M_L, M_S\rangle$.) Equation (6-27) is clearly symmetric in the two angular momenta. In the particular case of two equivalent electrons, the wave function will be symmetric in their spatial coordinates. Since J_- is a symmetric operator, $J_- = L_- + S_-$, $J_-|2L, 2L\rangle$ is again symmetric. Hence all the eigenkets with $J = 2L$ are symmetric in the two angular momenta. For $M = 2L - 1$ we have two possibilities $M_L = L$, $M_S = L - 1$; or $M_L = L - 1$, $M_S = L$. Out of the corresponding kets we form linear combinations corresponding to $J = 2L$, $J = 2L - 1$. Since

$$2^{-1/2}(|L - 1, L\rangle + |L, L - 1\rangle) \qquad (6\text{-}28)$$

is the only normalized, symmetric state that can be constructed, it must correspond to $J = 2L$. Accordingly,

$$2^{-1/2}(|L - 1, L\rangle - |L, L - 1\rangle) \qquad (6\text{-}29)$$

which is orthonormal to (6-28) corresponds to $J = 2L - 1$. By the same argument as before we see that all eigenvectors with $J = 2L - 1$ are antisymmetric.

For lower M the argument is similar. If $M = 2L - 2n$, where n is an integer, the possible pairs of M_L, M_S are

$$\begin{array}{llllll} M_L = L & L - 1 & \ldots & L - n & \ldots & L - 2n \\ M_S = L - 2n & L - 2n + 1 & \ldots & L - n & \ldots & L \end{array} \qquad (6\text{-}30)$$

In the cases where $M_S \neq M_L$, we can form a symmetric and an antisymmetric combination of kets, viz.,

$$2^{-1/2}(|M_L, M - M_L\rangle \pm |M - M_L, M_L\rangle) \tag{6-31}$$

while for $M_L = L - n$ we can form only a symmetric ket, $|L - n, L - n\rangle$. Thus there are n + 1 symmetric and n antisymmetric kets. On the other hand, if $M = 2L - 2n + 1$, there are n symmetric and n antisymmetric kets, since there is no state $M_L = M_S$. Applying the lowering operator $L_- + S_-$ to the antisymmetric kets for $M = 2L - 2n + 1$ will make n linearly independent, antisymmetric kets for $M = 2L - 2n$; this uses up all the antisymmetric kets available. Doing the same for the symmetric kets gives n linear combinations of symmetric kets of the type (6-31); but we know there are actually n + 1 symmetric kets for $M = 2L - 2n$, so that one is left over. This must belong to $J = 2L - 2n$, and we thus find that the wave function for $J = 2L - 2n$ must be symmetric. Conversely, going from $M = 2L - 2n$ to $2L - 2n - 1$, one antisymmetric wave function is added; therefore the wave function for $J = 2L - 2n - 1$ is antisymmetric. All states with $J = 2L, 2L - 2, \ldots, 0$ (or 1 if L is half-integral) will have eigenvectors symmetric in L and S; all states with $J = 2L - 1, 2L - 3, \ldots, 1$ (or 0) have antisymmetric ones.

This same result may be obtained from the symmetry properties of the Clebsch-Gordan coefficients which follow from the formula (6-21). One may readily prove that

$$C(LM_L\ L\ M_S, JM) = (-1)^{2L-J}\ C(LM_S\ LM_L, JM) \tag{6-32}$$

This then implies the symmetry under interchange as detailed above.

The rule can be applied, for example, to the compounding of the isotopic spins of two π-mesons. The isotopic spin of one π-meson is 1, hence the two-π states of total isotopic spin I = 2 and 0 are symmetric in the isotopic spin coordinates of the two π-mesons, while the state I = 1 is antisymmetric.

Similarly the addition of two spin $\frac{1}{2}$ angular momenta results in the symmetric S = 1 state and the antisymmetric S = 0 state, as we showed explicitly in Chapter 1.

ADDITION OF ANGULAR MOMENTA FOR IDENTICAL PARTICLES

In the previous section we constructed eigenfunctions of the total angular momentum $L^2 = (L_1 + L_2)^2$, $L_Z = L_{1Z} + L_{2Z}$ from linear combinations of simple products of single particle states, each of which is an eigenstate of $L_1^2, L_{1Z}; L_2^2, L_{2Z}$. For identical particles, we frequently wish to form total angular momentum eigen states by taking linear combinations of symmetrized states, rather than simple product states. For example, according to the program outlined at the beginning of this chapter, we need to take linear combinations of Slater determinants.

We now show that the same Clebsch-Gordan coefficients we used above, lead to total angular momentum eigenstates even when taking linear combinations of symmetrized product states. We again designate the two-particle state by $|\ell_1 m_1, \ell_2 m_2\rangle_\pm$. This now represents the unnormalized symmetrized state

$$u_{\ell_1 m_1}(1)\, u_{\ell_2 m_2}(2) \pm u_{\ell_2 m_2}(1)\, u_{\ell_1 m_1}(2)$$

This state is no longer an eigenstate of the unsymmetric operators L_1^2, L_{1Z}; L_2^2, L_{2Z}. However, it is an eigenstate of the symmetric operators $L_1^2 + L_2^2$, $L_{1Z} + L_{2Z} = L_Z$. We wish to take linear combinations of such states so as to obtain eigenstates of $L^2 = (L_1 + L_2)^2$. Analogous to (6-18), we form the state

$$|\psi\rangle_\pm = \sum_{\substack{m_1, m_2 \\ M = m_1 + m_2}} C(\ell_1 m_1 \ell_2 m_2, LM)\, |\ell_1 m_1, \ell_2 m_2\rangle_\pm$$

$$= |\psi\rangle_1 \pm |\psi\rangle_2 \tag{6-33}$$

$$|\psi\rangle_1 = \sum_{\substack{m_1, m_2 \\ M = m_1 + m_2}} C(\ell_1 m_1 \ell_2 m_2, LM)\, u_{\ell_1 m_1}(1)\, u_{\ell_2 m_2}(2) \tag{6-34a}$$

$$|\psi\rangle_2 = \sum_{\substack{m_1, m_2 \\ M = m_1 + m_2}} C(\ell_1 m_1 \ell_2 m_2, LM)\, u_{\ell_1 m_1}(2)\, u_{\ell_2 m_2}(1) \tag{6-34b}$$

Comparing (6-34) with (6-15) it is seen that (6-34a) is an eigenstate of $L^2 = (L_1 + L_2)^2$ with eigenvalue $L(L + 1)$, and (6-34b) is an eigenstate of $L^2 = (L_2 + L_1)^2$, also with eigenvalue $L(L + 1)$. Hence $|\psi\rangle_\pm$ is indeed identical with $|LM\rangle$, apart from an overall normalization factor, and the proper symmetry is preserved. An exceptional case occurs when $|\psi\rangle_1 = |\psi\rangle_2$. In this case $|\psi\rangle_-$ vanishes and the state $|LM\rangle$ cannot be reached. This will happen when $u_{\ell_1 m_1} = u_{\ell_2 m_2}$. This is just the Pauli principle in operation, and we discuss this in detail below.

As an example consider the addition of two spin $\frac{1}{2}$ angular momenta for two identical electrons, which we considered in Chapter 1. We saw there that the $M_S = 1$ is a symmetric combination. This state in our present notation is $|\frac{1}{2}\frac{1}{2}, \frac{1}{2}\frac{1}{2}\rangle_{\pm}$. It is seen that $|\frac{1}{2}\frac{1}{2}, \frac{1}{2}\frac{1}{2}\rangle_{-}$ vanishes, and only the symmetric state $|\frac{1}{2}\frac{1}{2}, \frac{1}{2}\frac{1}{2}\rangle_{+}$ survives.

ADDITION OF ANGULAR MOMENTA FOR EQUIVALENT ELECTRONS

In describing the possible resultant angular momenta for a collection of equivalent electrons (i.e., electrons described by orbitals with the same n and ℓ) the Pauli principle will limit the allowable combinations. Table 6-2 exhibits the addition of orbital and spin angular momenta for two equivalent electrons. The orbital angular momenta are $\ell_1 = \ell_2 = \ell$. The components, $m_{\ell 1}, m_{\ell 2}$ are listed in columns 1 and 3. The spin is $\frac{1}{2}$ for each electron. The components m_{s1}, m_{s2} are listed in columns 2 and 4; + signifying $\frac{1}{2}$, − signifying $-\frac{1}{2}$. $M_L = m_{\ell 1} + m_{\ell 2}$, $M_S = m_{s1} + m_{s2}$ are listed in columns 5 and 6. Columns 7 and 8 list the possible total orbital and total spin angular momenta that can be obtained by the linear combinations of states with a given M_L, M_S. The quantum numbers $n\ell$ are suppressed, since they are the same throughout. Each state, described by n_1, n_2; ℓ_1, ℓ_2; $m_{\ell 1}, m_{\ell 2}$; s_1, s_2; m_{s1}, m_{s2}; ($n_1 = n_2 = n$, $\ell_1 = \ell_2 = \ell$, $s_1 = s_2 = \frac{1}{2}$), is a Slater determinant of space-spin orbitals.

The entries in columns 1 to 4 of Table 6-2 are determined by the

Table 6-2
Addition of Angular Momenta for Two Equivalent Electrons

$m_{\ell 1}$	m_{s1}	$m_{\ell 2}$	m_{s2}	M_L	M_S	L	S
ℓ	+	ℓ	−	2ℓ	0	2ℓ	0
ℓ	+	$\ell - 1$	+	$2\ell - 1$	1	$2\ell - 1$	1
ℓ	−	$\ell - 1$	−	$2\ell - 1$	−1	$2\ell - 1$	1
ℓ	+	$\ell - 1$	−	$2\ell - 1$	0 }	2ℓ	0
ℓ	−	$\ell - 1$	+	$2\ell - 1$	0 }	$2\ell - 1$	1
ℓ	+	$\ell - 2$	+	$2\ell - 2$	1	$2\ell - 1$	1
ℓ	−	$\ell - 2$	−	$2\ell - 2$	−1	$2\ell - 1$	1
ℓ	+	$\ell - 2$	−	$2\ell - 2$	0 }	$2\ell - 1$	1
ℓ	−	$\ell - 2$	+	$2\ell - 2$	0 }	2ℓ	0
$\ell - 1$	+	$\ell - 1$	−	$2\ell - 2$	0 }	$2\ell - 2$	0

Pauli principle and by the requirement that electrons are indistinguishable. Thus for $m_{\ell 1} = m_{\ell 2} = \ell$, m_{s_1} and m_{s_2} must be different by the Pauli principle, and it does not matter whether we set $m_{s_1} = +$, $m_{s_2} = -$, or vice versa. On the other hand, for $m_{\ell 1} = \ell$ and $m_{\ell 2} = \ell - 1$ it is possible to have $m_{s_1} = m_{s_2}$. If $m_{s_1} \neq m_{s_2}$, it now matters whether $m_{s_1} = +$; $m_{s_2} = -$, or vice versa, because orbitals 1 and 2 are distinguished by having different values of m_ℓ.

The entries in columns 7 and 8 are determined by the following arguments. In the first line, L clearly is 2ℓ, since $M_L = 2\ell$. S in principle can be 1 or 0. In the case of line 1, S cannot be 1, because then the lowering operators S_-, operating on the $M_S = 0$ state, would generate a state with $M_S = -1$ with M_L still equal to 2ℓ. Similarly the raising operator S_+ on the $M_S = 0$ state would generate a $M_S = 1$ state with $M_L = 2\ell$. Thus if S = 1, we could arrive at three states $M_L = 2\ell$, $M_S = -1, 0, 1$. But only one $M_L = 2\ell$ state is available; hence S = 0.

In the next set there are four possibilities with $m_{\ell 1} = \ell$, $m_{\ell 2} = \ell - 1$ and various values of m_{s_1} and m_{s_2}. The first and second lines correspond to S = 1, since $M_S = \pm 1$. Since linear combinations are allowed only between states of the same M_S and M_L no linear combinations of these two are possible. Each one therefore is an eigenstate of L and S with the appropriate eigenvalues. The next two entries have the same M_L and M_S and linear combinations are possible. One linear combination will correspond to $L = 2\ell$, S = 0, the other to $L = 2\ell - 1$, S = 1.

The next set, with $M_L = 2\ell - 2$, has five possible states enumerated. Some linear combinations of these states must correspond to $L = 2\ell - 1$, S = 1. Immediately we conclude that the states in the first two lines are eigenstates of L and S with the indicated eigenvalues. The remaining three states can combine linearly, since they have the same M_L and M_S. One of the linear combinations must correspond to $L = 2\ell - 1$, S = 1; another to $L = 2\ell$, S = 0. The third state must then have a different L from the previous ones, and we conclude that it is $L = 2\ell - 2$. Since there is only one such state, S = 0.

The L and S values have been assigned by enumerating the states of given M_L M_S. Another method depends on considerations of symmetry. The entire wave function must be antisymmetric under interchange of electrons. Thus in the $\ell_1 m_{\ell 1} \ell_2 m_{\ell 2} s_1 m_{s_1} s_2 m_{s_2}$ representation the wave function is a Slater determinant:

$$R_{n\ell}(r_1)R_{n\ell}(r_2)Y_{\ell m_{\ell 1}}(\Omega_1)Y_{\ell m_{\ell 2}}(\Omega_2)\chi_1(1)\chi_2(2)$$

$$-R_{n\ell}(r_2)R_{n\ell}(r_1)Y_{\ell m_{\ell 1}}(\Omega_2)Y_{\ell m_{\ell 2}}(\Omega_1)\chi_1(2)\chi_2(1) \qquad (6\text{-}35)$$

where $R_{n\ell}$ is the radial function of an electron; $Y_{\ell m_{\ell i}}$ is the spherical harmonic and χ_i the spin function of the ith electron.

In the $LM_L SM_S$ representation, the wave function $|LM_L, SM_S\rangle$ is a linear combination of terms of the form (6-35), with $m_{\ell 1} + m_{\ell 2} = M_L$; $m_{s1} + m_{s2} = M_S$. We see that $R_{n\ell}(r_1)R_{n\ell}(r_2)$ will be a common factor of $|LM_L, SM_S\rangle$. Since $R_{n\ell}(r_1)R_{n\ell}(r_2)$ is symmetric the remaining part of $|LM_L, SM_S\rangle$, that is the angular and spin part, must be antisymmetric. If this angular and spin part is a product of two factors, one angular and the other spin, these two factors must have opposite symmetry under interchange. For $M_S = 1$ or $M_S = -1$ the desired factorization will occur, since then $\chi_1 = \alpha$, $\chi_2 = \alpha$ or $\chi_1 = \beta$, $\chi_2 = \beta$, respectively, and $\alpha_1\alpha_2$ or $\beta_1\beta_2$ can be factored out of the entire expression. For $M_S = 0$ we can have $\chi_1 = \alpha$, $\chi_2 = \beta$ $\chi_1 = \beta$, $\chi_2 = \alpha$. Then the angular-spin part of $|LM_L, SM_S\rangle$ can be written as

$$f(\Omega_1, \Omega_2)\alpha_1\beta_2 + g(\Omega_1, \Omega_2)\beta_1\alpha_2 \tag{6-36}$$

Here f and g are sums of products of spherical harmonics. Since (6-36) is an eigenfunction of L^2 and M_L, f and g independently must also be eigenfunctions. But we proved above that for $L = L_1 + L_2$ the spatial eigenfunctions and hence f and g, are either symmetric or antisymmetric, depending on the value of L. But (6-36) is antisymmetric, hence

$$\begin{aligned} f(\Omega_1, \Omega_2)\alpha_1\beta_2 &+ g(\Omega_1, \Omega_2)\beta_1\alpha_2 \\ &= -f(\Omega_2, \Omega_1)\alpha_2\beta_1 - g(\Omega_2, \Omega\ )\beta_2\alpha_1 \\ &= \pm(f(\Omega_1, \Omega_2)\alpha_2\beta_1 + g(\Omega_1, \Omega_2)\alpha_1\beta_2) \end{aligned}$$

$$f(\Omega_1, \Omega_2) = \pm g(\Omega_1, \Omega_2) \tag{6-37}$$

Therefore (6-36) can always be written as

$$f(\Omega_1, \Omega_2)(\alpha_1\beta_2 \pm \beta_1\alpha_2) \tag{6-38}$$

We conclude that for two equivalent electrons, $|LM_L, SM_S\rangle$ always factors into radial, angular, and spin parts, and the angular and spin parts have opposite symmetries under interchange. This means that whenever L is even, S = 0; when L is odd, S = 1. We see in Table 6-2 that this is obeyed. Indeed this could have been used to derive the entries in Table 6-2.

We remind the reader of standard spectroscopic notation and terminology.

A configuration of equivalent electrons is specified by writing ℓ^x, where x is the number of equivalent electrons of angular momentum ℓ. If it is desired to give the principal quantum number, this is written as $n\ell^x$. A configuration of nonequivalent electrons is specified by writing $\ell^x \ell'^y$. Thus two equivalent p electrons are specified by p^2, while two inequivalent p electrons are specified by pp′.

In adding up separately the individual orbital angular momenta and the spin angular momenta for electrons in a configuration, we arrive at states of definite L and S called *multiplets*. The multiplet is specified by writing ^{2S+1}L. The superscript 2S+1 is a numerical quantity called the *multiplicity*. The letter L takes on values S, P, D, etc., when the resultant L is 0, 1, 2, etc.

For three or more electrons it is much more difficult to give any general rules. Table 6-3 exhibits the addition of angular momenta for three equivalent p electrons. There are 20 allowed states (see p. 101). The table can, however, be abbreviated, because for every state with positive M_S, we can obtain a state with $M'_S = -M_S$ by simply changing the signs of all m_{si}. Similarly, we can independently change the signs of all $m_{\ell i}$. Thus we need only list the states of positive or zero M_L, and positive M_S, a total of 7 instead of 20.

The reasoning to establish L and S is again enumerating states. Symmetry arguments do not apply, since we no longer are adding two equal angular momenta. The curly brackets indicate that linear combinations are to be taken. It is seen that one obtains one ^4S, one ^2P, and one ^2D state. The ^4S state has 4 × 1 = 4 magnetic substates M_L, M_S; the ^2P state has 2 × 3 = 6; the ^2D state 2 × 5 = 10 magnetic substates. The total is 20, as expected. This is an important check of the completeness of our table of states.

Table 6-3

Addition of Angular Momenta for Three Equivalent p Electrons

$m_{\ell 1}$	m_{s1}	$m_{\ell 2}$	m_{s2}	$m_{\ell 3}$	m_{s3}	M_L	M_S	L	S
1	+	1	−	0	+	2	$\frac{1}{2}$	2	$\frac{1}{2}$
1	+	1	−	−1	+	1	$\frac{1}{2}$ }	2	$\frac{1}{2}$
1	+	0	+	0	−	1	$\frac{1}{2}$ }	1	$\frac{1}{2}$
1	+	0	+	−1	+	0	$\frac{3}{2}$	0	$\frac{3}{2}$
1	+	0	+	−1	−	0	$\frac{1}{2}$ }	2	$\frac{1}{2}$
1	+	0	−	−1	+	0	$\frac{1}{2}$ }	1	$\frac{1}{2}$
1	−	0	+	−1	+	0	$\frac{1}{2}$ }	0	$\frac{3}{2}$

In all cases we have so far discussed there is only one state for each combination LM_L, SM_S (or none). We shall see in the next chapter that this greatly simplifies the calculation of energy levels. The simplest configuration involving only equivalent electrons for which this is no longer true is d^3. In this case, two ^2D states occur. They cannot be distinguished except by their energy, and the calculation of their energy is more difficult than if there is only one state of given LS. For d^4 there are five pairs of states of the same LS, and for the f shell the occurrence of states of the same LS becomes very common.

PROBLEMS

1. (a) Derive the Clebsch-Gordan coefficient

$$C(LM_L\ SM_S;\ JM) \text{ with } L = S,\ J = 0$$

(b) Use the projection method to derive

$$C(LM_L\,SM_S;\ JM) \text{ with } J = 0,\ L = S = 1$$

(c) Calculate the Clebsch-Gordon coefficients for the addition of two angular momenta j and 1 using the lowering operator.

2. (a) What are the allowed multiplets in a d^3 configuration?
(b) Obtain the ^2P states for a p^3 configuration.
3. For two equivalent electrons show explicitly that the total number of states in the LS representation is equal to that in the $m_\ell m_s$ representation.
4. Prove (6-32).
5. (a) Find the highest multiplicity resulting from the configuration ℓ^x.
(b) Find the largest value of L for multiplets with the highest multiplicity resulting from the configuration ℓ^x.
6. Show explicitly that

$$(L + S)^2_{op}\,|\,J = L - \tfrac{1}{2}, M\rangle = J(J+1)\,|\,J = L - \tfrac{1}{2}, M\rangle$$

with $|\,J = L - \frac{1}{2}, M\rangle$ given by (6-26).

7. Consider an electron, with angular momentum ℓ and spin $\frac{1}{2}$, moving in a central potential. In the $\ell m_\ell s m_s$ representation the electron is described by the space function $R_{n\ell}Y_{\ell m}$, and the spin functions $\alpha = \binom{1}{0}$, $\beta = \binom{0}{1}$. Construct eigenfunctions of j^2 and j_z where $\mathbf{j} = \boldsymbol{\ell} + \mathbf{s}$.

Chapter 7

THEORY OF MULTIPLETS, ELECTROSTATIC INTERACTION

In this chapter, we carry out in detail the program outlined at the beginning of Chapter 6. We concern ourselves with an atom which has an electron configuration distributed over several complete and one incomplete shell. (The assumption that the atom has only one incomplete shell is for notational convenience. The theory is readily extended to atoms with several incomplete shells. We shall work explicitly an example where the electrons lie in different incomplete shells.) The central field approximation provides several — say k — degenerate solutions corresponding to the several different ways the electrons in the incomplete shell can be distributed. Evidently, these degenerate solutions differ among themselves by the assignment of the m_ℓ and m_s quantum numbers to the orbitals in the incomplete shell.

Each solution is a Slater determinant composed of orbitals for all the electrons in the configuration. The individual orbitals are of the form

$$[\mathcal{R}_{n\ell}(r)/r]Y_{\ell m_\ell}(\Omega)\,|\,m_s\rangle = |\,n\ell m_\ell m_s\rangle$$

since we are assuming the central field approximation. The entire Slater determinant may be specified uniquely in the following way. First, the complete shells are identified. Then the incomplete shell is named and a specific assignment of m_ℓ and m_s is exhibited for the orbitals in this incomplete shell.

For example phosphorus, with 3 electrons in the (incomplete) 3p shell has solutions which are specified by

$$1s^2,\ 2s^2,\ 2p^6,\ 3s^2,\ 3p^3:\ m_{\ell 1}\,m_{s1},\ m_{\ell 2}\,m_{s2},\ m_{\ell 3}\,m_{s3}$$

There are 20 such solutions corresponding to the 20 different ways the values for m_ℓ and m_s may be assigned. We call an assignment of m_ℓ and m_s values an *m set*. Phosphorus then has 20 m sets.

For notational economy, we shall represent each of these solutions by a ket in which only the m_ℓ and m_s of the orbitals in the incomplete shell are specified.

$$|m_{\ell 1}\, m_{s1},\ m_{\ell 2}\, m_{s2},\ \ldots,\ m_{\ell N}\, m_{sN}\,(\alpha)\rangle \qquad (7\text{-}1)$$

The index α ranges over all the k degenerate solutions and represents a Slater determinant with a certain m set, viz., a definite certain choice of m_ℓ and m_s for the orbitals in the incomplete shell. It is assumed that there are N electrons in the incomplete shell.

The degeneracy of the central field solutions is now partially removed by including the electrostatic interaction of the electrons as a perturbation. The single level now splits into several *multiplets* of levels, and the energy shift can be calculated by perturbation theory.

According to degenerate perturbation theory, the first-order correction to the energy is obtained by forming linear combinations of the degenerate solutions such that the perturbation Hamiltonian is diagonal in the subspace of the degenerate eigenstates. The first-order energy corrections are then simply the diagonal matrix elements of the perturbing Hamiltonian. Alternatively and equivalently, we may take linear combinations of the degenerate solutions, such that the total Hamiltonian is diagonal in the subspace; and the total energies, to first order, are then given by the diagonal matrix elements.

The problem at hand, therefore, is to obtain linear combinations of the Slater determinants (7-1) such that the total Hamiltonian (1-4) is diagonal. We proceed to solve this problem by first taking those linear combinations which diagonalize L^2, L_z; S^2, S_z. According to (6-7a) the total Hamiltonian has nonvanishing matrix elements only between states of equal LM_L, SM_S. Thus, except in those instances when more than one state of the same LM_L, SM_S can be formed from the individual orbitals, the Hamiltonian is diagonal. When more than one state with the same LM_L, SM_S is available, the Hamiltonian will have off-diagonal matrix elements between the different states of the same LM_L, SM_S. Such a submatrix must then be explicitly diagonalized.

We form the states

$$|LM_L,\ SM_S;\ \lambda\rangle$$
$$= \sum_\alpha C_\alpha(\lambda)\,|m_{\ell 1}\, m_{s1},\ \ldots,\ m_{\ell N}\, m_{sN}\,(\alpha)\rangle \qquad (7\text{-}2)$$

The C_α are the appropriate coefficients which transform from the individual orbital representation to the LM_L, SM_S representation. When two electrons are being combined, the C_α are Clebsch-Gordan coefficients. When more than two are present, the C_α are more general transformation coefficients. The index λ distinguishes between the different states of the same LM_L, SM_S. Such different states may arise when adding three or more electrons (see p. 116). The summation in α is a sum over all the k degenerate eigenstates; i.e., over the m sets.

The energy, correct to first order, is given by the eigenvalues of the matrix

$$\langle LM_L,\ SM_S;\ \lambda' | H | LM_L,\ SM_S;\ \lambda \rangle$$
$$= \sum_{\alpha\beta} C^*_\alpha(\lambda') C_\beta(\lambda)$$
$$\times \langle m_{\ell 1} m_{s1}, \ldots, m_{\ell N} m_{sN}(\alpha) | H | m_{\ell 1} m_{s1}, \ldots, m_{\ell N} m_{sN}(\beta) \rangle \tag{7-3}$$

The summation is over all possible $\alpha\beta$. Note that the assignment of $m_\ell m_s$ in the left-hand state α is in general different from the assignment of $m_\ell m_s$ in the right-hand state β.

DISCUSSION OF MATRIX ELEMENTS

In order to evaluate the matrix elements of the form

$$\langle m_{\ell 1} m_{s1}, \ldots, m_{\ell N} m_{sN}(\alpha) | H | m_{\ell 1} m_{s1}, \ldots, m_{\ell N} m_{sN}(\beta) \rangle$$

which occur in (7-3), we consider matrix elements of H between determinantal wave functions.

$$H = F_1 + F_2 \qquad F_1 = \sum_i f_i \qquad F_2 = \sum_{i<j} g_{ij}$$

$$f = -\frac{\hbar^2}{2m}\nabla^2 - \frac{Ze^2}{r} \qquad g_{12} = \frac{e^2}{r_{12}} \tag{7-4}$$

We note that

$$[f, \mathbf{L}] = 0 \qquad [g_{12}, \mathbf{L}_1 + \mathbf{L}_2] = 0$$
$$[f, \mathbf{S}] = 0 \qquad [g_{12}, \mathbf{S}] = 0 \tag{7-5}$$

Since f_i is a one-electron operator and g_{ij} is a two-electron operator, (4-13) through (4-18) give the appropriate matrix elements.

We consider first $\langle F_1 \rangle$ and examine

$$\sum_{\alpha\beta} C^*_\alpha(\lambda')C_\beta(\lambda)$$
$$\times \langle m_{\ell 1}\, m_{s1}, \ldots, m_{\ell N}\, m_{sN}(\alpha) \,|\, F_1 \,|\, m_{\ell 1}\, m_{s1}, \ldots, m_{\ell N}\, m_{sN}(\beta)\rangle \tag{7-6}$$

According to (4-13) and (4-14) the nonzero contributions in the series (7-6) arise only from two cases: (1) only one orbital different between the initial and final determinants, and (2) all orbitals the same.

In the first case, the contribution is zero for the following reasons. Suppose the ith orbital is different. In that instance according to (4-13) we need to evaluate

$$\int v_i^* f u_i \, d\tau \qquad v_i \neq u_i \tag{4-13}$$

We write this as

$$\int v_i^* f u_i \, d\tau = \langle n_i\, \ell_i\, m'_{\ell i}\, m'_{si} \,|\, f \,|\, n_i\, \ell_i\, m_{\ell i}\, m_{si}\rangle \tag{7-7}$$

where either $m'_{\ell i} \neq m_{\ell i}$ or $m'_{si} \neq m_{si}$ (or both). However since both L_z and S_z commute with f the matrix element vanishes.

Thus only the second case, when all orbitals are the same, between the initial and final states, survives.

$$\langle LM_L, SM_S;\, \lambda' \,|\, F_1 \,|\, LM_L, SM_S;\, \lambda\rangle$$
$$= \sum_\alpha C^*_\alpha(\lambda')C_\alpha(\lambda)$$
$$\times \langle m_{\ell 1}\, m_{s1}, \ldots, m_{\ell N}\, m_{sN}(\alpha) \,|\, F_1 \,|\, m_{\ell 1}\, m_{s1}, \ldots, m_{\ell N}\, m_{sN}(\alpha)\rangle \tag{7-8}$$

The determinantal matrix element in (7-8), according to (4-14), is $\Sigma_i \langle i|f|i\rangle$ where the sum over i goes over all the orbitals occupied in the Slater determinant with an m set α. However since **L** and **S** commute with f, $\langle i|f|i\rangle$ does not depend on $m_{\ell i}$ or m_{si} (see p. 104). Therefore the sum over i is the same for all Slater determinants (m sets) α arising from a given electron configuration.

$$\langle LM_L, SM_S; \lambda' | F_1 | LM_L, SM_S; \lambda \rangle$$

$$= \sum_\alpha C_\alpha^*(\lambda') C_\alpha(\lambda) \sum_i \langle i|f|i \rangle$$

$$= \left(\sum_i \langle i|f|i \rangle \right) \sum_\alpha C_\alpha^*(\lambda') C_\alpha(\lambda)$$

$$= \delta_{\lambda\lambda'} \sum_i \langle i|f|i \rangle \tag{7-9a}$$

This may also be written as

$$\delta_{\lambda\lambda'} \left(\sum_{\substack{i \\ \text{complete} \\ \text{shells}}} \langle i|f|i \rangle + \sum_{\substack{i \\ \text{incomplete} \\ \text{shell}}} \langle i|f|i \rangle \right)$$

$$= \delta_{\lambda\lambda'} \left(\sum_{\substack{i \\ \text{complete} \\ \text{shells}}} \langle i|f|i \rangle + N \langle i|f|i \rangle \right) \tag{7-9b}$$

The state i in the second term in the parentheses on the right-hand side is any state in the incomplete shell. It does not matter which one, since each state gives the same contribution. N is the number of electrons in the incomplete shell.

In much of the following we shall be concerned only with the energy splitting, namely we shall take differences of terms with different m_ℓ or m_s. For such applications we may ignore contributions like (7-9) which are independent of m_ℓ and m_s.

We now examine $\langle F_2 \rangle$.

$$\langle LM_L, SM_S; \lambda' | F_2 | LM_L, SM_S; \lambda \rangle$$

$$= \sum_{\alpha\beta} C_\alpha^*(\lambda') C_\beta(\lambda)$$

$$\times \langle m_{\ell 1} m_{s1}, \ldots, m_{\ell N} m_{sN}(\alpha) | F_2 | m_{\ell 1} m_{s1}, \ldots, m_{\ell N} m_{sN}(\beta) \rangle \tag{7-10}$$

According to (4-15), (4-17), and (4-18) the nonzero contributions to the series (7-10) arise from only three cases: (1) only two orbitals different between the initial and final determinants, (2) only one orbital different, and (3) all orbitals the same.

Case (2) is the simplest as it gives zero. In this instance, the determinantal matrix element in (7-10) is given by (4-17)

$$\sum_{j \neq i} [\langle ij | g | ij \rangle - \langle ij | g | ji \rangle] \tag{4-17}$$

where the ith orbital is different between the initial and final state. The differing orbitals must lie in the incomplete shell, since the Slater determinants have the same complete shells. By hypothesis we have

$$m'_{\ell i} \neq m_{\ell i} \quad \text{or} \quad m'_{si} \neq m_{si} \quad \text{or both} \tag{7-11}$$

where $m'_{\ell i}\, m'_{si}$ are the magnetic quantum numbers of the ith orbital in the initial state, and $m_{\ell i}\, m_{si}$ are the magnetic quantum numbers of the (different) i^{th} orbital in the final state.

Since $\mathbf{S}_1$, $\mathbf{S}_2$ and $\mathbf{L}_1 + \mathbf{L}_2$ commute with g, the direct matrix element $\langle ij | g | ij \rangle$ is zero unless $m'_{\ell i} + m_{\ell j} = m_{\ell i} + m_{\ell j}$ and $m'_{si} = m_{si}$. But this is inconsistent with (7-11), so the direct matrix element vanishes. Similarly the exchange term $\langle ij | g | ji \rangle$ vanishes unless $m'_{\ell i} + m_{\ell j} = m_{\ell j} + m_{\ell i}$ and $m'_{si} = m_{sj} = m_{si}$. This too contradicts (7-11). Thus the exchange term and the entire contribution from the second case vanish.[1]

Next, we consider the diagonal contribution of case (3) to (7-10).

$$\sum_{\alpha} C^*_\alpha(\lambda')\, C_\alpha(\lambda)$$
$$\times \langle m_{\ell 1} m_{s1}, \ldots, m_{\ell N} m_{sN}(\alpha) | F_2 | m_{\ell 1} m_{s2}, \ldots, m_{\ell N} m_{sN}(\alpha) \rangle \tag{7-12}$$

The determinantal matrix element in (7-12) is evaluated with (4-15). This gives

$$\sum_{\alpha} C^*_\alpha(\lambda') C_\alpha(\lambda) \langle m_{\ell 1} m_{s1}, \ldots, m_{\ell N} m_{sN}(\alpha) | F_2 | m_{\ell 1} m_{s1}, \ldots, m_{\ell N} m_{sN}(\alpha) \rangle$$
$$= \sum_{\alpha} C^*_\alpha(\lambda')\, C_\alpha(\lambda) \sum_{i<j} [\langle ij | g | ij \rangle - \langle ij | g | ji \rangle] \tag{7-13}$$

---

[1]The simplification that (4-13) and (4-17) vanish occurs only when there is just one incomplete shell. If there are two or more, the differing orbitals may lie in different incomplete shells. In that case they may have the same magnetic quantum numbers, as they already differ in the principal or azimuthal quantum numbers.

The internal summation is over all pairs of orbitals (i, j) occurring in the Slater determinant with the m set α.

We isolate the contribution to this summation over (i, j) of the orbitals lying in the complete shells. We first write

$$\sum_{i<j} [\langle ij|g|ij\rangle - \langle ij|g|ji\rangle]$$

$$= \frac{1}{2}\sum_{i,j} [\langle ij|g|ij\rangle - \langle ij|g|ji\rangle] \tag{7-14a}$$

We have made use of the fact that the summand is 0 for i = j and is symmetric in i and j. Then each of the individual sums over i and j can be separated into the part over the complete shells and the part over the incomplete shell

$$\frac{1}{2}\sum_{i,j} = \frac{1}{2}\sum_{\substack{i,j\\ \text{complete}\\ \text{shells}}} + \frac{1}{2}\sum_{\substack{i\\ \text{complete}\\ \text{shells}}}\sum_{\substack{j\\ \text{incomplete}\\ \text{shell}}} + \frac{1}{2}\sum_{\substack{i\\ \text{incomplete}\\ \text{shell}}}\sum_{\substack{j\\ \text{complete}\\ \text{shells}}} + \frac{1}{2}\sum_{\substack{i,j\\ \text{incomplete}\\ \text{shell}}}$$

$$= \sum_{\substack{i<j\\ \text{complete}\\ \text{shells}}} + \sum_{\substack{j\\ \text{complete}\\ \text{shells}}}\sum_{\substack{i\\ \text{incomplete}\\ \text{shell}}} + \sum_{\substack{i<j\\ \text{incomplete}\\ \text{shell}}} \tag{7-14b}$$

The first summation, over pairs of orbitals in complete shells, obviously is the same for all determinants α contributing to sum (7-12), and is independent of the magnetic quantum numbers of the incomplete shell.

The same is true for the second summation, where one orbital ranges over the complete shells and the other over the incomplete shell. To see this consider only the summation in j over complete shells, keeping i fixed at some orbital in the incomplete shell

$$\sum_{\substack{j\\ \text{complete}\\ \text{shells}}} \left[\left\langle ij\left|\frac{1}{r_{12}}\right|ij\right\rangle - \left\langle ij\left|\frac{1}{r_{12}}\right|ji\right\rangle\right] \tag{7-15a}$$

We first examine the direct term

$$\sum_{\substack{j\\ \text{complete}\\ \text{shells}}} \left\langle ij \left| \frac{1}{r_{12}} \right| ij \right\rangle$$

$$= \int d\tau_1 \, |u_i(1)|^2 \sum_{\substack{j\\ \text{complete}\\ \text{shells}}} \int d\tau_2 \, |u_j(2)| \frac{1}{r_{12}} \tag{7-15b}$$

The sum in j and the integral on Ω_2 have been performed in (4-44). There we called the result $V_{\text{Coulomb}}(r_1)$ and found that it depends only on r_1 and not on angles. (7-15b) is therefore

$$\int d\tau_1 \, |u_i(1)|^2 V_{\text{Coulomb}}(r_1) \tag{7-15c}$$

The integration over angles and the spin sum now eliminates the dependence on $m_{\ell i}$ and m_{si}.

Similarly, the sum of the exchange term for each complete shell is

$$\sum_{\substack{j\\ \text{complete}\\ \text{shells}}} \left\langle ij \left| \frac{1}{r_{12}} \right| ji \right\rangle$$

$$= \int d\tau_1 \, u_i^*(1) \sum_{m_{\ell j} m_{sj}} \int d\tau_2 \frac{u_j^*(2)\, u_j(1)\, u_i(2)}{r_{12}} \tag{7-15d}$$

This sum was performed in (4-46) and the Ω_2 integral was evaluated in (4-52). It was found that the entire angular dependence of the result is $Y_{\ell_i m_{\ell i}}(\Omega_1)$. When the Ω_1 integration is then performed in (7-15d) the $m_{\ell i}$ dependence disappears. It is obvious that the m_{si} dependence also disappears.

We conclude that the contribution of the second sum in the right-hand side of (7-14b) is the same for each orbital in the incomplete shell. Evidently, (7-13) becomes

$$\sum_{\alpha} C^*_{\alpha}(\lambda') C_{\alpha}(\lambda) \sum_{i<j} [\langle ij|g|ij\rangle - \langle ij|g|ji\rangle]$$

$$= \delta_{\lambda\lambda'} \left(\sum_{\substack{i<j \\ \text{complete} \\ \text{shells}}} [\langle ij|g|ij\rangle - \langle ij|g|ji\rangle] \right.$$

$$\left. + N \sum_{\substack{j \\ \text{complete} \\ \text{shells}}} [\langle ij|g|ij\rangle - \langle ij|g|ji\rangle] \right)$$

$$+ \sum_{\alpha} C^*_{\alpha}(\lambda') C_{\alpha}(\lambda) \sum_{\substack{i<j \\ \text{incomplete} \\ \text{shell}}} [\langle ij|g|ij\rangle - \langle ij|g|ji\rangle] \quad (7\text{-}16)$$

In the second sum in the parentheses, i is any orbital in the incomplete shell; it does not matter which one as the result is independent of $m_{\ell i}$ and m_{si}. N is the total number of electrons in the incomplete shell occurring in the Slater determinant α. The sum in the last term extends over all pairs of orbitals (i, j) contained in the incomplete shell occurring in the Slater determinant α.

Finally, we consider the first case in the contribution to (7-10), namely, just two orbitals different between initial and final states. The determinantal matrix element in (7-10) is given by (4-18)

$$\langle ij|g|ij\rangle - \langle ij|g|ji\rangle \quad (4\text{-}18)$$

where the ith and the jth orbitals are different between the initial and final states. No simplifications occur in this case. We note that both the differing orbitals must lie in the incomplete shell.

We summarize what we have learned. The matrix element of H between LM_L, SM_S eigenstates is

$$
\begin{aligned}
&\langle LM_L, SM_S; \lambda' | H | LM_L, SM_S; \lambda \rangle \\
&= \delta_{\lambda\lambda'} \Bigg(\sum_{\substack{i \\ \text{complete} \\ \text{shells}}} \langle i|f|i\rangle + \sum_{\substack{i<j \\ \text{complete} \\ \text{shells}}} [\langle ij|g|ij\rangle - \langle ij|g|ji\rangle] \\
&\qquad + N\langle i|f|i\rangle + N\sum_{\substack{j \\ \text{complete} \\ \text{shells}}} [\langle ij|g|ij\rangle - \langle ij|g|ji\rangle] \Bigg) \\
&\quad + \sum_{\alpha} C_\alpha^*(\lambda')C_\alpha(\lambda) \sum_{\substack{i<j \\ \text{incomplete} \\ \text{shell}}} [\langle ij|g|ij\rangle - \langle ij|g|ji\rangle] \\
&\quad + \sum_{\alpha\beta}{}' C_\alpha^*(\lambda')C_\beta(\lambda)[\langle ij|g|ij\rangle - \langle ij|g|ji\rangle] \qquad (7\text{-}17)
\end{aligned}
$$

The quantity multiplying $\delta_{\lambda\lambda'}$ is independent of the magnetic quantum numbers and may be ignored when calculating energy differences. In the expressions multiplying N, i refers to any orbital in the incomplete shell, it does not matter which one. The last term in (7-17) arises from the case discussed in the previous paragraph where the initial and final Slater determinants differ by exactly two orbitals in the incomplete shell. Hence the sum in α and β over the Slater determinants is restricted only to those which have the desired property: they differ by exactly two orbitals between the initial and final states. This is indicated by the prime on the summation sign.

Evidently within first-order perturbation theory, the splitting is determined just by the matrix elements of $g = e^2/r_{12}$ between orbitals lying in the incomplete shell. Also since complete shells have zero total L and S, the allowed states LM_L, SM_S are determined only by the orbitals of the incomplete shell.

Let us for the present ignore the possibility that more than one state with the same LM_L, SM_S can be formed from the individual orbitals. In this case a simplification in the method of calculation occurs which permits the evaluation of $\langle H \rangle$ without an explicit knowledge of the C_α and without calculating off-diagonal matrix elements. This is known as the *Slater sum rule* and is analogous to the step-down method of calculating Clebsch-Gordan coefficients.

SLATER SUM RULE

A convenient device for calculating the energy levels is the *Slater sum rule*. This rule is a statement of the well-known mathematical result that the trace of a matrix is invariant under similarity transformation. This implies, disregarding spin,

$$\sum_{\substack{m_1, m_2, \ldots, m_N \\ (\Sigma m_i = M)}} \langle m_1, \ldots, m_N | H | m_1, \ldots, m_N \rangle = \sum_{L \geq |M|} \langle LM | H | LM \rangle \tag{7-18a}$$

The sum on the left-hand side goes over all possible sets of single-electron angular-momentum quantum numbers which satisfy the relation $\Sigma_i m_i = M$. The left-hand side is thus the trace of the Hamiltonian matrix in the m_i representation, more accurately of the submatrix corresponding to a given M. The right-hand side is the sum of the energy eigenvalues for those L that are compatible with M. The transformation from the m_i representation to the LM representation is unitary; hence (7-18a) must hold. When spin is included we get instead

$$\sum_{\alpha} \langle m_{\ell 1} m_{s1}, \ldots, m_{\ell N} m_{sN} (\alpha) | H | m_{\ell 1} m_{s2}, \ldots, m_{\ell N} m_{sN} (\alpha) \rangle = \sum_{L \geq |M_L|} \sum_{S \geq |M_S|} \langle LM_L, SM_S | H | LM_L, SM_S \rangle \tag{7-18b}$$

where the magnetic quantum numbers of the m set α on the left-hand side of (7-18b) must fulfill the condition

$$\sum m_{\ell i} = M_L \qquad \sum m_{si} = M_S \tag{7-18c}$$

The diagonal determinantal matrix element on the left-hand side of (7-18b) is easily expressed using the results of the previous section, specifically (7-8), (7-9), (7-13), and (7-16),

$$\sum_{\alpha} \langle m_{\ell 1}\, m_{s1}, \ldots, m_{\ell N}\, m_{sN}(\alpha) \,|\, H \,|\, m_{\ell 1}\, m_{s1}, \ldots, m_{\ell N}\, m_{sN}(\alpha) \rangle$$

$$= N_1 \left(\sum_{\substack{i \\ \text{complete} \\ \text{shells}}} \langle i|f|i\rangle + \sum_{\substack{i<j \\ \text{complete} \\ \text{shells}}} [\langle ij|g|ij\rangle - \langle ij|g|ji\rangle] \right.$$

$$\left. + N\langle i|f|i\rangle + N\sum_{\substack{j \\ \text{complete} \\ \text{shells}}} [\langle ij|g|ij\rangle - \langle ij|g|ji\rangle] \right)$$

$$+ \sum_{\alpha} \sum_{\substack{i<j \\ \text{incomplete} \\ \text{shell}}} [\langle ij|g|ij\rangle - \langle ij|g|ji\rangle] \qquad \text{(7-19a)}$$

Here $N_1 = \Sigma_{\alpha} 1$ is the number of m sets α satisfying (7-18c). The expression multiplying N_1 is the same for all m sets and may be ignored when calculating energy splitting. The state i in the last two terms in the parentheses multiplying N is any state in the incomplete shell. The sum in the last term extends over all pairs of orbitals (i, j) contained in the incomplete shell occurring in the Slater determinant α.

We shall represent the term which does not contribute to the splitting by IN_1 or by $(I_1 + NI_2)N_1$ where I_1 stands for the first two terms in the parentheses of (7-19a) and NI_2 for the last two terms. I_1 is the total energy of the electrons in the complete shells and I_2 is the kinetic energy and nuclear interaction energy of one electron in the incomplete shell, as well as the interaction energy of this electron with all the electrons in the complete shells. Thus the sum rule becomes

$$\sum_{L \geq |M_L|} \sum_{S \geq |M_S|} \langle LM_L, SM_S | H | LM_L, SM_S \rangle$$

$$= (I_1 + NI_2)N_1 + \sum_{\alpha} \sum_{i<j} [\langle ij|g|ij\rangle - \langle ij|g|ji\rangle] \qquad \text{(7-19b)}$$

An example of an application of the sum rule follows. Consider a system with two nonequivalent electrons having orbital angular momenta ℓ_1 and ℓ_2 and having $m_{s1} = m_{s2} = \frac{1}{2}$. The spin dependence remains fixed at $S = M_S = 1$, hence we suppress it.

We enumerate states:

$m_{\ell 1}$	$m_{\ell 2}$	M_L	L
ℓ_1	ℓ_2	$\ell_1 + \ell_2$	$\ell_1 + \ell_2 = L_0$
$\ell_1 - 1$	ℓ_2	$\ell_1 + \ell_2 - 1$	L_0
ℓ_1	$\ell_2 - 1$	$\ell_1 + \ell_2 - 1$	$L_0 - 1$

On the left-hand side of (7-19b) only one term is present: $\langle LM_L | H | LM_L \rangle$ evaluated for $L = M_L = L_0$. In the right-hand sum of (7-19b), in the m_ℓ representation there is just one state $| m_{\ell 1} = \ell_1, m_{\ell 2} = \ell_2 \rangle$. Thus the sum rule gives

$$\langle L_0 L_0 | H | L_0 L_0 \rangle$$
$$= \left\langle m_{\ell 1} m_{\ell 2} \left| \frac{e^2}{r_{12}} \right| m_{\ell 1} m_{\ell 2} \right\rangle - \left\langle m_{\ell 1} m_{\ell 2} \left| \frac{e^2}{r_{12}} \right| m_{\ell 2} m_{\ell 1} \right\rangle + I$$
$$= \langle m_{\ell 1} = \ell_1, m_{\ell 2} = \ell_2 | H | m_{\ell 1} = \ell_1, m_{\ell 2} = \ell_2 \rangle \tag{7-20}$$

$\langle L_0 L_0 | H | L_0 L_0 \rangle$ is the energy of the state LM_L with $L = M_L = L_0$. However we showed before (p. 104) that $\langle LM_L | H | LM_L \rangle$ is independent of M_L. Hence (7-20) is the energy of all the states $L = L_0$, $|M_L| \leq L_0$, which we write $E(L_0)$.

We now compute the sum of the matrix elements for the two functions corresponding to $M = L_0 - 1$. The right-hand side of (7-19b) in the m_ℓ representation is

$$\langle m_{\ell 1} = \ell_1 - 1, m_{\ell 2} = \ell_2 | H | m_{\ell 1} = \ell_1 - 1, m_{\ell 2} = \ell_2 \rangle$$
$$+ \langle m_{\ell 1} = \ell_1, m_{\ell 2} = \ell_2 - 1 | H | m_{\ell 1} = \ell_1, m_{\ell 2} = \ell_2 - 1 \rangle \tag{7-21}$$

We know by the sum rule that this is equal to the left-hand side of (7-19b)

$$\langle L_0 \; L_0 - 1 | H | L_0 \; L_0 - 1 \rangle$$
$$+ \langle L_0 - 1 \; L_0 - 1 | H | L_0 - 1 \; L_0 - 1 \rangle$$
$$= E(L_0) + E(L_0 - 1) \tag{7-22}$$

Hence subtracting the result we obtained in (7-20) from (7-22) we get $E(L_0 - 1)$. This procedure can now be continued to get the remaining energies.

In the above we obtained the energies for the triplet states by taking $m_{s1} = m_{s2} = \frac{1}{2}$. To obtain the singlet energies we take $m_{s1} = \frac{1}{2}$, $m_{s2} = -\frac{1}{2}$, or $m_{s1} = -\frac{1}{2}$, $m_{s2} = \frac{1}{2}$. With different values of m_s the exchange term vanishes. Corresponding to given $M_L = L_0$, M_S will then equal 0 in the two instances. This then corresponds to $S = 1$ and $S = 0$. Knowing the triplet-state energy allows determination of the singlet-state energy from the sum rule. Since the exchange term which is positive definite (see below) appears with the negative sign in the triplet energy (7-20), and is absent in the diagonal matrix elements for m_{s1}, m_{s2}, it will appear with a positive sign in the singlet energy. This is in agreement with the result for parahelium.

It should be pointed out that the sum rule is not always a sufficient tool for determining the energies. For example, in the case of three nonequivalent s electrons we have states ^4S, ^2S, and ^2S. For $M_S = \frac{3}{2}$, arising from $m_{s1} = m_{s2} = m_{s3} = \frac{1}{2}$, we can get the energy of the ^4S term readily. However, there are three possibilities for $M_S = \frac{1}{2}$. Hence the sum rule will only give the sum of the energies of the two ^2S multiplets. Thus whenever more than one state with the same $LM_L SM_S$ can be formed from the individual orbitals, an explicit diagonalization of the expression (7-17) is necessary.

EQUIVALENCE OF HOLES AND ELECTRONS

We now prove the important and useful result that the level separation are the same for a shell lacking $N < 4\ell + 2$ electrons as for a shell with N electrons present. A shell lacking N electrons is said to contain N holes. This result, due to Heisenberg, removes the need for calculating atoms with shells more than half full.

First, a shell $n\ell$ containing N equivalent electrons has the same multiplets as a shell containing $4\ell + 2 - N$ electrons. This follows from our general procedure of obtaining resultant multiplets for equivalent electrons, as outlined at the end of Chapter 6. If we are dealing with a shell containing $4\ell + 2 - N$ electrons we can list the m_ℓ and m_s of the occupied orbitals as in Table 6-3. We can equally well describe the situation by listing the unoccupied orbitals and obtain a table like Table 6-3 for N electrons. The only difference is that M_L and M_S of a given state are the negatives of the sums of the m_ℓ and m_s for the unoccupied orbitals. But this does not change the permitted values of L and S.

We now examine the energy separations. The diagonal elements (7-13) for a set of $4\ell + 2 - N$ electrons can be written

$$\frac{1}{2}\left(\sum_{\substack{i \\ \text{complete} \\ \text{shell}}} \sum_{\substack{j \\ \text{holes}}} - \sum_{\substack{j \\ \text{complete} \\ \text{shell}}} \quad - \sum_{\substack{i \\ \text{complete} \\ \text{shell}}} \sum_{\substack{j \\ \text{holes}}} \quad + \sum_{\substack{j \\ \text{holes}}} \sum_{\substack{i \\ \text{holes}}}\right) \times (\langle ij|g|ij\rangle - \langle ij|g|ji\rangle) \qquad (7\text{-}23)$$

The first three terms can be neglected as they give a result independent of the magnetic quantum numbers. Hence the contribution of the diagonal elements to the level separation of a set of $4\ell + 2 - N$ electrons in an incomplete shell is the same as the contribution of N electrons.

We now discuss the off-diagonal elements. For one (two) orbitals different, we have an electron shifting from u_i' to u_i (and from u_j' to u_j). This is the same as a hole shifting from u_i to u_i' (and from u_j to u_j'), and will produce the same contribution whether we view it as an electron or as a hole. The desired result is now proved. It is seen that apart from a general displacement in energy due to the over-all number of electrons present, the level structure is quantitatively the same for N electrons as for $4\ell + 2 - N$ electrons.

This theorem can be extended to the case when more than one shell is incomplete. In that instance the result is that an atom with incomplete shells $n\ell$ and $n'\ell'$, containing N and N′ electrons, respectively, has the same level structure (apart from a general displacement) as an atom with $4\ell + 2 - N$ and $4\ell' + 2 - N'$ electrons in the incomplete shells. Note that level structure of an atom with $4\ell + 2 - N$ and N′ electrons in the incomplete shells is in general *not* the same.

EVALUATION OF INTEGRALS

Before proceeding with a specific calculation of the multiplet energies, we evaluate the integrals that we shall need.

The one-electron operator contribution to (7-17) involves $\langle i|f|i\rangle$. This is

$$\langle i|f|i\rangle = \langle n\ell|f|n\ell\rangle$$

$$= \int_0^\infty \left\{\frac{\hbar^2}{2m}\mathcal{R}_{n\ell}\left[-\frac{d^2\mathcal{R}_{n\ell}}{dr^2} + \frac{\ell(\ell+1)}{r^2}\mathcal{R}_{n\ell}\right] - \frac{Ze^2}{r}\mathcal{R}_{n\ell}^2\right\} dr \qquad (7\text{-}24)$$

The general two-electron operator matrix element that we shall need is $\langle ij|g|k\ell\rangle$.

$$\langle 12|g|34\rangle$$

$$= \langle n_1 \ell_1 m_{\ell 1} m_{s_1}, n_2 \ell_2 m_{\ell 2} m_{s_2} \left| \frac{e^2}{r_{12}} \right| n_3 \ell_3 m_{\ell 3} m_{s_3}, n_4 \ell_4 m_{\ell 4} m_{s_4} \rangle \tag{7-25a}$$

The angular and spin parts of this are

$$\delta(m_{s_1} m_{s_3})\, \delta(m_{s_2} m_{s_4}) \int Y^*_{\ell_1 m_{\ell 1}}(\Omega_1) Y^*_{\ell_2 m_{\ell 2}}(\Omega_2) \frac{1}{r_{12}}$$
$$\times Y_{\ell_3 m_{\ell 3}}(\Omega_1) Y_{\ell_4 m_{\ell 4}}(\Omega_2)\, d\Omega_1\, d\Omega_2 \tag{7-25b}$$

Expanding $1/r_{12}$ in terms of spherical harmonics and considering the Ω_1 integral we obtain

$$\int Y^*_{\ell_1 m_{\ell 1}}(\Omega_1) Y_{\ell_3 m_{\ell 3}}(\Omega_1) Y_{k\mu}(\Omega_1)\, d\Omega_1$$

$$= \sqrt{\frac{2k+1}{4\pi}}\, c^k(\ell_1 m_{\ell 1}, \ell_3 m_{\ell 3})\, \delta(\mu, m_{\ell 1} - m_{\ell 3}) \tag{7-25c}$$

This is the definition of the c^k; compare (4-48). The Ω_2 integration gives

$$\sqrt{\frac{2k+1}{4\pi}}\, c^k(\ell_4 m_{\ell 4}, \ell_2 m_{\ell 2})\, \delta(\mu, m_{\ell 4} - m_{\ell 2}) \tag{7-25d}$$

(Note that the subscripts of ℓ and m_ℓ in the argument of c^k are reversed in comparison to the previous equation.) Therefore including the radial part, the entire matrix element is

$$\langle 12|g|34\rangle = \delta(m_{s_1}, m_{s_3})\, \delta(m_{s_2}, m_{s_4})\, \delta(m_{\ell 1} + m_{\ell 2}, m_{\ell 3} + m_{\ell 4})$$
$$\times \sum_{k=0}^{\infty} c^k(\ell_1 m_{\ell 1}, \ell_3 m_{\ell 3})\, c^k(\ell_4 m_{\ell 4}, \ell_2 m_{\ell 2})\, R^k(12, 34) \tag{7-26}$$

where

$$R^k(12, 34)$$

$$= e^2 \int_0^\infty \int_0^\infty \mathcal{R}_{n_1 \ell_1}(r_1) \mathcal{R}_{n_2 \ell_2}(r_2) \mathcal{R}_{n_3 \ell_3}(r_1) \mathcal{R}_{n_4 \ell_4}(r_2) \frac{r_<^k}{r_>^{k+1}}\, dr_1\, dr_2 \tag{7-27}$$

We anticipated the presence of the Kronecker deltas by noting that g commutes with $\mathbf{L}_1 + \mathbf{L}_2$, and $\mathbf{S}_1$, $\mathbf{S}_2$. The sum over k does not go to infinity since, as previously stated (4-49), $k + \ell + \ell'$ must be even and $|\ell - \ell'| \leq k \leq \ell + \ell'$.

We introduce further notation for the case when the initial and final states are the same:

$$F^k(n_i \ell_i, n_j \ell_j) = R^k(ij, ij)$$

$$G^k(n_i \ell_i, n_j \ell_j) = R^k(ij, ji) \tag{7-28}$$

$$a^k(\ell_i m_{\ell i}, \ell_j m_{\ell j}) = c^k(\ell_i m_{\ell i}, \ell_i m_{\ell i}) c^k(\ell_j m_{\ell j}, \ell_j m_{\ell j})$$

$$b^k(\ell_i m_{\ell i}, \ell_j m_{\ell j}) = [c^k(\ell_i m_{\ell i}, \ell_j m_{\ell j})]^2 \tag{7-29}$$

We may express the diagonal contribution which is the relevant quantity in applications of the Slater sum rule as

$$\langle ij|g|ij\rangle - \langle ij|g|ji\rangle$$
$$= \sum_k \{a^k(\ell_i m_{\ell i}, \ell_j m_{\ell j}) F^k(n_i \ell_i, n_j \ell_j)$$
$$- \delta(m_{si} m_{sj}) b^k(\ell_i m_{\ell i}, \ell_j m_{\ell j}) G^k(n_i \ell_i, n_j \ell_j)\} \tag{7-30}$$

We note that from its definition F^k is positive definite. Although a similar statement cannot be made for G^k, it can be shown that the entire contribution from the exchange term $\Sigma_k b_k G^k$ is also positive definite (Slater,[2] Vol. II, p. 286).

SAMPLE CALCULATION

We now work out an example for two equivalent p electrons ($\ell_1 = \ell_2 = 1$). Evidently, k in (7-26) can take on values 0 and 2. The relevant c^k we need are given in Table (7-1). From these the a^k and b^k can be determined by equation (7-29).

The angular momentum states that can be reached are ^1D, ^3P, ^1S. There is one combination which leads to the ^1D, $M_L = 2$ state: $m_{\ell 1} = m_{\ell 2} = 1$; $m_{s1} = -m_{s2} = \frac{1}{2}$. According to the Slater sum rule, we have

---

[2] J. C. Slater, *Quantum Theory of Atomic Structure*, McGraw-Hill, New York, 1960.

Table 7-1
$c^k(1m_1, 1m_2)$ for k = 0 and k = 2

$c^0(1m_1, 1m_2)$

m_2 \ m_1	1	0	−1
1	1	0	0
0	0	1	0
−1	0	0	1

$c^2(1m_1, 1m_2)$

m_2 \ m_1	1	0	−1
1	$-1/5$	$-\sqrt{3}/5$	$-\sqrt{6}/5$
0	$\sqrt{3}/5$	$\sqrt{4}/5$	$\sqrt{3}/5$
−1	$-\sqrt{6}/5$	$-\sqrt{3}/5$	$-1/5$

$$E(^1D) = \langle L = 2 \;\; M_L = 2, \;\; S = 0 \;\; M_S = 0 | H | L = 2 \;\; M_L = 2, S = 0 \;\; M_S = 0\rangle$$

$$= \langle m_{\ell 1} = 1, m_{\ell 2} = 1 \left| \frac{e^2}{r_{12}} \right| m_{\ell 1} = 1, m_{\ell 2} = 1\rangle + I \qquad (7\text{-}31a)$$

The exchange term vanishes since $m_{s1} \neq m_{s2}$. Using (7-30)

$$E(^1D) = a^0(11, 11) F^0(nl, nl) + a^2(11, 11) F^2(nl, nl) + I$$

$$= F^0 + \tfrac{1}{25} F^2 + I \qquad (7\text{-}31b)$$

For the ^3P state there also is only one combination which reaches the ^3P, $M_L = 1$, $M_S = 1$ state: $m_{\ell 1} = 1$, $m_{\ell 2} = 0$, $m_{s2} = m_{s2} = \frac{1}{2}$. A calculation similar to the above gives

$$E(^3P) = F^0 - \tfrac{5}{25} F^2 + I \qquad (7\text{-}32)$$

Finally, to calculate the ^1S state, we note that the three different $m_\ell m_s$ states: $m_{\ell 1} = -m_{\ell 2} = 1$, $m_s = -m_{s2} = \frac{1}{2}$; $m_{\ell 1} = -m_{\ell 2} = 1$, $m_{s1} = -m_{s2} = -\frac{1}{2}$; $m_{\ell 1} = m_{\ell 2} = 0$, $m_{s1} = -m_{s2} = \frac{1}{2}$; all reach the $M_L = 0$, $M_S = 0$ component of the ^1D, ^3P, and ^1S states. According to the sum rule, the sum of the energies of these states equals the sum of the diagonal elements in the $m_\ell m_s$ representation. Evaluating the expressions, we get

$$E(^1D) + E(^3P) + E(^1S) = 3F^0 + \tfrac{6}{25} F^2 + 3I \qquad (7\text{-}33)$$

subtracting out the known values for $E(^2D)$ and $E(^3P)$ leaves

$$E(^1S) = F^0 + \tfrac{10}{25} F^2 + I \qquad (7\text{-}34)$$

Table 7-2
Multiplet Energies for Two Equivalent p Electrons

$m_{\ell 1}$	m_{s1}	$m_{\ell 2}$	m_{s2}	State	a^2	$\delta(m_{s1}, m_{s2})b^2$	Energy
1	+	1	−	^{1}D	1/25	0	1/25 F^2
1	+	0	+	^{3}P	−2/25	3/25	−5/25 F^2
1	+	0	−	^{1}D + ^{3}P	−2/25	0	−4/25 F^2
1	−	0	+				
1	+	−1	+	^{3}P	1/25	6/25	−5/25 F^2
1	+	−1	−	^{1}D + ^{3}P + ^{1}S	1/25	0	6/25 F^2
1	−	−1	+		1/25	0	
0	+	0	−		4/25	0	

Table 7-2 summarizes this calculation. Since k = 0 always contributes the same term F^0 we omit this, as well as the contribution I which arises from the one-electron operators and the (unspecified) complete shells. From Table 7-2 we see that not only can we determine all the energy levels from the sum rules, but also we get three independent determinations of the ^{3}P state which are in agreement. This also happens in many other cases. Without evaluating the radial integrals (which must of course be positive) Table 7-2 or similar tables give the order of the various energy levels arising from a configuration and the ratio of their spacings. The results for several configurations are given in Table 7-3.

Table 7-3
Multiplet Structure for Various Electron Configurations

p^2	d^2	p^3
^{1}D: 1 $F^{2\prime}$	^{1}S: 14 $F^{2\prime\prime}$ + 126 $F^{4\prime\prime}$	^{2}D: − 6 $F^{2\prime}$
^{3}P: − 5 $F^{2\prime}$	^{3}P: 7 $F^{2\prime\prime}$ − 84 $F^{4\prime\prime}$	^{2}P: 0
^{1}S: 10 $F^{2\prime}$	^{1}D: −3 $F^{2\prime\prime}$ + 36 $F^{4\prime\prime}$	^{4}S: − 15 $F^{2\prime}$
	^{3}F: −8 $F^{2\prime\prime}$ − 9 $F^{4\prime\prime}$	
	^{1}G: 4 $F^{2\prime\prime}$ + 1 $F^{4\prime\prime}$	

$$F^{2\prime} = \frac{1}{25} F^2, \; F^{2\prime\prime} = \frac{1}{49} F^2, \; F^{4\prime\prime} = \frac{1}{441} F^4$$

COMPARISON WITH EXPERIMENTAL RESULTS

In Condon and Shortley,[3] pp. 197–207, and Slater,[2] Vol. I, pp. 339–342, are tabulated many comparisons between this theory and experiments. In Table 7-4 we list a typical comparison. It is seen that the agreement is qualitatively good, but quantitatively far from perfect. It is felt that the D and P states probably lie fairly close to the calculated energies but that the S states are displaced to lower energies, owing to interactions with other configurations (see p. 145). The magnitude of the splitting is of the order of 0.1 Rydberg or 1 eV.

A different kind of comparison is to attempt to find a consistent empirical set of values for the radial integrals so that the energy levels can be computed. Slater has deduced for F^2 the value 0.35 Ry

Table 7-4
Theoretical and Experimental Energy Separation Ratios for Various Terms

p^2			p^3		
Atom	Configuration	$\frac{^1S - {}^1D}{^1D - {}^3P}$	Atom	Configuration	$\frac{^2P - {}^2D}{^2D - {}^4S}$
Theory	np^2	1.50	Theory	np^3	0.667
C	$2p^2$	1.13	N	$2p^3$	0.500
N^+	$2p^2$	1.14	O^+	$2p^3$	0.509
O^{++}	$2p^2$	1.14	S^+	$3p^3$	0.651
Si	$3p^2$	1.48	As	$4p^3$	0.715
Ge	$4p^2$	1.50	Sb	$5p^3$	0.908
Sn	$5p^2$	1.39	Bi	$6p^3$	1.121

p^4		
Atom	Configuration	$\frac{^1S - {}^1D}{^1D - {}^3P}$
Theory	np^4	1.50
O	$2p^4$	1.14
Te	$5p^4$	1.50

[3] E. U. Condon and G. H. Shortley, *The Theory of Atomic Spectra*, Cambridge University Press, Cambridge, 1959.

for C and 0.88 for O^{+++}. Increasing Z and degree of ionization q tends to increase F^2. This might be expected because F^k is fundamentally the average of $1/r$, albeit with complicated weighting factors. As Z or q increases, the distance r of the electrons from the nucleus tends to decrease. In interpreting the results in the d^2 case, F^4/F^2 has been assigned a value of 0.75 by Condon and Shortley and 0.55 to 0.6 by Slater. The ^3F state is found to be lowest in energy.

From the many calculations that have been performed, several rules have emerged. (1) The lowest energy goes with the highest spin. Since the highest spin corresponds to a symmetric combination of the individual spin functions, the space function will be antisymmetric and produces the least Coulomb repulsion (see p. 34). (2) Among the multiplets with the highest spin, the highest L gives the lowest energy. High M_L implies the orbits are near the equatorial plane, which permits the electrons to be far apart in the mean, and makes their interaction energy small.

These two rules are known as *Hund's rule*. They have been tested and confirmed in studying many spectra, including those of the rare earths which involve f electrons. The rules only apply to the lowest energy state. It is *not* true that *all* states of maximum S have lower energy than all states of the next smaller S: E.g., in the d^2 configuration, the ^1D state usually lies lower than ^3P; in the d^4, the state ^1I is always lower than the higher of two ^3F states and

Table 7-5

Multiplet Energies for the p^n Configuration

n	Lowest multiplet	Energy	Difference
1	^2P	$I_1 + 1I_2$	
			$I_2 + F^0 - \frac{5}{25}\ F^2$
2	^3P	$I_1 + 2I_2 + F^0 - \frac{5}{25} F^2$	
			$I_2 + 2F^0 - \frac{10}{25} F^2$
3	^4S	$I_1 + 3I_2 + 3F^0 - \frac{15}{25} F^2$	
			$I_2 + 3F^0$
4	^3P	$I_1 + 4I_2 + 6F^0 - \frac{15}{25} F^2$	
			$I_2 + 4F^0 - \frac{5}{25} F^2$
5	^2P	$I_1 + 5I_2 + 10F^0 - \frac{20}{25} F^2$	
			$I_2 + 5F^0 - \frac{10}{25} F^2$
6	^1S	$I_1 + 6I_2 + 15F^0 - \frac{30}{25} F^2$	

also lower than the ^3D state. Similarly, for S less than its maximum value, the state of highest L is not always the lowest: In the simple case of d^2, the lowest singlet state is ^1D, not ^1G. (For the nuclear shell structure, Hund's rule gives the opposite pattern, since the force is attractive.)

Applying this rule we can find the lowest state for a p^n configuration. Table 7-5 lists the lowest multiplet for p^n configurations, interaction between the electrons in the p^n shell for this multiplet, and the difference in energy between n and n − 1 electrons.

Apart from the general increase in the interaction (the $I_2 + nF^0$ terms) due to the addition of more electrons, there is an increase in binding (more negative contribution from F^2) from the first to second and from the second to third. Then there is a drop, and a repetition of the pattern. Thus we have shown that a half-filled p shell (3 electrons) gives a particularly high ionization potential. The same can be shown for half-complete d- and f-shells. This behavior was observed in the ionization potential of various atoms (see p. 81).

AVERAGE ENERGY

In what has been done above, the contribution which does not depend on m_ℓ or m_s, that is, the contribution that is the same for all multiplets of a given configuration, has been ignored. We now calculate a quantity which is closely related to this common energy of the configuration.

The object that will concern us here is the ***average energy*** W, which is defined as the average of the energies of the multiplets, weighted by the multiplicities.

$$W = \frac{\Sigma_{LS}(2L+1)(2S+1)E(L,S)}{\Sigma_{LS}(2L+1)(2S+1)} \qquad (7\text{-}35)$$

The denominator of this expression is simply the total number of states

$$\sum_{LS}(2L+1)(2S+1) = \binom{4\ell+2}{N} \qquad (7\text{-}36)$$

The numerator is the sum of the energy eigenvalues in the LS representation, which according to the Slater sum rule is the sum of the diagonal elements of the Hamiltonian in the $m_\ell m_s$ representation

$$\sum_{LS} (2L+1)(2S+1)E(L,S) = \sum_{\alpha} \langle m_{\ell 1} m_{s1}, \ldots, m_{\ell N} m_{sN}(\alpha) | H | m_{\ell 1} m_{s1}, \ldots, m_{\ell N} m_{sN}(\alpha) \rangle \tag{7-37}$$

The concept of average energy is useful because the right-hand side of (7-37) can be evaluated explicitly, which we proceed to do below, and because W has a simple physical significance.

According to the usual analysis of diagonal determinantal matrix elements, which we have presented before (see (7-19)), the right-hand side of (7-37) is

$$\left\{ \sum_{\substack{i \\ \text{complete} \\ \text{shells}}} \langle i|f|i\rangle + \sum_{\substack{i<j \\ \text{complete} \\ \text{shells}}} [\langle ij|g|ij\rangle - \langle ij|g|ji\rangle] + N\langle i|f|i\rangle + N \sum_{\substack{j \\ \text{complete} \\ \text{shells}}} [\langle ij|g|ij\rangle - \langle ij|g|ji\rangle] \right\} \sum_{\alpha} 1 + \sum_{\alpha} \sum_{\substack{i<j \\ \text{incomplete} \\ \text{shell}}} [\langle ij|g|ij\rangle - \langle ij|g|ji\rangle] \tag{7-38}$$

Here as before the index i in the expressions multiplied by N refers to any orbital in the incomplete shell. The sum in the last term extends over all pairs of orbitals (i, j) contained in the incomplete shell occurring in the Slater determinant α. The sum over α extends over all states, hence

$$\sum_{\alpha} 1 = \binom{4\ell+2}{N} \tag{7-39}$$

and the average energy is

$$W = \sum_{\substack{i \\ \text{complete} \\ \text{shells}}} \langle i|f|i\rangle + \sum_{\substack{i<j \\ \text{complete} \\ \text{shells}}} [\langle ij|g|ij\rangle - \langle ij|g|ji\rangle]$$

$$+ N\langle i|f|i\rangle + N \sum_{\substack{j \\ \text{complete} \\ \text{shells}}} [\langle ij|g|ij\rangle - \langle ij|g|ji\rangle]$$

$$+ \binom{4\ell+2}{N}^{-1} \sum_{\alpha} \sum_{\substack{i<j \\ \text{incomplete} \\ \text{shell}}} [\langle ij|g|ij\rangle - \langle ij|g|ji\rangle] \qquad (7\text{-}40)$$

We now simplify the last term in the above.

The sum over ij in the incomplete $n\ell$ in the last term in (7-40), runs over all pairs occurring in the determinant α. Since we also sum over α, the overall sum runs several times through *all* the states which can lie in the shell; i.e.,

$$\sum_{\alpha} \sum_{\substack{i<j \\ \text{incomplete} \\ \text{shell } n\ell}} = C \sum_{\substack{i<j \\ \text{all states of} \\ \text{shell } n\ell}} \qquad (7\text{-}41)$$

The constant C, which in general is different from one, is present since each term $\langle ij|g|ij\rangle - \langle ij|g|ji\rangle$ of the summand on the left-hand side is counted more than once as α and ij run over their ranges. To evaluate C, we need to know how many times a specific pair ij occurs in the left-hand sum of (7-41). Since there are $\binom{4\ell+2}{N}$ determinantal wave functions, each containing N orbitals of the incomplete shell, a given pair ij occurs in $\binom{4\ell}{N-2}$ of these, as this is the number of ways we can chose $N - 2$ additional orbitals to supplement the specific pair ij. This is obviously the same for all pairs since every pair is equally likely to occur. Thus we conclude that C is $\binom{4\ell}{N-2}$. The last term of (7-40) is

$$\binom{4\ell+2}{N}^{-1} \binom{4\ell}{N-2} \sum_{i<j} = \frac{N(N-1)}{(4\ell+2)(4\ell+1)} \tfrac{1}{2} \sum_{i,j} \qquad (7\text{-}42)$$

The sum over ij in (7-42) is now over all the orbitals that can lie in the incomplete shell or we may say that the sum runs over the complete shell $n\ell$. If we fix i, the sum over j is, as we have shown before, independent of $m_\ell m_s$; thus it gives the same contribution for every i. As there are $4\ell + 2$ states through which i runs, we obtain finally for (7-42)

$$\frac{N(N-1)}{2}\,\frac{1}{4\ell+1}\sum_{\substack{j\\ \text{complete}\\ \text{shell } n\ell}}\left[\langle ij|g|ij\rangle - \langle ij|g|ji\rangle\right] \tag{7-43}$$

This has an obvious physical meaning. The orbital i has $4\ell + 1$ interactions with the other orbitals in a complete shell. Therefore $(1/4\ell + 1)\Sigma_j$ is the average electrostatic interaction of one orbital. In an incomplete shell of N orbitals, there are $N(N - 1)/2$ interacting pairs. Thus (7-43) is simply the number of interacting pairs times the average interaction.

Our final result for the average energy of a configuration with several complete shells and an incomplete shell $n\ell$ occupied by N orbitals is

$$W = \sum_{\substack{i\\ \text{complete}\\ \text{shells}}}\langle i|f|i\rangle + \sum_{\substack{i<j\\ \text{complete}\\ \text{shells}}}\left[\langle ij|g|ij\rangle - \langle ij|g|ji\rangle\right]$$

$$+ N\langle i|f|i\rangle + N\sum_{\substack{j\\ \text{complete}\\ \text{shells}}}\left[\langle ij|g|ij\rangle - \langle ij|g|ji\rangle\right]$$

$$+ \frac{N(N-1)}{2}w \tag{7-44a}$$

$$w = \frac{1}{4\ell+1}\sum_{\substack{j\\ \text{complete}\\ \text{shell } n\ell}}\left[\langle ij|g|ij\rangle - \langle ij|g|ji\rangle\right] \tag{7-44b}$$

An explicit expression for w may be obtained in terms of the radial integrals F^k defined in (7-28). The direct term is

$$\sum_{\substack{j \\ \text{complete} \\ \text{shell } n\ell}} \langle ij|g|ij\rangle$$

$$= e^2 \int d\tau_1 \, |u_{n\ell m_{\ell i}}(\mathbf{r}_1)|^2 \sum_{\substack{m_{\ell j} \\ m_{sj}}} \int d\tau_2 \, |u_{n\ell m_{\ell j}}(\mathbf{r}_2)|^2 \frac{1}{r_{12}} \qquad \text{(7-45a)}$$

The sum and the Ω_2 integral were performed in (4-44). Inserting that result here gives

$$\sum_{\substack{j \\ \text{complete} \\ \text{shell } n\ell}} \langle ij|g|ij\rangle$$

$$= 2(2\ell + 1)\, e^2 \int d\tau_1 \, |u_{n\ell m_{\ell i}}(\mathbf{r}_1)|^2 \int_0^\infty dr_2 \, \mathcal{R}^2_{n\ell}(r_2) \frac{1}{r_>} \qquad \text{(7-45b)}$$

Performing the Ω_1 integral leaves

$$\sum_{\substack{j \\ \text{complete} \\ \text{shell } n\ell}} \langle ij|g|ij\rangle$$

$$= 2(\ell + 1)\, e^2 \int_0^\infty dr_1 \, dr_2 \, \mathcal{R}^2_{n\ell}(r_1)\, \mathcal{R}^2_{n\ell}(r_2) \frac{1}{r_>}$$

$$= (4\ell + 2)\, F^0(n\ell, n\ell) \qquad \text{(7-45c)}$$

The exchange term is

$$\sum_{\substack{j \\ \text{complete} \\ \text{shell } n\ell}} \langle ij|g|ji\rangle$$

$$= e^2 \int d\tau_1 \, u^*_{n\ell m_{\ell i}}(\mathbf{r}_2) \sum_{\substack{m_{\ell j} \\ m_{sj}}} \delta(m_{si}, m_{sj})$$

$$\times \int d\tau_2 \, u^*_{n\ell m_{\ell j}}(\mathbf{r}_2)\, u_{n\ell m_{\ell j}}(\mathbf{r}_1)\, u_{n\ell m_{\ell i}}(\mathbf{r}_2) \frac{1}{r_{12}} \qquad \text{(7-46a)}$$

The sum and the Ω_2 integral were given in (4-46) and (4-49). Inserting that result here yields

$$\sum_{\substack{j \\ \text{complete} \\ \text{shell } n\ell}} \langle ij|g|ji\rangle = \sum_{k=0}^{2\ell} c^k(\ell 0, \ell 0)\, e^2 \int d\tau_1\, u^*_{n\ell m_{\ell i}}(\mathbf{r}_1)$$

$$\times \int_0^\infty dr_2\, \mathcal{R}^2_{n\ell}(r_2) \frac{\mathcal{R}_{n\ell}(r_1)}{r_1} \frac{r_<^k}{r_>^{k+1}} Y_{\ell m_{\ell i}}(\Omega_1) \qquad (7\text{-}46b)$$

Performing the Ω_1 integral leaves

$$\sum_{\substack{j \\ \text{complete} \\ \text{shell } n\ell}} \langle ij|g|ji\rangle$$

$$= \sum_{k=0}^{2\ell} c^k(\ell 0, \ell 0)\, e^2 \int_0^\infty dr_1\, dr_2\, \mathcal{R}^2_{n\ell}(r_1)\, \mathcal{R}^2_{n\ell}(r_2) \frac{r_<^k}{r_>^{k+1}}$$

$$= F^0(n\ell, n\ell) + \sum_{k=2}^{2\ell} c^k(\ell 0, \ell 0)\, F^k(n\ell, n\ell) \qquad (7\text{-}46c)$$

We have separated the $k = 0$ term and used $c^0(\ell 0, \ell 0) = 1$.
Combining (7-45c) and (7-46c) gives

$$w = F^0(n\ell, n\ell) - \frac{1}{4\ell + 1} \sum_{k=2}^{2\ell} c^k(\ell 0, \ell 0)\, F^k(n\ell, n\ell) \qquad (7\text{-}47)$$

Using the results in Tables 7-2 or 7-3, we can easily verify the expression (7-44a) for W, with w given by (7-47) and W defined by (7-35). In the literature one frequently finds the multiplet energies given in terms of W. This will obviously change the expressions for the energy, since the reference point is not as before $I_1 + NI_2$, but $I_1 + NI_2 + [N(N-1)/2]w$. The energy differences remain, of course, the same.

The physical significance of the average energy is that it is just the Hartree-Fock energy

$$E = \sum_i \langle i|f|i\rangle + \sum_{i<j} [\langle ij|g|ij\rangle - \langle ij|g|ji\rangle] \qquad (7\text{-}48)$$

when the central field approximation has been made. This may be verified by calculating (7-48) with the help of the central field equations (4-55) and (4-56). Thus W is the unperturbed energy, and it is quite natural to express the energy levels of an atom, correct through first order, in terms of W.

We may apply Koopmans' argument (Chapter 4) to obtain the average removal energy of an electron in the incomplete shell. Using the Koopmans' approximation, viz., taking the eigenfunctions of the ion to be the same as those of the atom, we get from (7-44) for minus the removal energy of the ith electron in the incomplete shell the expression

$$\langle i|f|i\rangle + \sum_{\substack{j \\ \text{complete} \\ \text{shells}}} [\langle ij|g|ij\rangle - \langle ij|g|ji\rangle] + (N-1)w \qquad (7\text{-}49)$$

It can be easily verified that this is just the Hartree-Fock eigenvalue for an electron in an incomplete shell in the central field approximation (see (4-58)).

CONFIGURATION INTERACTION

By ignoring all spin effects in the Hamiltonian we found that **L** and **S** commute with the Hamiltonian. We saw that **L** and **S** can be separately quantized, and this is called Russell-Saunders, or LS, coupling. Strictly speaking, L, S, M_L, and M_S are the only good quantum numbers. Parity is determined by the sum of the individual ℓ's. Hence even though the individual ℓ's are not good quantum numbers, the evenness or oddness of their sum remains a good quantum number. However, since the difference between the energy levels arising from different configurations is in general large compared to the electrostatic interaction energy, we assumed that the Hartree-Fock equations yield suitable zero-order wave functions from which to find the interaction energies. More precisely, we should allow for mixing, because several configurations may lead to a multiplet of the same LS. This is the second approximation, the zeroth being the Hartree-Fock, the first, the electrostatic interaction.

We are led to construct the submatrix of the Hamiltonian which connects the configurations contributing to a given LS. We label the rows and columns by the different contributing configurations. In the discussion above we have considered only diagonal elements of this matrix, i.e., elements between the same configuration. Now we take into account the entire Hamiltonian and diagonalize.

If only two configurations contribute, the Hamiltonian is

$$\begin{pmatrix} H_{aa} & H_{ab} \\ H_{ab}^* & H_{bb} \end{pmatrix} \tag{7-50}$$

The energy eigenvalues become

$$\epsilon = \tfrac{1}{2}(H_{aa} + H_{bb}) \pm \sqrt{\tfrac{1}{4}(H_{bb} - H_{aa})^2 + |H_{ab}|^2} \tag{7-51}$$

The square root is greater than $\frac{1}{2}(H_{bb} - H_{aa})$. Hence H_{ab}, which represents the degree of mixing, will spread the eigenvalues farther apart. A singlet S state, for example, is depressed by its interaction with higher singlet S states.

ELECTRON CORRELATION

The Hartree-Fock wave function is of course not accurate because it treats each electron separately. In reality, the electrons are correlated. Due to their mutual repulsion, any pair of electrons tend to be somewhat more distant from each other than the Hartree-Fock wave function would indicate.

The basic theory of electron correlation was developed by R. K. Nesbet.[4] His theory is patterned on that of Brueckner, Bethe and Goldstone for the treatment of the correlation energy of nucleons in the nucleus. The correlation between any pair of electrons ij is treated as independent of all other electrons. Nesbet[4] starts from a Hartree-Fock wave function Φ_0, and then adds perturbations in which either one or both the electrons ij are excited, so that the wave function has the form

$$\Psi_{ij} = \Phi_0 + \sum_a c_i^a \Phi_i^a + \sum_b c_j^b \Phi_j^b + \sum_{ab} \Phi_{ij}^{ab} \ . \tag{7-52}$$

---

[4]R. K. Nesbet, *Phys. Rev.* **175**, 2(1968). This also contains references to the work of others.

Here Φ_i^a means a wave function of the entire atom in which electron i is promoted to the excited state a, while all other electrons remain in their Hartree-Fock states. This is considered a trial function, and the expectation value of the Hamiltonian is minimized, by choice of the coefficients c_i^a and the functions Φ_{ij}^{ab}. In this calculation, one must not exclude excited states which violate the Pauli principle; this was proved by Goldstone for the nuclear case.

One very important result is that the sum of pair correlations is sufficient to give the total energy of the atom very accurately. Correlations between three electrons give very little contribution; the total is less than 0.002 Ry for all atoms up to Ne. Since the interaction is only between pairs of electrons it is reasonable that the correlation in the wave function is also essentially only between pairs.

Another interesting result is that the correlation energy between any pair of electrons is very insensitive to Z, changing only by a few percent between Be and Ne (Z = 4 and Z = 10). Correlation energies are given in Table 7.6 for atoms between Z = 3 and 10, the first full line of the

Table 7-6

Correlation Energies (all are negative) for Electrons in Atoms from Be to Ne, in Rydbergs

Electron pair	State	Energy
$1s^2$	^1S	.0799
1s2s	^1S	.0075
	^3S	.0009
1s2p	^1P	.0044
	^3P	.0030
$2s^2$	^1S	.0217
2s2p	^3P	.0066
	^1P	.0346
$2p^2$	^3P	.0218
	^1D	.0330
	^1S	.0882

Table 7-7

Total Correlation Energies
Comparison with Observations
(in Rydbergs, minus sign omitted)

	Calculated	Observed
Be	.186	.188
B	.245	.248
C	.306	.310
N	.369	.372
O	.500	.508
F	.630	.632
Ne	.764	.762

periodic table. For the interaction of a pair like $2p^2$, three arrangements are given, as discussed in Table 7-2, viz., ^3P ^1D ^1S. The correlation energy for ^1S is largest, because in the Hartree-Fock wave function the p electrons are closest together in the ^1S state. Correlation pushes them apart, and makes more difference for ^1S than, e.g., for ^3P. The correlation energy for a given state of an atom, e.g., the ^3P state of oxygen, is obtained by adding the correlation energies of all the 2p pairs contained in that state of O.

All correlation energies are of course negative. They are given in Table 7-6 in Ry units. The correlation between electrons in the same shell, e.g., both in the 2p shell, are larger than those for two electrons in different shells, e.g., one in the 1s and the other in the 2s or 2p. The largest correlation energy is for two 2p electrons in the ^1S state; this is .088 Ry = 1.19 eV. On the other hand, that between a 1s and a 2s electron in the ^3S state, is only .0009 Ry, i.e., about .01 eV. Since there are many electron pairs in an atom like Ne, the total correlation energy is quite substantial, .76 Ry = 10.3 eV. But in the ionization energy, i.e., the difference between the energy of Ne and that of Ne^+, the contribution of the correlation energy is of course much smaller.

Table 7-7 gives the total correlation energies, as calculated by Nesbet and as observed. The observed energy is obtained by taking the spectroscopic energy level, and subtracting the Hartree-Fock energy. The agreement is remarkably close, generally within .04 eV. The problem of calculating accurate atomic energies can therefore be considered as solved, at least for atoms up to Ne.

Also for heavier atoms the method is very promising because the correlation energy of a given pair of electron orbitals is so insensitive to Z. It should therefore be satisfactory to use the correlation energy for 2p electrons, as calculated for $Z < 10$, also for larger Z. Then, for such an atom, one would need only to calculate the correlation energies of the electrons in orbitals with $n > 2$. Of course complications will enter when relativity corrections become important.

Correlation energies have also been calculated for molecules. In N_2 for instance, the correlation energy contributes half the binding energy as compared to two atoms of N. The accuracy achievable with Nesbet's independent pair method is good enough to give useful results for molecular binding energies.

PROBLEMS

1. Calculate the electrostatic energy of the various multiplets of highest multiplicity for the configuration d^7, in terms of F^2 and F^4.
2. Find the electrostatic energy of multiplets in a d^2 configuration in terms of F^2 and F^4, thus verifying column 2 of Table 7-3.
3. Verify the entries in Table 7-2.
4. Using the results summarized in Tables 7-2 and 7-3, calculate the average energy W for two equivalent p electrons from the definition (7-35), as well as from the formulas (7-44), (7-47). Retain only the terms involving F^2.
5. Show explicitly the connection between the average energy W, and the Hartree-Fock energy given by (7-48) in the central field approximation.

Chapter 8

THEORY OF MULTIPLETS, SPIN-ORBIT INTERACTION, AND INTERACTIONS WITH EXTERNAL FIELDS

In our study of the level structure of atoms, we have been neglecting the spin-orbit interaction by assuming that the energy of the atomic levels is due only to the Coulomb electrostatic interaction. In this chapter we examine this spin-orbit interaction, as well as the interaction of the electron configuration with prescribed external fields. The effects arising from the orbital angular momentum of the electrons are readily handled by the theory as presented so far. The interactions involving the electron spin, however, can be dealt with only in an *ad hoc* fashion; the more satisfactory treatment, requiring the Dirac theory, we postpone to Chapter 23.

INTERACTION WITH A CONSTANT EXTERNAL MAGNETIC FIELD

The quantum mechanical (nonrelativistic) description of a charged (spinless) mass point in a general external magnetic field, described by a vector potential **A**, can be effected by adding to the Hamiltonian the terms

$$H_{mag} = \frac{iq\hbar}{mc}\mathbf{A}\cdot\nabla + \frac{q^2}{2mc^2}A^2 \qquad (8\text{-}1)$$

(Schiff,[1] p. 177. We are in a gauge where $\nabla\cdot\mathbf{A}$ and the scalar potential are zero. This is always possible if the field has no sources. The charge of the particle is taken to be q.)

A solenoidal vector potential for a constant magnetic field $\mathcal{H}$ can be chosen to be

[1] L. I. Schiff, *Quantum Mechanics*, 3rd ed., McGraw-Hill, New York, 1968.

$$\mathbf{A} = \tfrac{1}{2}\mathcal{H} \times \mathbf{r} \tag{8-2}$$

The first term of (8-1) then reads

$$-\frac{q}{2mc}\,\mathcal{H} \times \mathbf{r}\cdot\mathbf{p} = -\frac{q}{2mc}\,\mathcal{H}\cdot\mathbf{r}\times\mathbf{p} = -\frac{q}{2mc}\,\mathcal{H}\cdot\boldsymbol{\ell}$$

where $\boldsymbol{\ell}$ is the orbital momentum operator for the particle. Equation (8-1) becomes

$$H_{mag} = -\frac{q}{2mc}\,\mathcal{H}\cdot\boldsymbol{\ell} + \frac{q^2}{8mc^2}(\mathcal{H}\times\mathbf{r})^2 \tag{8-3}$$

We shall estimate the magnitude of the terms in (8-3) for an atom; using atomic units: $r \approx 1$, ℓ of an electron is of order 1, and q is $-e = -1$. $\mathcal{H}$ is at most 30 kilogauss in Zeeman effect measurements, or 1.8×10^{-3} atomic units. The first term is then $\sim 0.65 \times 10^{-5}$ atomic units, $\sim 1.5\ \text{cm}^{-1}$, which is easily measurable. If $\mathcal{H}$ is as high as 200 kilogauss, the second term is $\sim 0.0002\ \text{cm}^{-1}$ and is clearly negligible. Therefore we neglect the quadratic term for the present. We discuss the quadratic term on p. 171.

The effect of the magnetic field can now be described by ascribing to the particle a magnetic dipole moment

$$\mathbf{M}_\ell = g\boldsymbol{\ell} \tag{8-4a}$$

where

$$g = \frac{q}{2mc} \tag{8-4b}$$

is the *gyromagnetic ratio*. If we orient our coordinates such that the z axis is along $\mathcal{H}$, the contribution to the Hamiltonian from the first term in (8-3) becomes $-g\mathcal{H}(\hbar/i)(\partial/\partial\varphi)$. We write the Schrödinger equation

$$H^0\Psi - g\mathcal{H}\frac{\hbar}{i}\frac{\partial}{\partial\varphi}\Psi = E\Psi \tag{8-5}$$

If the potential occurring in H^0 is spherically symmetric, we set $\Psi = R_{n\ell}Y_{\ell m_\ell}$ and chose Ψ to be an eigenfunction of H^0

$$H^0\Psi = E^0\Psi \tag{8-6a}$$

Then this Ψ also solves (8-5) if we chose E to be

$$E = E^0 - m_\ell g\hbar\mathcal{H} \tag{8-6b}$$

where m_ℓ is the magnetic quantum number. Thus the interaction with a constant magnetic field shifts the energy by an amount $m_\ell g_m \hbar \mathcal{H}$, $-\ell \leq m_\ell \leq \ell$, and removes the m_ℓ degeneracy in energy. The last term in (8-6b) is obviously the energy which we should expect for the interaction of the magnetic field with the magnetic moment (8-4), viz.,

$$E_{mag} = -\mathcal{H} \cdot \mathbf{M}_\ell \tag{8-7a}$$

with $\mathbf{M}_\ell$ given by (8-4). We may consider (8-7a) as the definition of the magnetic moment if E is linear in $\mathcal{H}$; otherwise

$$\mathbf{M}_\ell = -\frac{\partial E}{\partial \mathcal{H}} \tag{8-7b}$$

If the particle is an electron it has spin $\mathbf{s}$, and the spin magnetic moment is found experimentally to be

$$\mathbf{M}_s = 2g_m \mathbf{s} \times (1.00116)$$
$$g_m = \frac{-e}{2mc} \tag{8-8}$$

where the magnetic moment is again defined from the magnetic energy by (8-7). The statements (8-7) and (8-8) are verified by Stern-Gerlach experiments or by analysis of the Zeeman effect.

The minus sign is present in g_m since the charge of the electron is taken to be -e. The factor 2 in M_s is the so-called *Dirac moment* and follows naturally from the Dirac theory; see Chapter 23. The factor 1.00116 is a field theoretic effect, called a *radiative correction*. It has been observed and can be predicted by field theory. The part 0.00116 is called the *anamalous moment*.

SPIN-ORBIT INTERACTION FOR ATOMS

The spin-orbit interaction is due to the interaction of the magnetic moment of the electron with the magnetic field set up by its motion. Since this effect is entirely relativistic we can expect that only the Dirac theory can give a complete analysis. However, we can give here a pseudo-derivation which gives the correct result.

If one electron is moving with a velocity $\mathbf{v}$ relative to the entire atomic configuration, then viewed from the electron's rest fame, the nucleus and the other electrons are moving with a velocity $-\mathbf{v}$ and so is their effective field $\mathcal{E}$. Associated with the moving electric field is a magnetic field arising from the relativistic transformation equation, which to first order in v/c is (Panofsky and Phillips,[2] p. 330).

$$\mathcal{H} = \frac{1}{c}\,\boldsymbol{\mathcal{E}} \times \mathbf{v} = -\frac{1}{c}\,\nabla\phi \times \mathbf{v} \tag{8-9a}$$

(We keep only terms in first-order v/c, since we shall fit this into the nonrelativistic Schrödinger theory.) If ϕ, the effective potential, is spherically symmetric (8-9a) gives, with the substitution $-e\phi = V$ and further rearrangement:

$$\mathcal{H} = \frac{1}{mec}\,\frac{dV}{dr}\,\frac{\mathbf{r}}{r} \times \mathbf{p}$$

$$= \frac{1}{mecr}\,\frac{dV}{dr}\,\boldsymbol{\ell} \tag{8-9b}$$

Combining (8-4), (8-7), (8-8), and (8-9) we obtain the interaction energy,

$$\frac{1}{m^2c^2r}\,\frac{dV}{dr}\,\boldsymbol{\ell}\cdot\mathbf{s} \tag{8-10}$$

(The radiative correction is ignored. Notice that in its rest frame the electron has no orbital magnetic moment; hence no ℓ^2 term occurs.) Equation (8-10) also holds in the rest frame of the nucleus, since to first order in v/c the energy is the same. However, there is a further numerical factor. If one views the electron classically as a spinning top, then the relativistic kinematical transformation laws involve a factor of ½, an effect known as *Thomas precession*. Therefore the correct spin-orbit interaction energy is given by

$$H_{SO} = \frac{1}{2m^2c^2}\left(\frac{1}{r}\,\frac{dV}{dr}\right)\boldsymbol{\ell}\cdot\mathbf{s} \tag{8-11a}$$

This formula will be derived from the Dirac theory in Chapter 23. For several electrons (8-11a) becomes

$$H_{SO} = \frac{1}{2m^2c^2}\sum_i\left(\frac{1}{r}\,\frac{dV}{dr}\right)_i\boldsymbol{\ell}_i\cdot\mathbf{s}_i \tag{8-11b}$$

---

[2] W. K. H. Panofsky and M. Phillips, *Classical Electricity and Magnetism*, 2nd ed., Addison-Wesley, Reading, MA, 1962.

As long as the spin-orbit interaction is much smaller than the level separation of the various LS terms, i.e., much smaller than the electrostatic interaction, we can consider (8-11) as a perturbation and evaluate its diagonal elements in a scheme labeled by SLJM, even though **S** and **L** no longer commute with the Hamiltonian. **J**, of course, is still a constant of motion, hence we may quantize J^2 and J_z. Accordingly the energy states are described by L and S and then for a given LS further splitting occurs corresponding to different resultant J.

Conventionally, we say that a definite *multiplet* ${}^{2S+1}L$ splits into several *terms* of different J. A term is represented by ${}^{2S+1}L_J$ and J varies between L + S and |L − S|.

The replacement of the actual electric field at the site of the ith electron by $(-\mathbf{r}/r)(dV/dr)$ is of course an approximation. In addition to the evident assumption of spherical symmetry, this approximation involves the assumption that the electric forces on the ith electron can be described by a potential. This is not rigorously true in a Hartree-Fock theory due to the presence of the nonlocal exchange potential.

We shall now show that the contribution to (8-11b) arising from the electrons in complete shells may be ignored. In calculating matrix elements of H_{so}, we may work in the $m_\ell m_s$ representation. An arbitrary matrix element in this representation has the form

$$\langle m_{\ell 1} m_{s1}, \ldots, m_{\ell N} m_{sN}(\alpha) | H_{so} | m_{\ell 1} m_{s1}, \ldots, m_{\ell N} m_{sN}(\beta)\rangle \tag{8-12}$$

Here the notation for the state $|m_{\ell 1} m_{s1}, \ldots, m_{\ell N} m_{sN}(\alpha)\rangle$ is described in Chapter 7; see equation (7.1). $|m_{\ell 1} m_{s1}, \ldots, m_{\ell N} m_{sN}(\alpha)\rangle$ represents a determinantal wave function composed of orbitals for all the electrons in the configuration. The $m_\ell m_s$ assignment to the N electrons in the incomplete shell has been explicitly exhibited, while the specification of the complete shells has been suppressed. The index α labels various determinants which have the same complete shells, but differ in the $m_\ell m_s$ assignment of the electrons in the incomplete shell. Thus for the arbitrary matrix element (8-12), the $m_\ell m_s$ of the initial state α may in general be different from the $m_\ell m_s$ of the final state β.

Noting that H_{so} is a one-electron operator in the sense of Chapter 4, equations (4-13) and (4-14) indicate that (8-12) is nonzero only in two instances: (1) only one orbital different between the initial and final determinants; (2) all orbitals the same. In the first case, the differing orbital must lie in the incomplete shell, since the determinants α and β have the same complete shells. Therefore, the complete shells obviously play no role in this case. According to equation (4-13), the contribution in this case is

$$\int v_i^* f u_i \, d\tau \qquad v_i \neq u_i \,;\; v_i,\, u_i \text{ in incomplete shell} \quad (4\text{-}13)$$

We write this as

$$\int v_i^* f u_i \, d\tau = \langle n\ell m'_{\ell i}\, m'_{si} \,|\, f \,|\, n\ell m_{\ell i}\, m_{si} \rangle \tag{8-13a}$$

where either $m'_{\ell i} \neq m_{\ell i}$ or $m'_{si} \neq m_{si}$ (or both). We have represented the orbital u_i lying in the incomplete shell $n\ell$ by

$$|\, n\ell m_{\ell i}\, m_{si} \rangle = \frac{\mathcal{R}_{n\ell}}{r} Y_{\ell m_{\ell i}}(\Omega)\,|\, m_{si} \rangle$$

$$= \frac{\mathcal{R}_{n\ell}}{r}\,|\, \ell m_{\ell i}\, m_{si} \rangle$$

and similarly for v_i. Inserting the expression for f, we conclude that in the case that the initial and final determinants differ by exactly one orbital, (8-12) evaluates to

$$\frac{1}{2m^2c^2} \int_0^\infty \frac{1}{r} \frac{dV}{dr} \mathcal{R}_{n\ell}^2 \, dr \, \langle \ell m'_{\ell i}\, m'_{si} \,|\, \boldsymbol{\ell} \cdot \mathbf{s} \,|\, \ell m_{\ell i}\, m_{si} \rangle$$

$$= \frac{1}{2m^2c^2} \xi_{n\ell} \langle \ell m'_{\ell i}\, m'_{si} \,|\, \boldsymbol{\ell} \cdot \mathbf{s} \,|\, \ell m_{\ell i}\, m_{si} \rangle \tag{8-13b}$$

$$\xi_{n\ell} = \int_0^\infty \frac{1}{r} \frac{dV}{dr} \mathcal{R}_{n\ell}^2 \, dr \tag{8-13c}$$

It is only in the second case, initial and final states the same, that the complete shells *may* contribute. In this case, equation (4-14) gives

$$\langle m_{\ell 1} m_{s1}, \ldots, m_{\ell N} m_{sN}(\alpha) \,|\, H_{so} \,|\, m_{\ell 1} m_{s1}, \ldots, m_{\ell N} m_{sN}(\alpha) \rangle$$

$$= \sum_i \langle i \,|\, f \,|\, i \rangle \tag{8-14}$$

The sum on i extends over all the electrons, viz., over the complete shells as well as over the N electrons of the incomplete shell which are present in the determinant α. The contribution to the above sum from the complete shells is zero. The reason for this is that the complete shells sum to a spherically symmetric state, with zero total angular momentum. To see this explicitly, we isolate the contribution to (8-14) of the complete shells $n_i\,\ell_i$.

$$\sum_{\substack{i\\ \text{complete}\\ \text{shells}}} \langle i|f|i\rangle = \frac{1}{2m^2c^2}\sum_i \left\langle n_i\,\ell_i\,m_{\ell i}\,m_{si}\left|\frac{1}{r}\frac{dV}{dr}\,\boldsymbol{\ell}\cdot\mathbf{s}\right|n_i\,\ell_i\,m_{\ell i}\,m_{si}\right\rangle$$

$$= \frac{1}{2m^2c^2}\sum_i \xi_{n_i\,\ell_i}\langle \ell_i\,m_{\ell i}\,m_{si}\,|\,\boldsymbol{\ell}\cdot\mathbf{s}\,|\,\ell_i\,m_{\ell i}\,m_{si}\rangle \tag{8-15a}$$

$$\xi_{n_i\,\ell_i} = \int_0^\infty \frac{1}{r}\frac{dV}{dr}\,\mathcal{R}^2_{n_i\,\ell_i}\,dr \tag{8-15b}$$

Only the diagonal matrix element of $\ell_z\,s_z$ is nonzero in (8-15a), as $\ell_x\,s_x$ and $\ell_y\,s_y$ have no diagonal components. Hence (8-15a) is

$$\sum_{\substack{i\\ \text{complete}\\ \text{shells}}} \langle i|f|i\rangle = \frac{\hbar^2}{2m^2c^2}\sum_i \xi_{n_i\,\ell_i}\,m_{\ell i}\,m_{si} \tag{8-15c}$$

Consider now a specific complete shell $n_i\,\ell_i$ contributing to (8-15c).

$$\sum_{\substack{i\\ \text{complete shell}\\ n_i\,\ell_i}} \langle i|f|i\rangle = \frac{\hbar}{2m^2c^2}\,\xi_{n_i\,\ell_i}\sum_i m_{\ell i}\,m_{si} \tag{8-15d}$$

This is seen to vanish since the summation over the magnetic quantum numbers of a complete shell extends over an equal number of positive and negative quantities of the same magnitude. Thus we conclude that only the incomplete shell $n\ell$ contributes to (8-14)

$$\langle m_{\ell 1}\,m_{s1},\ldots,m_{\ell N}\,m_{sN}(\alpha)\,|\,H_{so}\,|\,m_{\ell 1}\,m_{s1},\ldots,m_{\ell N}\,m_{sN}(\alpha)\rangle$$

$$= \frac{1}{2m^2c^2}\,\xi_{n\ell}\sum_{\substack{i\\ \text{incomplete}\\ \text{shell } n\ell}}\langle \ell m_{\ell i}\,m_{si}\,|\,\boldsymbol{\ell}\cdot\mathbf{s}\,|\,\ell m_{\ell i}\,m_{si}\rangle \tag{8-15e}$$

In summary, we have learned that the complete shells do not contribute to the spin orbit energy. Thus we may ignore them from the outset by replacing H_{SO} of equation (8-11b) by

$$\text{'}H\text{'}_{SO} = \frac{1}{2m^2c^2}\,\xi_{n\ell}\sum_{i=1}^{N}\boldsymbol{\ell}_i\cdot\mathbf{s}_i \tag{8-16a}$$

where the summation now extends only over the N electrons in the incomplete shell $n\ell$. The quantity $(1/r)\,dV/dr$ has been replaced by $\xi_{n\ell}$, which is its average over the incomplete shell. This is the only way that $(1/r)(dV/dr)$ contributes [see (8-13c), (8-15e)]. In order that 'H'$_{SO}$ have the proper matrix elements, (8-13c) and (8-15e), we also pretend that the determinantal wave functions are composed only of orbitals of the incomplete shell. As we shall presently need to form states of definite angular momentum, this is no restriction since the complete shells have zero L, S, and J.

The parameter $\xi_{n\ell}$, occurring in the simplified Hamiltonian (8-16a), is defined in (8-13c).

$$\xi_{n\ell} = \int_0^\infty \frac{1}{r}\frac{dV}{dr}\,\mathcal{R}^2_{n\ell}\,dr \tag{8-16b}$$

where $\mathcal{R}_{n\ell}/r$ is the radial wave function of the incomplete shell. Since dV/dr may be written as $Z(r)e^2/r$, where $Z(r)$ is the charge inside a sphere of radius r,

$$\xi_{n\ell} = \left\langle \frac{Z(r)e^2}{r^3}\right\rangle_{n\ell} > 0 \tag{8-16c}$$

To obtain the first-order correction to the energy arising from the spin-orbit interaction, we shall want to evaluate the diagonal matrix elements of (8-16a) in the LSJM representation. To do this, we need to know the matrix elements of $\boldsymbol{\ell}_i$ and $\mathbf{s}_i$. We now prove a general theorem which will enable us to evaluate these matrix elements, and which is generally important in quantum mechanics.

A THEOREM ABOUT MATRIX ELEMENTS

We consider any operator **A** which satisfies the following commutation relations with **J**, the angular momentum

$$[A_a, J_b] = [J_a, A_b] = i\hbar\varepsilon_{abc}A_c$$

$$[A_a, J_a] = 0 \tag{8-17}$$

where a, b, c are cyclic permutations of x, y, z. Such an operator is called a *vector operator*. (The reason for this nomenclature is that under rotations **A** transforms as a vector.)

There are many operators which satisfy the commutation relations (8-17). The most important type for us is that of an angular momentum **J**, which is the sum of mutually commuting components,

$$\mathbf{J} = \mathbf{J}_1 + \cdots + \mathbf{J}_n \tag{8-18}$$

Then each $\mathbf{J}_i$ may be substituted for **A** in (8-17). This follows from the fact that the components of $\mathbf{J}_i$ satisfy

$$[J_{ia}, J_{ib}] = i\hbar\epsilon_{abc} J_{ic} \tag{8-19}$$

since $\mathbf{J}_i$ is an angular momentum; and that J_{ia} commutes with J_{kb} for $k \neq i$. In particular we shall use the theorem to be derived for $\boldsymbol{\ell}_i$ and $\mathbf{s}_i$ occurring in (8-16a). We know that

$$\mathbf{L} = \sum_i \boldsymbol{\ell}_i \qquad \mathbf{S} = \sum_i \mathbf{s}_i \tag{8-20}$$

Then we may choose any $\boldsymbol{\ell}_i$ to be **A** in (8-17) if we identify **L** with **J**, and similarly for $\mathbf{s}_i$ and **S**.

Moreover, **A** in (8-17) may be the position vector of an electron and **J** its orbital momentum. Or **J** may be the sum of the orbital momenta of all electrons, or the total angular momentum of the atom, with **A** being either the position vector of one electron, or the sum of the position vectors of all electrons, or the sum of their linear momenta. In this form, the theorem will be applicable to the calculation of optical transition probabilities.

The theorem we shall prove states that in a representation that diagonalizes $\mathbf{J}^2$ and J_z, the matrix elements of **A** between states of the same J are proportional to the matrix elements of **J**, and the proportionality constant is independent of the magnetic quantum numbers.

To prove this theorem, it is convenient to work with different combinations of the operators. We set

$$J_\pm = J_x \pm iJ_y \qquad J_0 = J_z$$
$$A_\pm = A_x \pm iA_y \qquad A_0 = A_z \tag{8-21}$$

A vector operator resolved into the components $A_\pm$, A_0 as in (8-21) is said to be in a *spherical basis* in contradistinction to the usual xyz or *Cartesian basis*. The commutation relations for $A_\pm$, A_0 follow from (8-17).

$$[A_\pm, J_0] = \mp\hbar A_\pm \tag{8-22a}$$

$$[A_\pm, J_\pm] = 0 \tag{8-22b}$$

$$[A_\pm, J_\mp] = \pm 2\hbar A_0 \tag{8-22c}$$

$$[A_0, J_0] = 0 \tag{8-22d}$$

$$[A_0, J_\pm] = \pm\hbar A_\pm \tag{8-22e}$$

In equations (8-22b) and (8-22c), the upper and lower signs are to be taken together.

All the information that we have about **A** is contained in (8-22), and we shall use these relations to obtain our results. (Equations (8-22d) and (8-22e) can be shown to follow from the previous three, hence we shall concern ourselves only with (8-22a), (8-22b), and (8-22c).)

We take matrix elements of (8-22) with respect to states, $|\lambda J\, M\rangle$, in which J^2 and J_Z are diagonal, in particular elements leading from a state J to another state with the same J. We indicate that there may be other quantum numbers λ (such as the energy) which may be different in initial and final states. We assume that **J** commutes with whatever variables are represented in λ, e.g., the Hamiltonian, so that **J** has only matrix elements diagonal in λ; and furthermore that these do not depend on λ but only on J and M. That is we take

$$J_\pm|\lambda JM\rangle = \langle J\, M\pm 1|J_\pm|JM\rangle|\lambda J\, M\pm 1\rangle \tag{8-23a}$$

$$\langle\lambda JM|J_\pm = \langle JM|J_\pm|J\, M\mp 1\rangle\langle\lambda J\, M\mp 1| \tag{8-23b}$$

$$J_0|\lambda JM\rangle = \langle JM|J_0|JM\rangle|\lambda JM\rangle$$

$$= \hbar M|\lambda JM\rangle \tag{8-23c}$$

The coefficients appearing in the right hand of (8-23a) and (8-23b) are of course known [see (1-29)].

We now take matrix elements of (8-22a), and use (8-23c)

$$\mp\langle\lambda' JM'|A_\pm|\lambda JM\rangle = (M - M')\langle\lambda' JM'|A_\pm|\lambda JM\rangle \tag{8-24a}$$

Therefore $M' = M \pm 1$, otherwise the matrix element vanishes. Thus, the matrix elements of $A_\pm$ are proportional to those of $J_\pm$, and the proportionality constant may be written as

$$\frac{\langle \lambda' J\ M \pm 1 | A_\pm | \lambda JM \rangle}{\langle J\ M \pm 1 | J_\pm | JM \rangle} \tag{8-24b}$$

Next, we take the same matrix element of (8-22b) and use (8-23a) and (8-23b)

$$\langle \lambda' JM' | A_\pm | \lambda J\ M \pm 1 \rangle \langle J\ M \pm 1 | J_\pm | JM \rangle$$
$$= \langle \lambda' J\ M' \mp 1 | A_\pm | \lambda JM \rangle \langle JM' | J_\pm | J\ M' \mp 1 \rangle \tag{8-25a}$$

Using the previous result, we get $M' = M \pm 2$.

$$\langle \lambda' J\ M \pm 2 | A_\pm | \lambda J\ M \pm 1 \rangle \langle J\ M \pm 1 | J_\pm | JM \rangle$$
$$= \langle \lambda' J\ M \pm 1 | A_\pm | \lambda JM \rangle \langle J\ M \pm 2 | J_\pm | J\ M \pm 1 \rangle$$

or

$$\frac{\langle \lambda' J\ M \pm 2 | A_\pm | \lambda\ J\ M \pm 1 \rangle}{\langle J\ M \pm 2 | J_\pm | J\ M \pm 1 \rangle} = \frac{\langle \lambda' J\ M \pm 1 | A_\pm | \lambda JM \rangle}{\langle J\ M \pm 1 | J_\pm | JM \rangle} \tag{8-25b}$$

Equation (8-25b) indicates that the proportionality constant (8-24b) is independent of M, and we may call it $\langle \lambda' J || A || \lambda J \rangle_\pm$. Thus we have obtained the result that

$$\langle \lambda' JM' | A_\pm | \lambda\ JM \rangle = \langle \lambda' J || A || \lambda\ J \rangle_\pm \langle JM' | J_\pm | JM \rangle \tag{8-25c}$$

We next obtain a similar relation for A_0 and show that $\langle \lambda' J || A || \lambda J \rangle_+ = \langle \lambda' J || A || \lambda J \rangle_-$. We take matrix elements of (8-22c) and use (8-23a) and (8-23b)

$$\pm 2 \langle \lambda' JM' | A_0 | \lambda JM \rangle$$
$$= \langle \lambda' JM' | A_\pm | \lambda\ J\ M \mp 1 \rangle \langle J\ M \mp 1 | J_\mp | JM \rangle$$
$$- \langle \lambda' J\ M' \pm 1 | A_\pm | \lambda JM \rangle \langle JM' | J_\mp | J\ M' \pm 1 \rangle \tag{8-26a}$$

Inserting the value of the matrix element of $A_\pm$ from (8-25c) we obtain for the right-hand side of (8-26a)

$$\langle \lambda' J || A || \lambda J \rangle_\pm [\langle JM' | J_\pm | J\ M \mp 1 \rangle \langle J\ M \mp 1 | J_\mp | JM \rangle$$
$$- \langle J\ M' \pm 1 | J_\pm | JM \rangle \langle JM' | J_\mp | J\ M' \pm 1 \rangle] \tag{8-26b}$$

Evidently A_0 has nonvanishing matrix elements only when $M' = M$. In that instance the bracketed term in (8-26b) is simply

$$|\langle J\,M \mp 1 | J_{\mp} | JM\rangle|^2 - |\langle J\,M \pm 1 | J_{\pm} | JM\rangle|^2$$

$$= \pm 2\hbar^2 M = \pm 2\hbar \langle JM | J_0 | JM\rangle \qquad (8\text{-}26c)$$

Combining (8-26a), (8-26b), and (8-26c) yields

$$\langle \lambda' JM' | A_0 | \lambda JM\rangle = \langle \lambda' J || A || \lambda J\rangle_{\pm} \langle JM' | J_0 | JM\rangle \qquad (8\text{-}27)$$

Evidently the proportionality coefficient must be the same regardless whether we take the + or − signs. Hence we obtain as a final result

$$\langle \lambda' JM' | \mathbf{A} | \lambda JM\rangle = \langle \lambda' J || A || \lambda J\rangle \langle JM' | \mathbf{J} | JM\rangle \qquad (8\text{-}28)$$

which proves the theorem.

A simple corollary of (8-28) is

$$\langle \lambda' JM' | \mathbf{A}\cdot\mathbf{J} | \lambda JM\rangle = \langle \lambda' J || A || \lambda J\rangle \langle JM' | J_{op}^2 | JM\rangle$$

$$= \delta(M, M') \langle \lambda' J || A || \lambda J\rangle \hbar^2 J(J+1) \qquad (8\text{-}29)$$

Thus the scalar product $\mathbf{A}\cdot\mathbf{J}$ is diagonal in M. Frequently, J(J + 1) easier to calculate than any component of **A**, and this then serves to determine the constant $\langle \lambda' J || A || \lambda J\rangle$.

Dirac proved a more general theorem which permits us to calculate matrix elements of a vector operator between states of different J. Some intricate algebra (see Problem 10) yields the result

$$[J^2, [J^2, A]] = 2\hbar^2(J^2A + AJ^2) - 4\hbar^2(A\cdot J)J \qquad (8\text{-}30)$$

Taking matrix elements between states of the same J, the theorem (8-28) can be recovered. Taking matrix components of (8-30) between states JM; J′M′; J ≠ J′, the last term on the right will be zero, since **J** does not have elements between different J values. A little algebra yields

$$[(J + J' + 1)^2 - 1][(J - J')^2 - 1] \times \langle \lambda' J'M' | \mathbf{A} | \lambda JM\rangle = 0 \qquad (8\text{-}31)$$

Hence **A** has nonzero matrix elements only for J′ = J, J ± 1. The case J′ = J we have studied above. When J′ = J ± 1, we may still conclude that A_z has nonzero matrix elements only for M = M′

since A_z commutes with J_z, (8-22d). Similarly, from (8-22e) we may infer that $A_\pm$ has nonzero matrix elements only for $M' = M \pm 1$.

The elements of **A** are given in Condon and Shortley,[3] pp. 59–73; and Slater,[4] Vol. II, Appendix 31. We shall use them below.

If **J** = **L** + **S** we see that L and S each satisfy (8-17). In this case (8-28) and (8-29) give the old vector model result that the time average of **L** (or **S**) can be replaced by its component along **J**, which is $[(\mathbf{L}\cdot\mathbf{J})\mathbf{J}]/\mathbf{J}^2$, since, in the time average, the perpendicular component cancels out, owing to the "precession" of **L** (or **S**) about **J**.

It is seen that the commutation relations (8-17) of an operator with the angular momentum are sufficient to determine the M dependence of the matrix elements of this operator. Thus the "geometrical" characteristics of the operator, viz., the M dependence of its matrix elements, are completely determined while the "dynamical" characteristics are contained in the M independent quantity $\langle \lambda' J' \| A \| \lambda J \rangle$. This constant is called the *reduced matrix element.*

We have encountered a similar situation before, see p. 104. There we saw that the fact that an operator commutes with **J** also determines the M dependence of its matrix elements. In that instance the result was trivial. The matrix elements did not depend on M.

Both these situations are special cases of a general theorem, known as the *Wigner-Eckart* theorem, which gives the M dependence of the matrix elements of an operator which satisfies definite commutation relations with the angular momentum (see Schiff[5], p. 223).

EVALUATION OF THE SPIN-ORBIT INTERACTION

Returning now to (8-16) we wish to evaluate

$$\langle \mathrm{LSJM} \,|\, \text{`}H'_{\mathrm{so}} \,|\, \mathrm{LSJM} \rangle \tag{8-32a}$$

Expressing this in the $LM_L SM_S$ representation, by the usual Clebsch-Gordon series, we have

---

[3]E. U. Condon and G. H. Shortley, *The Theory of Atomic Spectra*, Cambridge University Press, Cambridge, 1959.

[4]J. C. Slater, *Quantum Theory of Atomic Structure*, McGraw-Hill, New York, 1960.

[5]L. I. Schiff, *Quantum Mechanics*, 3rd ed., McGraw-Hill, New York, 1968.

$$\langle LSJM | \text{'}H'_{so} | LSJM\rangle = \sum_{\substack{M_L M_S M'_L M'_S \\ M_L + M_S = M = M'_L + M'_S}} C^*(LM'_L\, SM'_S,\, JM)$$
$$\times\, C(LM_L\, SM_S,\, JM)\langle LM'_L,\, SM'_S | \text{'}H'_{so} | LM_L,\, SM_S\rangle \tag{8-32b}$$

We now use the Wigner-Eckart theorem. In (8-17) we identify **J** with **L**(**S**) and **A** with $\boldsymbol{\ell}_i$ ($\mathbf{s}_i$) where $\mathbf{L} = \Sigma_i\, \boldsymbol{\ell}_i$ ($\mathbf{S} = \Sigma_i\, \mathbf{s}_i$). We see that $\boldsymbol{\ell}_i$ ($\mathbf{s}_i$) satisfies (8-17). Hence from (8-28)

$$\langle LM'_L,\, SM'_S | \boldsymbol{\ell}_i \cdot \mathbf{s}_i | LM_L,\, SM_S\rangle$$
$$= \langle L || \ell_i || L\rangle\langle S || s_i || S\rangle$$
$$\times \langle LM'_L,\, SM_S | \mathbf{L}\cdot\mathbf{S} | LM_L,\, SM_S\rangle \tag{8-32c}$$

and

$$\langle LM'_L,\, SM'_S | \text{'}H'_{so} | LM_L,\, SM_S\rangle$$
$$= \frac{1}{2m^2c^2}\,\xi_{n\ell}\sum_i \langle LM'_L,\, SM'_S | \boldsymbol{\ell}_i \cdot \mathbf{s}_i | LM_L,\, SM_S\rangle$$
$$= \frac{1}{2m^2c^2}\,\xi_{n\ell}\sum_i \alpha(i)\beta(i)\langle LM'_L,\, SM'_S | \mathbf{L}\cdot\mathbf{S} | LM_L,\, SM_S\rangle \tag{8-32d}$$

where α and β are the reduced matrix elements

$$\alpha(i) = \langle L || \ell_i || L\rangle$$
$$\beta(i) = \langle S || s_i || S\rangle \tag{8-32e}$$

These constants depend on the multiplet LS, but not on $M_L\, M_S\, M'_L\, M'_S$ Substituting (8-32d) into (8-32b) and performing the Clebsch-Gordan sum yields

$$\langle LSJM | \text{'}H'_{so} | LSJM\rangle = \frac{1}{2m^2c^2}\,\xi_{n\ell}\sum_i \alpha(i)\beta(i)$$

$$\times \sum C^*(LM'_L\, SM'_S,\, JM)C(LM_L\, SM_S,\, JM)\;\langle LM'_L\, SM'_S | \mathbf{L}\cdot\mathbf{S} | LM_L\, SM_S\rangle$$

$$= \frac{1}{2m^2c^2}\,\xi_{n\ell}\sum_i \alpha(i)\beta(i)\ \langle LSJM\,|\,\mathbf{L}\cdot\mathbf{S}\,|\,LSJM\rangle \qquad (8\text{-}32f)$$

Thus within first-order perturbation theory, the spin-orbit interaction 'H'_{so} given by (8-16) may be replaced by

$$\text{“}H\text{''}_{so} = \frac{1}{2m^2c^2}\,\xi_{n\ell}\sum_i \alpha(i)\beta(i)\,\mathbf{L}\cdot\mathbf{S} \qquad (8\text{-}33)$$

The matrix element (8-32f) is now easily evaluated by making use of the operator identity

$$2\mathbf{L}\cdot\mathbf{S} = (J^2 - L^2 - S^2)_{op} \qquad (8\text{-}34)$$

L^2_{op} and S^2_{op} remain diagonal in the LSJM representation, but L_z and S_z no longer are. We get, for (8-32f),

$$E_{so} = \langle LSJM\,|\,\text{“}H\text{''}_{so}\,|\,LSJM\rangle$$

$$= \frac{1}{4}\left(\frac{\hbar}{mc}\right)^2 \xi_{n\ell}\gamma_{LS}[J(J+1) - L(L+1) - S(S+1)] \qquad (8\text{-}35)$$

where

$$\gamma_{LS} = \sum_i \alpha(i)\beta(i) \qquad (8\text{-}36)$$

is a numerical coefficient which can be calculated if the wave function of the $LM_L\,SM_S$ state is known in terms of the $m_{\ell i}\,m_{si}$ representation. We discuss methods for evaluating γ_{LS} below.

As the only J dependence of E_{so} is through the term in brackets, we can evaluate the difference

$$E_{so}(J) - E_{so}(J-1) = \left(\frac{1}{2}\,\frac{\hbar}{mc}\right)^2 \xi_{n\ell}\gamma_{LS}\,2J \qquad (8\text{-}37)$$

This is *Landé's interval rule* which states that the separation of two J states belonging to the same LS is proportional to the larger value of J.

Whenever at most one state of given LM_L, SM_S can be formed from the individual oribitals, a diagonal sum rule, such as the Slater sum rule discussed in the previous chapter, may be used to evaluate the constant γ_{LS}. The sum rule is used to connect the trace of H_{so} in the $LM_L\,SM_S$ representation to the trace in the $m_\ell\,m_s$ representation. Recall the statement of the sum rule (7-18b) which when applied to H_{so} reads

$$\sum_{\alpha} \langle m_{\ell 1} m_{s1}, \ldots, m_{\ell N} m_{sN}(\alpha) | H_{so} | m_{\ell 1} m_{s1}, \ldots, m_{\ell N} m_{sN}(\alpha) \rangle$$
$$= \sum_{L \geq |M_L|} \sum_{S \geq |M_S|} \langle LM_L, SM_S | H_{so} | LM_L, SM_S \rangle \qquad (8\text{-}38a)$$

The sum is α is over all determinantal wave functions which have an $m_\ell m_s$ assignment for the electrons in the incomplete shell, satisfying

$$\sum_{i=1}^{N} m_{\ell i} = M_L \qquad \sum_{i=1}^{N} m_{si} = M_S \qquad (8\text{-}38b)$$

The diagonal matrix element on the left-hand side of (8-38a) has been evaluated in (8-15e)

$$\sum_{\alpha} \langle m_{\ell 1} m_{s_1}, \ldots, m_{\ell N} m_{sN}(\alpha) | H_{so} | m_{\ell 1} m_{s_1}, \ldots, m_{\ell N} m_{sN}(\alpha) \rangle$$
$$= \frac{\hbar^2}{2m^2 c^2} \xi_{n\ell} \sum_{\alpha} \sum_{i} m_{\ell i} m_{si} \qquad (8\text{-}38c)$$

The internal sum in i extends over the N electrons in the incomplete shell $n\ell$ occurring in the determinantal wave function α. To evaluate the right-hand side of (8-38a) we replace H_{so} by "H''_{so}", which is permissible, as we have argued before, for matrix diagonal elements.

$$\langle LM_L, SM_S | H_{so} | LM_L, SM_S \rangle$$
$$= \langle LM_L, SM_S | \text{“}H''_{so}\text{”} | LM_L, SM_S \rangle$$
$$= \frac{\hbar^2}{2m^2 c^2} \xi_{n\ell} \gamma_{LS} M_L M_S \qquad (8\text{-}38d)$$

Thus the sum rules give

$$\sum_{\alpha} \sum_{i=1}^{N} m_{\ell i} m_{si} = M_L M_S \sum_{\substack{L \geq |M_L| \\ S \geq |M_S|}} \gamma_{LS} \qquad (8\text{-}38e)$$

As an application, consider N electrons in an incomplete shell $n\ell$, with $N \leq 2\ell + 1$. We assume maximum S, M_S; i.e., all $m_{si} = \frac{1}{2}$, $M_S = S$. Then (8-38e) becomes

$$\tfrac{1}{2}\sum_{\alpha}\sum_{i=1}^{N} m_{\ell i} = \tfrac{1}{2}\sum_{\alpha} M_L = \frac{M_L}{2}\sum_{\alpha} 1 = M_L S \sum_{L \geq |M_L|} \gamma_{LS}$$

$$\frac{1}{2S}\sum_{\alpha} 1 = \sum_{L \geq |M_L|} \gamma_{LS} \qquad (\text{maximum } S,\ M_S) \qquad (8\text{-}39a)$$

If we also assume maximum M_L; i.e., $m_{\ell 1} = \ell$, $m_{\ell 2} = \ell - 1$, etc., then there is exactly one determinantal wave function contributing to the left-hand side. On the right-hand side there also is just one value L contributing, the largest L. Thus

$$\gamma_{LS} = \frac{1}{2S} \qquad (\text{maximum } LM_L,\ SM_S) \qquad (8\text{-}39b)$$

But γ_{LS} is independent of M_L, M_S; hence the above holds for maximum L, S independent of M_L, M_S. In fact, (8-39b) holds for all L, with maximum S. To see this consider the application of (8-39a) to the case when the determinants α have $m_\ell m_s$ assignments leading to maximum S, M_S; and to two values of L: L_0 the maximum, and L_1 the next lower. For the case $M_L = L_1$ there will be two determinants satisfying $\Sigma_i m_{\ell i} = M_L$, since by hypothesis two LM_L states are to be formed from them. Then (8-39a) becomes

$$\frac{1}{2S}(2) = \gamma_{L_0 S} + \gamma_{L_1 S} \qquad (8\text{-}39c)$$

But $\gamma_{L_0 S}$ has been found to be 1/2S, hence $\gamma_{L_1 S}$ also is 1/2S. This process can be continued step by step, down through the available values of L. We conclude thus

$$\gamma_{LS} = \frac{1}{2S} \qquad (\text{maximum } S) \qquad (8\text{-}39d)$$

For multiplicities lower than the maximum, the sum rule again may be used to evaluate γ_{LS}.

For more than half-full shells, the following argument is useful. We can write

$$\sum_k \boldsymbol{\ell}_k \cdot \mathbf{s}_k = \left(\sum_{\substack{\text{whole}\\\text{shell}}} - \sum_{\substack{\text{empty}\\\text{states}}}\right) \boldsymbol{\ell}_k \cdot \mathbf{s}_k$$

$$= -\sum_{\substack{\text{empty}\\\text{states}}} \boldsymbol{\ell}_k \cdot \mathbf{s}_k \tag{8-40}$$

Hence the spin-orbit energies for a configuration of $4\ell + 2 - k$ electrons are negatives of those for a configuration of k electrons. In particular, for more than half-filled shells,

$$\gamma_{LS} = -\frac{1}{2S} \quad \text{(highest S)} \tag{8-41}$$

If $\gamma_{LS} > 0$, highest energies correspond to highest J; if $\gamma_{LS} < 0$, the reverse is true. The former is called a *regular multiplet*, the latter an *inverted multiplet*. For less than $2\ell + 1$ electrons in the shell, most multiplets are regular, but there are some exceptions, e.g., the ^2F state arising from a d^3 configuration, which has $\gamma_{LS} = -\frac{1}{6}$. For more than half-filled shells, most (but not all) multiplets are inverted.

The following generalization of Hund's rule (p. 138) may be given. To obtain the term with the lowest energy for a regular multiplet (less than half-filled shells) take the highest S, then the highest L, and finally the lowest J. For an inverted multiplet (more than half-filled shells) take the highest S, then the highest L, and finally the highest J.

For exactly half-filled shells there is no first-order spin-orbit interaction, since a half-filled shell can be viewed two ways, either as $2\ell + 1$ electrons or $2\ell + 1$ holes. The energies of these two cases by (8-40) must be equal in magnitude and opposite in sign. Hence the energy vanishes. The state of highest multiplicity in a half-filled shell has $L = 0$, so no spin-orbit splitting occurs. To obtain the spin-orbit energy for other multiplets for half-filled shells we must use a second-order perturbation theory.

H_{so} is of the order $10^{-4}\ \xi_{n\ell}\gamma_{LS}$-atomic units as can be seen from (8-38). For the Fe group 3d electrons $(\hbar/2mc)^2\ \xi_{n\ell}$ is found to be 50 to 1000 cm^{-1}, the value increasing from the beginning to the end of the shell. We recall that the configuration energies were usually of the order of 10^5 cm^{-1}; the electrostatic interaction, 10^4 cm^{-1}; the spin orbit term now is of order 10^2 cm^{-1}; and the effect of an external magnetic field, 1 cm^{-1}. Hence our assumptions about quantum numbers are usually correct.

However, it may happen that the spin-orbit interaction is much larger than the electrostatic interaction. This occurs in X-ray spectra. (Although comparatively rare in atomic theory, in nuclear theory this is quite common.) Each electron is then characterized by $n\ell jm$ rather than $n\ell m_\ell m_s$. A configuration $(n\ell)^k$ then splits up first into subconfigurations, characterized by the number k_1 of electrons having $j = \ell + \frac{1}{2}$; this may be written $(n\ell, \ell + \frac{1}{2})^{k_1} \times (n\ell, \ell - \frac{1}{2})^{k-k_1}$. The electrostatic energy then splits each subconfiguration into states of different J, with

$$J = \sum_i j_i \tag{8-42}$$

This is called jj coupling. The calculation of electrostatic energy is more complicated than for LS coupling and is treated in Condon and Shortley,[3] Chapter 10. They also treat the case of intermediate coupling when spin-orbit and electrostatic interaction are of the same order of magnitude. In either case, L and S are no longer good quantum numbers but J is.

COMPARISON WITH EXPERIMENT

The first-order energy correction due to the spin-orbit interaction (8-35) can be readily compared with experiment. We consider Fe which has the configuration $1s^2 2s^2 2p^6 3s^2 3p^6 3d^6 4s^2$. Evidently, there is just one incomplete shell, the 3d, which leads to inverted multiplets since it is more than half-full. According to Hund's rule we expect the ground state to be the term $^5D_4$. In Table 8-1, we list after Condon and Shortley[3] p. 195, the observed energy relative to the ground state of the various terms of the ^5D multiplet. If Landé's rule were obeyed exactly, the last column should show no variation with J, as according to (8-37) and (8-41) it is

Table 8-1
Spin-orbit Effects in the ^5D Multiplet of Fe. Energy is Given in cm^{-1} Relative to the Ground State.

J	^5D_J	E(J)	E(J) − E(J − 1)	$\frac{E(J) - E(J-1)}{J}$
4	$^5D_4$	0.000		
			−415.934	−104.0
3	$^5D_3$	415.934		
			−288.067	− 96.0
2	$^5D_2$	704.001		
			−184.125	− 92.1
1	$^5D_1$	888.126		
			− 89.942	− 89.9
0	$^5D_0$	978.068		

$$\frac{1}{2}\left(\frac{\hbar}{mc}\right)^2 \xi_{n\ell}\gamma_{LS} = -\frac{1}{8}\left(\frac{\hbar}{mc}\right)^2 \xi_{n\ell}$$

ZEEMAN EFFECT

The interaction of an electron with a homogenous magnetic field $\mathcal{H}$ along the z axis is given by

$$H_z = \frac{e\mathcal{H}}{2mc}(L_z + 2S_z) \tag{8-43}$$

The total Hamiltonian commutes with J_z but not with J^2. Since we now have an externally defined orientation in space, we do not expect the Hamiltonian to be invariant under rotations. The quantity $e\hbar/2mc$ is called the Bohr magneton μ_0 and is equal to 9×10^{-21} cgs units. When the field is sufficiently weak so that the effect of H_Z is small, we can consider (8-43) as a perturbation in the SLJM scheme. This is known as the *Zeeman effect*.

We need to evaluate the diagonal elements $\langle JM | L_z + 2S_Z | JM \rangle$. Writing this as $\langle JM | J_z + S_z | JM \rangle$, we apply (8-28)

$$\langle LSJM | S_z | LSJM \rangle = \langle LSJ || S || LSJ \rangle \langle JM | J_z | JM \rangle \tag{8-44}$$

The reduced matrix element may be found with the help of (8-29)

$$\langle LSJM | \mathbf{S}\cdot\mathbf{J} | LSJM \rangle = \langle LSJ || S || LSJ \rangle \hbar^2 J(J+1) \tag{8-45a}$$

The operator $\mathbf{S}\cdot\mathbf{J}$ may be rewritten as

$$2\mathbf{S}\cdot\mathbf{J} = (J^2 + S^2 - L^2)_{op} \tag{8-45b}$$

Therefore

$$\langle LSJ || S || LSJ \rangle = \frac{J(J+1) + S(S+1) - L(L+1)}{2J(J+1)} \tag{8-45c}$$

and

$$\langle LSJM | H_z | LSJM \rangle = \frac{e}{2mc}\mathcal{H}\langle JM | J_z | JM \rangle g$$

$$= \mu_0 M \mathcal{H} g \tag{8-46}$$

where g is the *Landé g factor* and is given by

$$g = 1 + \langle LSJ || S || LSJ \rangle = 1 + \frac{J(J+1) - L(L+1) + S(S+1)}{2J(J+1)}$$

$$= \frac{3}{2} + \frac{(S-L)(S+L+1)}{2J(J+1)} \qquad (8\text{-}47)$$

Special values of g are

$$L = 0, J = S: g = 2 \qquad J = L + S: g = 1 + \frac{S}{J}$$

$$J = L - S: g = 1 - \frac{S}{J+1} \qquad J = S - L: g = 1 + \frac{S+1}{J+1} \qquad (8\text{-}48)$$

For one-electron spectra

$$J = L + \tfrac{1}{2}: g = 1 + \frac{1}{2L+1}$$

$$J = L - \tfrac{1}{2}: g = 1 - \frac{1}{2L+1} \qquad (8\text{-}49)$$

The splitting of the spectral lines is given by

$$E_i - E_f = \hbar\omega_0 + \mathcal{H}\mu_0 [g_i M_{L_i} - g_f M_{L_f}] \qquad (8\text{-}50)$$

where $\hbar\omega_0$ is the energy difference without magnetic field, and i and f refer to initial and final states, respectively.

PASCHEN-BACK EFFECT

If the external magnetic field is sufficiently strong, the Zeeman term in the Hamiltonian may dominate the spin-orbit interaction. This is known as the *Paschen-Back effect*. In this case we can consider the entire magnetic interaction $H_{so} + H_z$ as a perturbation in the $LM_L SM_S$ scheme. The diagonal matrix elements now become

$$H_{mag} = \frac{1}{2}\left(\frac{\hbar}{mc}\right)^2 \xi_{n\ell}\gamma_{LS} M_L M_S + \mu_0 \mathcal{H}(M_L + 2M_S) \qquad (8\text{-}51)$$

In first order, M_S cannot change in a transition, and M_L can change only by ±1 or 0 (see Chapter 11). The spectral line splitting is given by

$$E_i - E_f = \hbar\omega_0 + \mu_0\mathcal{H}(M_{L_i} - M_{L_f}) + \frac{M_S}{2}\left(\frac{\hbar}{mc}\right)^2$$

$$\times (\xi_{n\ell_i}\gamma_{LS_i} M_{L_i} - \xi_{n\ell_f}\gamma_{LS_f} M_{L_f}) \qquad (8\text{-}52)$$

where $\hbar\omega_0$ is the energy change without magnetic effects. The main splitting is, by assumption, given by the term with $\mathcal{H}$, which gives simply the Lorentz triplet, $M_{L_i} - M_{L_f} = +1, 0$, and -1. The last term gives the spin-orbit interaction. For $\Delta M_L = 0$ and $\xi_{n\ell_i} = \xi_{n\ell_f}$, this becomes

$$\Delta E = \hbar\omega_0 + \frac{M_S}{2}\left(\frac{\hbar}{mc}\right)^2 \xi_{n\ell} M_L (\gamma_{LS_i} - \gamma_{LS_f}) \tag{8-53}$$

This gives a splitting into $(2L + 1)(2S + 1)$ components. If L is known, simple counting of the components will determine the spin S. As early as 1929, the nuclear spin of Bi^{209} was determined this way and found to be $\frac{9}{2}$.

For intermediate magnetic field strength, the interaction with the external field and the spin-orbit interaction become comparable. In this case, the secular equation has to be solved explicitly for the energy. M is still a good quantum number but neither M_L, M_S, nor J are. In the case $S = \frac{1}{2}$ (e.g., alkali atoms), there are two values of M_S, $\frac{1}{2}$ and $-\frac{1}{2}$, for each value of M, except for $M = \pm(L + \frac{1}{2})$. For $M = L + \frac{1}{2}$ we must have $M_S = \frac{1}{2}$, and (8-51) is exact at all values of the field strength. For other M, the higher energy value in the limit of high field, (8-51), is given by $M_S = \frac{1}{2}$, $M_L = M - \frac{1}{2}$. In the limit of low field, and for a regular doublet, the level $J = L + \frac{1}{2}$ has the higher energy. Thus the state $J = L + \frac{1}{2}$, M at low field goes over into the state $M_S = \frac{1}{2}$, M at high field, and $J = L - \frac{1}{2}$, M into $M_S = -\frac{1}{2}$, M. This transition can easily be followed in detail by explicit solution of the eigenvalue problem for intermediate fields (Schiff,[5] pp. 440-443).

QUADRATIC ZEEMAN EFFECT

For very strong magnetic fields and large values of n (corresponding to large $\langle r^2 \rangle$) the quadratic term in (8-3) may become important. Since the magnitude of the spin-orbit interaction is characterized by $\xi_{n\ell} \sim \langle 1/r^3 \rangle \sim 1/n^3$ for large n, we can neglect it. In this case, electron spin becomes a constant of the motion and can be ignored in the discussion. With these approximations the entire magnetic interaction is given by

$$\frac{e}{2mc}\mathcal{H}\ell_z + \frac{e^2}{8mc^2}\mathcal{H}^2 r^2 \sin^2\theta \tag{8-54}$$

θ being the angle between the radius vector and the z axis. We have seen that the effect of the first term is to displace the energy by $m_\ell g_m \hbar\mathcal{H}$. Thus the problem of evaluating the *quadratic* Zeeman effect (for one electron) reduces to evaluating the effect of the perturbation $(e^2/8mc^2)\mathcal{H}^2 r^2 \sin^2\theta$. [For many electrons the perturbation is $(e^2/8mc^2)\mathcal{H}^2 \Sigma_i r_i^2 \sin^2\theta_i$.] Methods for doing this are discussed by Schiff,[5] pp. 296–298. We shall not pursue the general problem any further.

The diamagnetism of atoms receives its explanation from the quadratic Zeeman effect. In particular, for noble gases L_z and $S_z = 0$, and our approximations become exact. The only magnetic effect is the quadratic Zeeman effect, since there even is no shift in levels with magnetic quantum number. Then for helium, for example,

$$\Delta E = \frac{e^2\,\mathcal{H}^2}{8mc^2}\langle r_1^2 \sin^2\theta_1 + r_2^2 \sin^2\theta_2\rangle \qquad (8\text{-}55)$$

Using Hartree's wave functions,[6] which are symmetric in r_1 and r_2 and are spherically symmetric, we obtain

$$\Delta E = 1.05 \times 10^{-5}\mathcal{H}^2 \quad \text{(atomic units)} \qquad (8\text{-}56)$$

(To translate into cgs units, note that $\mathcal{H}^2$ has the dimension energy/volume, so the number must be multiplied by a_0^3.) The magnetic susceptibility per mole χ is defined by

$$N_0\,\Delta E = -\tfrac{1}{2}\chi\mathcal{H}^2 \qquad (8\text{-}57)$$

where N_0 is Avogadro's number. Inserting (8-56) we obtain $\chi = -1.87 \times 10^{-6}$ cm^3/mol. The measured value of the magnetic susceptibility for helium is $\chi = -1.88 \times 10^{-6}$. The agreement is excellent.

For heavier noble gases available wave functions are less good, so the theory cannot be checked to this accuracy. The term in $\mathcal{H}^2$ is also responsible for the diamagnetism of diatomic molecules. For atoms with a resultant $J \neq 0$, the $\mathcal{H}^2$ term gives a diamagnetic contribution to the susceptibility χ which subtracts from the dominant, paramagnetic contribution.

---

[6] See for example H. A. Bethe and E. E. Salpeter, *Quantum Mechanics of One- and Two-Electron Atoms*, Plenum, New York, 1977, p. 227.

STARK EFFECT

When an atom is placed in an external electric field one observes the *Stark effect*. The Hamiltonian then has a term

$$H_{\mathcal{E}} = e\mathcal{E}z = e\mathcal{E}\sum_k z_k \tag{8-58}$$

added to it, where $\mathcal{E}$ is the external field, assumed constant and pointing along the z axis and the summation in k is taken over all the electrons. Since (8-58) is an odd operator, its diagonal elements will vanish because of the definite parity of the eigenfunctions. Therefore the first-order perturbation is always 0. An exception exists for the excited states of hydrogen. Owing to the accidental ℓ degeneracy, linear combinations of the $n\ell$ wave functions can be taken, which no longer possess definite parity and give a nonzero result for the Stark effect (Schiff,[5] pp. 252-253).

In general, however, we must go to second-order perturbation theory. The shift in energy ΔE_a will be given by

$$\Delta E_a = \sum_b{}' \frac{|H_{\mathcal{E}ab}|^2}{E_a - E_b} = e^2\mathcal{E}^2 \sum_b{}' \frac{|z_{ab}|^2}{E_a - E_b}$$

$$z = \sum_k z_k \tag{8-59}$$

where the prime in the sum sign indicates that the state b = a is omitted in the sum. This is called the *quadratic* Stark effect.

For hydrogen in the ground state, the sum in (8-59) may be evaluated owing to the special symmetries of the Coulomb problem. The result is (Schiff,[5] pp. 263-265)

$$\Delta E_0 = -\frac{9}{4}E^2a_0^3, \quad a_0 = \frac{\hbar^2}{me^2} \tag{8-60}$$

For a many-electron atom, the results will depend on the mode of angular momentum coupling. We can nevertheless say some general things about (8-59).

Equation (8-59) holds only with the usual stipulation that the energy shift must be small compared to the energy separation of the unperturbed levels. z_{ab} will be nonzero only between states of opposite parity. Since z is the z component of the position operator $r = \Sigma_i r_i$ which satisfies (8-17), the theorem (8-31) will hold, and z will have nonzero elements only for $J_b = J_a$, $J_a \pm 1$, and $M_a = M_b$.

In the case of LS coupling, L is a good quantum number; then the theorem (8-31) holds also for L and we must have $L_b = L_a, L_a \pm 1$. In this case, of course, $S_b = S_a$, since z does not depend on S.

Let us calculate the pattern observed in a state described by quantum numbers αJM, where α stands for the other quantum numbers which specify the state. Therefore z_{ab} is nonvanishing only in the following three instances:

$$\langle \alpha JM | z | \alpha' J-1\, M\rangle = A(J, \alpha, \alpha')\sqrt{J^2 - M^2}$$

$$\langle \alpha JM | z | \alpha' JM\rangle = B(J, \alpha, \alpha')M$$

$$\langle \alpha JM | z | \alpha' J+1\, M\rangle = C(J, \alpha, \alpha')\sqrt{(J+1)^2 - M^2} \qquad (8\text{-}61)$$

where A, B, and C are some functions of J, α, and α'. The JJ element follows from (8-27) and the off-diagonal elements from formulas given in Condon and Shortley,[5] p. 63. Equation (8-59) now reads

$$\Delta E_a = e^2 \mathcal{E}^2 \Bigg[\left(\sum_{\alpha'}{}' \frac{|A(J\alpha\alpha')|^2}{E_{\alpha J} - E_{\alpha' J-1}}\right)(J^2 - M^2) + \left(\sum_{\alpha'}{}' \frac{|B(J\alpha\alpha')|^2}{E_{\alpha J} - E_{\alpha' J}}\right)(M^2) + \left(\sum_{\alpha'}{}' \frac{|C(J\alpha\alpha')|^2}{E_{\alpha J} - E_{\alpha' J+1}}\right)((J+1)^2 - M^2)\Bigg] \qquad (8\text{-}62)$$

Hence we see that the quadratic Stark effect does not depend on the sign of M. This is because the electric field acts on the charge-probability density, which does not depend on the sign of M.

The ground state Stark effect will always be negative, this being the general characteristic of any second-order perturbation effect on the lowest state.

For sufficiently excited states any atomic configuration becomes hydrogenic. Then $z \sim n^2$ and $1/(E_a - E_b)$ certainly has a n^3 dependence (see p. 40). Therefore $\Delta E \sim n^7$, and the perturbation treatment no longer applies. It has been shown that for very large n the Stark effect is linear in E, as in hydrogen.

For extremely high electric fields the Stark effect is capable of ionizing the atom. Looking at the one-electron potential energy $-(Ze^2/r) + eEz$, we see that the atomic center is not the only place where the potential is at a relative minimum. In the direction of the negative z, i.e., of the anode, the magnitude of the term $+e\mathcal{E}z$ will eventually make the potential even lower than in the atom. It is a well-known quantum mechanical result that whenever two potential troughs exist, an electron can pass from one trough (atom) to the other (anode) by the tunnel effect. Furthermore, once the electron has passed through the potential barrier it will be accelerated to the anode and the atom will be left ionized. Some discussion of this, ionization in hydrogen can be found in Landau and Lifshitz,[7] pp. 292-297.

PROBLEMS

1. An atom in a ^3P state is put into a magnetic field. What are the quantum numbers of the resulting states (a) in a weak, (b) in a strong magnetic field.
2. Calculate the Stark effect of the 3p and 3d triplet states of the He atom. The energy of an unperturbed level is given by the Rydberg formula

$$E = -\frac{Ry}{(n-\delta)^2}$$

and the Rydberg corrections are

for	3s	3p	3d
δ	0.296	0.0677	0.0029

The matrix element of z between two states can be written as

$$\langle n\ell m\,|\,z\,|\,n'\ \ell-1\ m\rangle = \langle n\ell\,|\,r\,|\,n'\ \ell-1\rangle\left(\frac{\ell^2-m^2}{4\ell^2-1}\right)^{1/2}$$

and we have approximately

$$\langle 3s\,|\,r\,|\,3p\rangle = 12.7a_0$$

$$\langle 3p\,|\,r\,|\,3d\rangle = 10.6a_0 \qquad a_0 = \text{Bohr radius}$$

What field strength is required to shift the most displaced component of the 3p state by 10 percent of the difference in energy between 3p and 3d? What is peculiar about the substate $m = 2$ of the 3d level, and how is it displaced?

---

[7] L. D. Landau and E. M. Lifshitz, *Quantum Mechanics*, 3rd ed., Pergamon, Oxford, 1977.

3. Calculate the energy levels, including electrostatic and spin-orbit energy, for a p^2 configuration. Assume the spin-orbit interaction to be

$$2\zeta(\boldsymbol{\ell}_1 \cdot \mathbf{s}_1 + \boldsymbol{\ell}_2 \cdot \mathbf{s}_2) \tag{1}$$

Procedure: Write down the Hamiltonian matrix for each J in the LS representation, such as for J = 2:

$$\begin{pmatrix} {}^3P + a\zeta & b\zeta \\ b\zeta & {}^1D + c\zeta \end{pmatrix} \tag{2}$$

where a, b, c are as yet unknown coefficients, and ^3P, ^1D are the electrostatic energies for these two states. Think carefully which of the coefficients a, b, c (and others to be defined for J = 1 and 0) must be zero. Then write down the eigenvalue equation in the limit when ζ is large compared with the electrostatic energies, and compare the result with that of the j-j coupling. This determines the coefficients a, b, c. Then solve for the eigenvalues for arbitrary ratio ζ/F^2. [F^2 is the radial integral defined in (7-28).]

(a) Plot the resulting eigenvalues as functions of ζ/F^2.

(b) Observe that states of the same J never cross. States of different J do cross; where?

(c) Show that in the LS limit, the ^3P states obey the Landé interval rule, and obtain the coefficient a in the matrix (2) directly from an eigenfunction of the $^3P_2$ state.

4. The ^2P state of an alkali atom is split into two states $j = \frac{1}{2}$ and $\frac{3}{2}$ by the spin-orbit interaction

$$V_2 = 2\zeta\boldsymbol{\ell}\cdot\mathbf{s} \qquad (\zeta > 0)$$

The atom is placed in a magnetic field $\mathcal{H}$. Calculate the six energy levels as functions of the parameter

$$X = \mathcal{H}\frac{\mu_0}{\zeta}$$

Find the levels for large X including the terms independent of X. How are the levels for X = 0 connected with those for large X? Plot the result. For which levels does the slope become steeper as X increases? Find the crossings of energy curves from the plot and algebraically.

Hint: Write the Hamiltonian in the m_ℓ, m_s representation in which both V_2 and

$$V_3 = \mathcal{H}\mu_0(\ell_z + 2s_z)$$

are easily expressed (though not both diagonal).

5. An atom in a state of angular momentum J is in a strong electric and weak magnetic field, in the same direction. In what manner is the level split? Give a qualitative formula for the level shift terms of M.
6. The different term energies $\langle LSJM | H_{SO} | LSJM \rangle = E_{SO}(J)$ of a given multiplet ${}^{2S+1}L$ satisfy

$$\sum_{J=|L-S|}^{J=L+S} (2J+1)E_{SO}(J) = X \qquad (1)$$

Find X by (a) explicitly evaluating the above sum, using the expression for $E_{SO}(J)$ given in the text (b) formulating an appropriate diagonal sum rule and evaluating (1) in a different representation.

7. Find the splitting of the lowest state of an Fe atom in a magnetic field of 20,000 gauss. Take $F^2 = 50{,}000\ \text{cm}^{-1}$, $F^4 = 30{,}000\ \text{cm}^{-1}$, $(\hbar^2/2m^2c^2)\zeta_{3d} = 400\ \text{cm}^{-1}$. Justify your use of LS coupling and determine whether the external field effects are Zeeman, Paschen-Back, or quadratic Zeeman.
8. The experimental spectrum of hydrogen in the n = 2 levels is

	${}^2s_{1/2}$	${}^2p_{1/2}$	${}^2p_{3/2}$
ΔE	0	0.1	1

| ← Lamb shift →|

| ←―――― Fine Structure ――――→ |

The ${}^2s_{1/2}$, ${}^2p_{3/2}$ splitting of $0.36\ \text{cm}^{-1}$ is a fine structure effect arising from the spin-orbit interaction and relativistic motion of the electron. This is predicted by the Dirac theory; see Chapter 23. The ${}^2p_{1/2}$, ${}^2s_{1/2}$ splitting of about $0.036\ \text{cm}^{-1}$, is the Lamb shift, and is not given by the Dirac equation but is a field theoretic effect. Without inquiring as to the dynamical origin of the Lamb shift, we may include this shift in the Schrödinger equation by introducing a Lamb shift operator L whose only effect is to shift the energy of the ${}^2s_{1/2}$ state relative to ${}^2p_{1/2}$ state.

Calculate the Stark effect in hydrogen when the field strength in units of energy) is: (a) large compared to the fine structure; (b) of the order of the fine structure; (c) much smaller than the fine structure but greater than the Lamb shift; and finally (d) very weak.

The wave functions in the njm representation, $j = \ell \pm \frac{1}{2}$, are given in Problem 7 of Chapter 6.

9. Derive (8-22d) and (8-22e) using (8-22a, b, c) and the commutation relations of J.
10. Derive (8-30).
11. Although (8-59) may be evaluated exactly for the hydrogen atom ground state, here is an estimate for that quantity.
    (a) Obtain a *lower* estimate by the following method.

$$E = E_0 + \langle 0|H_I|0\rangle + \sum_{n>0} \frac{|\langle 0|H_I|n\rangle|^2}{E_0 - E_n}$$

$$E \geq \tilde{E}$$

$$\tilde{E} = E_0 + \langle 0|H_I|0\rangle + \sum_{n>0} \frac{|\langle 0|H_I|0\rangle|^2}{E_0 - E_1}$$

    Evaluate $\tilde{E}$. (Here $|n\rangle$ is the nth energy eigenstate of hydrogen with energy eigenvalue E_n.)

    (b) Obtain an *upper* estimate by the variation method. The trial wave function $\phi(r)$ taken to be

$$\phi(r) = C[1 + Ae\,\mathcal{E}\,Z]\,u_0(r)$$

    where $u_0(r)$ is the ground state hydrogen wave function, A is the variational parameter and C is the normalization constant so chosen that $\int d^3r\,\phi^2(r) = 1$. Procedure: Find C, compute $\langle\phi|H_0+H_I|\phi\rangle$ as a function of A, vary A and determine minimum. You may approximate the minimization problem by keeping terms of order $\mathcal{E}^2$ and ignoring terms of order $\mathcal{E}^3$, $\mathcal{E}^4$, etc.

    (c) Compare the above bounds to the exact result (8-60).

Chapter 9

MOLECULES

In this chapter we shall briefly consider the quantum mechanical treatment of molecules. Here the physical situation is much more complex than that of atoms. For one thing, the electrons move in a field which can no longer be considered spherically symmetric since there are two or more nuclei acting as sources of the field. However, one simplifying feature exists, and allows for separate calculation of the energy associated with nuclear motion and of the energy of electronic motion. This separation occurs as a consequence of the large ratio of nuclear mass to electron mass. Therefore, as we shall see below, the kinetic energy of the nuclei, E_n, is much smaller than the kinetic energy of the electrons, E_e. Since period of motion is of the order of $\hbar$ divided by energy, the nuclear periods are much larger than the electron periods. Hence it is expected that to a good approximation the nuclei can be considered fixed in calculating electron motion. The nuclear motion is calculated with the approximation that the motion can be classified into translations, vibrations, and rotations of the nuclei. This forms the basis of the *Born-Oppenheimer approximation*, which we now discuss.

BORN-OPPENHEIMER APPROXIMATION

We examine the relative magnitude of the energy of the various modes of motion of the molecule.

If R is a molecular dimension, then the electronic energies are of the order of

$$E_e \sim \frac{\hbar^2}{mR^2} \sim \text{several eV} \tag{9-1}$$

where m is the electronic mass. This follows from the uncertainty principle, which requires a momentum $\hbar/R$ to fix the electron in a distance R.

The translational motion of the system as a whole is the same as that of a free particle. As this has no nonclassical features we shall not consider it any further.

The vibrational energy of the nuclei E_v, for a fairly low mode of vibration, is $\hbar(K_0/M)^{1/2}(v + \frac{1}{2})$, where M is the molecular mass, K_0 is an appropriate stiffness constant, and v is the vibrational quantum number. One can estimate K_0 by noting that the displacement in a normal mode must be of the order of R if the molecule is to dissociate. The energy of vibration is then about $K_0 R^2$, by definition of the stiffness constant. On the other hand, the energy of dissociation must be of the order of E_e; hence $K_0 R^2 \sim E_e$. Thus the energy difference between two neighboring vibrational levels is about

$$E_v \sim \hbar\left(\frac{E_e}{MR^2}\right)^{1/2} \sim \left(\frac{m}{M}\right)^{1/2} E_e \sim 0.1 \text{ eV} \qquad (9\text{-}2)$$

m/M is usually in the range 10^{-3} to 10^{-4}.

The rotational energy of the nuclei E_r is calculated from the moment of inertia $\sim MR^2$ of the molecule:

$$E_r \sim \frac{\hbar^2}{MR^2} \sim \frac{m}{M} E_e \sim 0.001 \text{ eV} \qquad (9\text{-}3)$$

We thus see that the energy of nuclear motions indeed satisfies the Born-Oppenheimer assumption: it is much smaller than the electronic energy.

The time-independent Schrödinger equation for the molecule is

$$\left[-\frac{\hbar^2}{2m}\sum_{i=1}^{n}\nabla_i^2 - \sum_{j=1}^{N}\frac{\hbar^2}{2M_j}\nabla_j^2 + V(\mathbf{R}_1, \ldots, \mathbf{R}_N; \mathbf{r}_1, \ldots, \mathbf{r}_n) - E\right] \times \Psi(\mathbf{R}_1, \ldots, \mathbf{R}_N; \mathbf{r}_1, \ldots, \mathbf{r}_n) = 0 \qquad (9\text{-}4)$$

where the R's and M's refer to the positions and masses of the nuclei and the lower case letters refer to electronic coordinates and masses. The kinetic energy of nuclear motion is of the order m/M and can be neglected according to the Born-Oppenheimer approximation. Then, in the remaining equation, the wave function Ψ depends on the (fixed) $\mathbf{R}_i$ only parametrically. Therefore Ψ is approximated by

$$\Psi(\mathbf{r}_i, \mathbf{R}_j) = u_{\mathbf{R}_j}(\mathbf{r}_i)\, w(\mathbf{R}_j) \tag{9-5}$$

where the u's satisfy the equation

$$\left[-\frac{\hbar^2}{2m}\sum_i \nabla_i^2 + V(\mathbf{R}_j, \mathbf{r}_i)\right] u_{\mathbf{R}_j}(\mathbf{r}_i) = U(\mathbf{R}_j)\, u_{\mathbf{R}_j}(\mathbf{r}_i) \tag{9-6}$$

(The symbols R_j and $\mathbf{r}_i$ are abbreviations for the sets $\mathbf{R}_1, \ldots, \mathbf{R}_N$; and $\mathbf{r}_1, \ldots, \mathbf{r}_n$.) $U(\mathbf{R}_j)$ in (9-6) is the energy eigenvalue of (9-6) and depends parametrically on $\mathbf{R}_j$. That is, to each arrangement of the nuclei indexed by $\mathbf{R}_j$, there corresponds an electron distribution $|u_{\mathbf{R}_j}(\mathbf{r}_i)|^2$ with energy $U(\mathbf{R}_j)$. Substituting (9-5) in (9-4) gives, with the aid of (9-6) and some algebra,

$$u_{\mathbf{R}_j}(\mathbf{r}_i)\left[-\frac{\hbar^2}{2}\sum_j \frac{1}{M_j}\nabla_j^2 + U(\mathbf{R}_j) - E\right] w(\mathbf{R}_j)$$

$$= \frac{\hbar^2}{2}\sum_j \frac{1}{M_j}\left[w(\mathbf{R}_j)\nabla_j^2 u_{\mathbf{R}_j}(\mathbf{r}_i)\right.$$

$$\left. + 2(\nabla_j w(\mathbf{R}_j))\cdot(\nabla_j u_{\mathbf{R}_j}(\mathbf{r}_i))\right] \tag{9-7}$$

Upon multiplication of both sides of (9-7) by $u_{\mathbf{R}_j}(\mathbf{r}_i)$, and integration over $d\tau_i$, the left-hand side acquires the factor $\int u_{\mathbf{R}_j}^2(\mathbf{r}_i)\, d\tau_i$. This integral equals one, if we assume that $u_{\mathbf{R}_j}(\mathbf{r}_i)$ is real. The second term on the right-hand side in (9-7) is proportional to

$$\nabla_j w(\mathbf{R}_j)\cdot\nabla_j \int u_{\mathbf{R}_j}^2(\mathbf{r}_i)\, d\tau_i = \nabla_j w(\mathbf{R}_j)\cdot\nabla_j(1) = 0$$

The first term on the right-hand side of (9-7) involves

$$\int u_{\mathbf{R}_j}(\mathbf{r}_i)\nabla_j^2 u_{\mathbf{R}_j}(\mathbf{r}_i)\, d\tau_i$$

$$= \tfrac{1}{2}\nabla_j^2 \int u_{\mathbf{R}_j}^2(\mathbf{r}_i)\, d\tau_i - \int (\nabla_j u_{\mathbf{R}_j}(\mathbf{r}_i))^2\, d\tau_i$$

$$= -\int (\nabla_j u_{\mathbf{R}_j}(\mathbf{r}_i))^2\, d\tau_i$$

Hence (9-7) can be written

$$\left[-\sum_{j=1}^{N}\frac{\hbar^2}{2M_j}\nabla_j^2 + U(\mathbf{R}_j) + W(\mathbf{R}_j)\right] w(\mathbf{R}_j) = Ew(\mathbf{R}_j) \tag{9-8}$$

where

$$W(\mathbf{R}_j) = \sum_{j=1}^{N} \frac{\hbar^2}{2M_j} \int (\nabla_j u_{\mathbf{R}_j}(\mathbf{r}_i))^2 \, d\tau_i \tag{9-9}$$

Equation (9-9) can be evaluated and gives a small correction to the nuclear motion potential $U(\mathbf{R}_j)$. More important is the approximation we made of considering only the projection of (9-7) on $u_{\mathbf{R}_j}(\mathbf{r}_i)$. In an exact solution, it would be necessary to fulfill (9-7) for all $\mathbf{R}_j$ and $\mathbf{r}_i$. Born and Oppenheimer have shown that, owing to the smallness of the amplitudes of the nuclear motion in comparison with the equilibrium internuclear distances, the correction terms arising from this can be neglected as long as high vibrational and rotational modes are not excited.

HYDROGEN MOLECULE

As an example, we outline an approximate treatment of the hydrogen molecule. The only nuclear coordinate that occurs in the electronic equation is R, the magnitude of the internucleon distance. The Hamiltonian occurring in (9-6) is

$$-\frac{\hbar^2}{2m}[\nabla_1^2 + \nabla_2^2]$$

$$+ e^2\left[\frac{1}{R} + \frac{1}{r_{12}} - \frac{1}{r_{1A}} - \frac{1}{r_{2A}} - \frac{1}{r_{1B}} - \frac{1}{r_{2B}}\right] \tag{9-10a}$$

A, B refer to the two nuclei, 1, 2 to the two electrons. Explicitly if the position of the two nuclei relative to the origin is $\mathbf{r}_A$ and $\mathbf{r}_B$, and of the two electrons, $\mathbf{r}_1$ and $\mathbf{r}_2$; then we have

$$R = |\mathbf{r}_A - \mathbf{r}_B| \qquad r_{12} = |\mathbf{r}_1 - \mathbf{r}_2|$$

$$\mathbf{r}_{1A} = \mathbf{r}_1 - \mathbf{r}_A \qquad \mathbf{r}_{2A} = \mathbf{r}_2 - \mathbf{r}_A$$

$$\mathbf{r}_{1B} = \mathbf{r}_1 - \mathbf{r}_B \qquad \mathbf{r}_{2B} = \mathbf{r}_2 - \mathbf{r}_B \tag{9-10b}$$

The procedure then is to solve the Schrödinger equation of the electronic motion, where the Hamiltonian is given by (9-10a) with R considered to be a fixed parameter. The energy eigenvalue U(R) then depends on R and can be used as the nuclear potential for equation (9-8).

We solve the Schrödinger equation for the electronic motion by perturbation theory. As we wish to maintain proper symmetry of the electron wave function, we apply the unsymmetric perturbation theory as developed in Chapter 3.

$$H_a^0 = -\frac{\hbar^2}{2m}[\nabla_1^2 + \nabla_2^2] - e^2\left[\frac{1}{r_{1A}} + \frac{1}{r_{2B}}\right]$$

$$\lambda H_a^1 = e^2\left[\frac{1}{r_{12}} - \frac{1}{r_{1B}} - \frac{1}{r_{2A}} + \frac{1}{R}\right]$$

$$U_a^0 = u_A(\mathbf{r}_{1A})\,u_B(\mathbf{r}_{2B})$$

$$H_b^0 = -\frac{\hbar^2}{2m}[\nabla_1^2 + \nabla_2^2] - e^2\left[\frac{1}{r_{1B}} + \frac{1}{r_{2A}}\right]$$

$$\lambda H_b^1 = e^2\left[\frac{1}{r_{12}} - \frac{1}{r_{1A}} - \frac{1}{r_{2B}} + \frac{1}{R}\right]$$

$$U_b^0 = \pm u_A(\mathbf{r}_{1B})\,u_B(\mathbf{r}_{2A})$$

$$E^0 = -\frac{e^2}{2a_0}\left[\frac{1}{n_A^2} + \frac{1}{n_B^2}\right] \qquad (9\text{-}11)$$

Here $u_A(\mathbf{r})$ is a hydrogenic wave function with principal quantum number n_A. The nuclear motion is then described by the potential

$$U(R) = E^0 + \lambda E^1(R) \qquad (9\text{-}12a)$$

where $\lambda E^1(R)$ is given by [see (3-17)]

$$(1+\gamma)\lambda E^1 = \tfrac{1}{2}\int d\tau\,(U_a^0 + U_b^0)^*(\lambda H_a^1 U_a^0 + \lambda H_b^1 U_b^0)$$

$$= \int d\tau\,(U_a^0 + U_b^0)^*\,\lambda H_a^1 U_a^0 \qquad (9\text{-}12b)$$

$$\gamma = \tfrac{1}{2}\int d\tau\,(U_b^{0*} U_a^0 + U_a^{0*} U_b^0)$$

$$= \int U_b^{0*} U_a^0\, d\tau \qquad (9\text{-}12c)$$

In both equations, the second line follows from the first by symmetry.[1]

---

[1] Equation (9-12b) differs from (3-17) by the presence of the term γ. In Chapter 3 we assumed U_a^0 and U_b^0 to be orthogonal. Here we do not make this assumption.

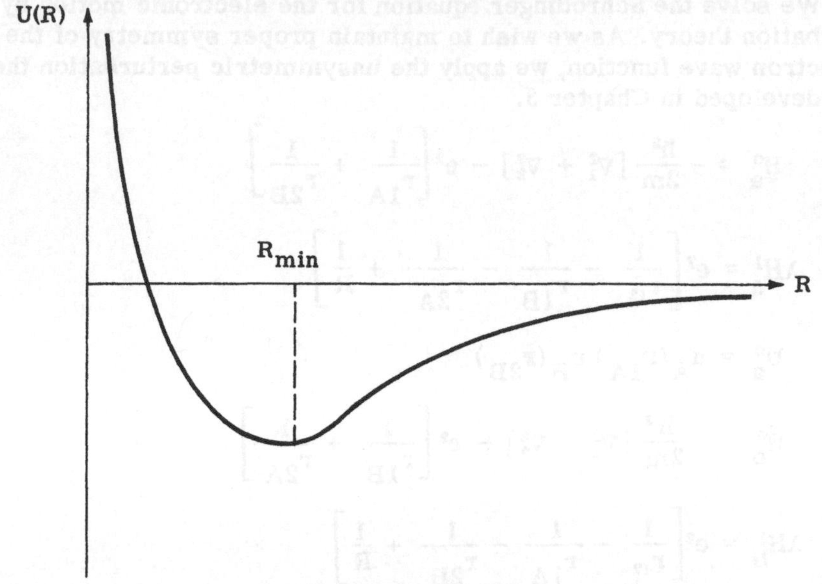

Figure 9-1
Internucleon potential for the hydrogen molecule.

It happens that the exchange term is negative, therefore the energy is lower for the combination

$$u_A(r_{1A})\,u_B(r_{2B}) + u_A(r_{1B})\,u_B(r_{2A})$$

The result that is found for U(R) is represented qualitatively in Figure 9-1. A knowledge of U(R) gives the equilibrium position of the nuclei, R_{min}. $d^2U/dR^2\,|_{R_{min}}$ gives the vibration frequency of the nuclei (see below).

For the antisymmetric combination of spatial wave functions the potential function U(R) is found to be repulsive[2] at all R. Hence, remembering the Pauli principle, we conclude that binding can occur only for the antisymmetric combination of spin-wave functions, which corresponds to S = 0 (see p. 16). This is called *homopolar binding*

---

[2]There is, however, a weak attraction at very large R, owing to Van der Waals' forces. These forces, which are described, for example, in L. I. Schiff *Quantum Mechanics*, 3rd ed., McGraw-Hill, New York, 1968, pp. 259 ff., arise from second-order perturbation theory and hence are not included in our present theory.

Table 9-1

Theoretical and Experimental Results for the Hydrogen Molecule

	Theory	Experiment
Dissociation energy	3.2 - 4.69 eV	4.73 eV
R_{min}	0.72 Å	0.74 Å
Vibrational frequency	4290 cm^{-1}	4270 cm^{-1}

and is frequently described by saying that the electrons must have "opposite spin." Detailed results of calculations based on this theory as well as on various refinements are listed in Table 9-1. The highest dissociation energy, 4.69 eV, is derived from a variational calculation.

DIATOMIC MOLECULES

For general diatomic molecules, with nuclear masses M_1 and M_2, the Schrödinger equation for the relative motion of the nuclei follows from (9-8),

$$\left[-\frac{\hbar^2}{2M}\nabla^2 + U(R)\right] w(R, \theta, \varphi) = Ew(R, \theta, \varphi) \tag{9-13}$$

$M = (M_1 M_2)/(M_1 + M_2)$, and W(R) in (9-8) has been included in U(R). The usual central force separation is possible:

$$w = \frac{w(R)}{R} Y_{Km}(\Omega)$$

$$\left(-\frac{\hbar^2}{2M}\frac{d^2}{dR^2} + U(R) + \frac{\hbar^2}{2M}\frac{K(K+1)}{R^2}\right) w(R) = Ew(R) \tag{9-14}$$

It has been found that for the lowest electronic states the internucleon potential can be represented quite well by the *Morse potential*

$$U(R) = U_0\left[e^{-2(R-R_0)/a} - 2e^{-(R-R_0)/a}\right] \tag{9-15}$$

Equation (9-14) can be rewritten in terms of the effective one-dimensional potential

$$U'(R) = U(R) + \frac{\hbar^2}{2M}\,\frac{K(K+1)}{R^2}$$

$U'(R)$ can be expanded about its minimum $U_0' = U'(R_1)$ to give

$$U'(R) = U_0' + \tfrac{1}{2}K_0(R - R_1)^2$$

$$+\, b(R - R_1)^3 + c(R - R_1)^4 \tag{9-16}$$

The third and fourth terms in (9-16) can be considered to be perturbations of a simple harmonic oscillator. By the usual perturbation method we then obtain to second order in v,

$$E = U_0' + \hbar\left(\frac{K_0}{M}\right)^{1/2}(v + \tfrac{1}{2}) - \frac{\hbar^2 b^2}{MK_0^2}\left[\tfrac{15}{4}(v + \tfrac{1}{2})^2 + \tfrac{7}{16}\right]$$

$$+\, \frac{3\hbar^2 c}{2MK_0}\left[(v + \tfrac{1}{2})^2 + \tfrac{1}{4}\right] \tag{9-17}$$

where v = 0, 1, 2, ... is the vibrational quantum number.

If U(R) is of the form (9-15), then

$$R_1 = R_0 + \frac{\hbar^2 K(K+1)a^2}{2MR_0^3 U_0}$$

$$U_0' = -U_0 + \frac{\hbar^2 K(K+1)}{2MR_0^2} - \frac{\hbar^4 K^2(K+1)^2 a^2}{4M^2 R_0^6 U_0}$$

$$K_0 = \frac{2U_0}{a^2} - \frac{3\hbar^2 K(K+1)}{MR_0^2 a^2}\,\frac{a}{R_0}\left(1 - \frac{a}{R_0}\right)$$

$$b = -\frac{U_0}{a^3} \qquad c = \frac{7U_0}{12a^4} \tag{9-18}$$

In (9-18) only enough terms were kept to give E correctly to second order in v and K^2.

The first equation in (9-18) shows that the molecule stretches, due to rotation. The second equation gives the equilibrium energy $-U_0$ and the rotational energy to second order in K^2. It is seen that first-order rotational energy agrees with that of a rigid rotator. The frequency of vibration, $\omega_0 = (K_0/M)^{1/2}$, is found to decrease with $M = (M_1 M_2)/(M_1 + M_2)$. For H_2, $\omega_0 \sim 4000\ \text{cm}^{-1}$; for HX, where X is an arbitrary, heavier atom, $\omega_0 \sim 3000\ \text{cm}^{-1}$; for N_2, $\omega_0 \sim 1000\ \text{cm}^{-1}$. It is found that for all heavy, tightly bound molecules, K_0 is approximately the same.

Using the expressions for b and c in (9-18), and $K = 0$, the last two terms in (9-17) can be combined to give

$$-\frac{\hbar^2}{2Ma^2}\,(v + \tfrac{1}{2})^2$$

neglecting a small term independent of v. Then, for $K = 0$, all terms in (9-17) can be combined to give

$$E = -\left[\sqrt{U_0} - \frac{\hbar}{a\sqrt{2M}}\,(v + \tfrac{1}{2})\right]^2 \tag{9-19}$$

The pure vibrational spectrum (9-19) is very simple. The spacing between vibrational levels decreases with increasing vibrational quantum number v, and goes to zero when $E = 0$. The number of vibrational levels, however, is finite, and can easily be seen to be equal to

$$N = \frac{2U_0}{\hbar\omega_0} \tag{9-20}$$

where $\hbar\omega_0$ is the spacing between levels at low v.

SYMMETRY FOR HOMONUCLEAR DIATOMIC MOLECULES

The wave function describing a ***homonuclear diatomic molecule***, viz., a molecule with two identical nuclei, factors into a product of electronic and nuclear spatial and spin wave functions [see (9-5)].

$$\Psi = u_{el}\,\chi_{el}\,u_N\,\chi_N \tag{9-21}$$

Ψ must be symmetric (antisymmetric) under nuclear interchange if the spin of the nucleus is integral (half-integral). χ_{el} is symmetric under nuclear interchange, as it does not depend on the nuclei. The electronic spatial wave function u_{el} will depend on the distances r_{Ai} and r_{Bi} of each electron from the nuclei A and B. Interchange of the two nuclei means interchange of r_{Ai} with r_{Bi}, simultaneously for all i. Geometrically, this interchange means a reflection of the entire electron system at the midplane between the two nuclei. This reflection leaves the Hamiltonian invariant, and therefore u_{el} will have a definite symmetry with respect to it, either symmetric or antisymmetric. (In the simple case of the H_2 molecule, u_{el} is symmetric with respect to the two nuclei, and in this case the symmetry with respect to the nuclei is related to the symmetry of the spatial function between the two electrons. In general, this relation does not hold. In fact, the spatial electron function usually has a complicated symmetry with respect to electron interchange, whereas its symmetry with respect to the two nuclei is always simple.) In most molecular ground states, u_{el} is symmetric in the two nuclei.

Whenever u_{el} is symmetric with respect to the two nuclei, $u_N \chi_N$ must be symmetric (antisymmetric) under nuclear interchange if the nuclear spin is an integer (half-integer). In u_N, nuclear interchange is equivalent to a change in sign of $\mathbf{R}$, the relative position vector. (The coordinate system is assumed to be chosen centered at the midpoint of the line joining the two nuclei.) Hence the symmetry of u_N is determined by $(-1)^K$. Thus we conclude that for nuclei of zero or integer spin, the nuclear spin function must be symmetric for even K, antisymmetric for odd K. For nuclei with half-integer spin, the nuclear spin function must be antisymmetric for even K and symmetric for odd K.

From the discussion of addition of equal angular momenta, i.e., addition of the two nuclear spins $I\hbar$, it is seen that the total of $(2I + 1)^2$ spin states can be divided into $(I + 1)(2I + 1)$ symmetric states and $I(2I + 1)$ antisymmetric states. It is seen that the ratio of the number of symmetric spin states to antisymmetric spin states is $(I + 1)/I$. Thus in a gas of homonuclear diatomic molecules in statistical equilibrium, the ratio of the number of molecules with even K to the number with odd K will be $(I + 1)/I$ if I is 0 or an integer, and $I/(I + 1)$ if I is a half-integer. This gives rise to alternating intensities in the rotational spectra of molecules. We repeat that this analysis depends on u_{el} being symmetric under interchange of the nuclei. When this is not so, the results are modified in an obvious fashion.

Because of the extremely weak interaction of nuclear spins with the electrons, the probability of changing the spin orientations is very small. Hence gases of molecules differing in total nuclear spin act almost as different gases. For example, H_2 can have two states: resultant nuclear spin 1 (ortho) or resultant nuclear spin 0 (para). The statistical weight ratio of ortho to para states is therefore 3 to 1.

PROBLEMS

1. Supply the missing algebra in the derivation of (9-7).
2. Solve the one-dimensional Schrödinger equation, for bound states, with the Morse potential: $V(x) = U_0[e^{-2x/a} - 2e^{-x/a}]$, $U_0 > 0$. Hint: Solve the differential equation by defining the parameters

$$s = \left[\frac{2m\,|\,E\,|\,a^2}{\hbar^2}\right]^{1/2} \qquad v = \left[\frac{2m\,U_0\,a^2}{\hbar^2}\right]^{1/2} - s - \frac{1}{2}$$

by changing the independent variable x to y

$$y = 2\left[\frac{2m\,U_0\,a^2}{\hbar^2}\right]^{1/2} \exp(-x/a)$$

and by changing the dependent variable ψ to w

$$\psi = e^{-1/2y}\, y^s\, w(y)$$

The resulting differential equation can be solved with confluent hypergeometric functions. Show that when $[2m\,U_0 a^2/\hbar^2]^{1/2} > \frac{1}{2}$, the energy levels are given by (9-19); while when $[2m\,U_0 a^2/\hbar^2]^{1/2} < \frac{1}{2}$, no discrete energy spectrum exists.
3. Verify (9-16), (9-17), and (9-18).
4. Give the Hamiltonian for a rigid rotator with a moment of inertia I. What are the energy eigenvalues? (A rigid rotator approximates a diatomic molecule when vibrational motion is ignored.)
5. Analyze the lowest electronic states of a diatomic molecule where the internuclear potential is given by

$$U(R) = -2U_0\left(\frac{a}{R} - \frac{a^2}{2R^2}\right)$$

Carry out the analysis in a fashion similar to the analysis in the text, equations (9-15) to (9-20).

Part II

SEMICLASSICAL RADIATION THEORY

SO FAR in our discussion of atoms, we have dealt only with stationary states. We shall now treat transitions between these states. A proper theoretical framework for discussing this phenomenon is given by quantum electrodynamical field theory, in which the electromagnetic field is quantized. We shall give here a semiclassical treatment, in which the electromagnetic field is described classically. The semi-classical approach cannot replace the complete theory. Yet it is of interest since lowest order results, which are physically relevant, are so simply derived. Moreover, in the modern field theoretical approach to particle physics, semiclassical methods have produced much insight and frequently are the only ones available. Hence it is useful to present them in atomic physics, especially since the fully quantized theory is adequately covered in other texts, e.g., Gottfried[1] and Messiah[2].

---

[1]K. Gottfried, *Quantum Mechanics*, Vol. I, Benjamin/Cummings, Reading MA, 1968.

[2]A. Messiah, *Quantum Mechanics*, Vols. I and II, Wiley, New York, 1961, 1962.

Chapter 10

SEMICLASSICAL THEORY OF RADIATION

SCHRÖDINGER EQUATION

We study the interaction of an atomic system with an electromagnetic radiation field. The Schrödinger equation for a mass point of charge q in an electromagnetic field described by a vector potential A is

$$i\hbar \frac{\partial \Psi}{\partial t} = \left[-\frac{\hbar^2}{2m} \nabla^2 + \frac{iq\hbar}{mc} \mathbf{A}\cdot\nabla + V \right] \Psi \qquad (10\text{-}1)$$

(Schiff[1], p. 177). We are in the gauge $\nabla \cdot \mathbf{A} = 0$, and the scalar potential $\phi = 0$, which is always possible when no sources of the electromagnetic field are present. We have suppressed the term in A^2 which is negligible.

The problem will be treated semiclassically in the sense that although the motion of the particle is quantized, the electromagnetic field will be considered classically. Therefore it is assumed that the vector potential can be specified without any uncertainty at each point of space time by means of the classical Maxwell equations in a vacuum.[2]

---

[1] L. I. Schiff, *Quantum Mechanics*, 3rd ed., McGraw-Hill, New York, 1968.

[2] The classical electromagnetic theory needed for our development is described, for example, in W. K. H. Panofsky and M. Phillips, *Classical Electricity and Magnetism*, 2nd ed., Addison-Wesley, Reading MA, 1962.

$$\nabla \times \mathcal{E} = -\frac{1}{c}\frac{\partial \mathcal{H}}{\partial t}$$

$$\nabla \times \mathcal{H} = \frac{1}{c}\frac{\partial \mathcal{E}}{\partial t}$$

$$\nabla \cdot \mathcal{H} = 0$$

$$\nabla \cdot \mathcal{E} = 0 \tag{10-2}$$

Then

$$\mathcal{H} = \nabla \times \mathbf{A}$$

$$\mathcal{E} = -\frac{1}{c}\frac{\partial \mathbf{A}}{\partial t} \tag{10-3}$$

$$\nabla^2 \mathbf{A} - \frac{1}{c^2}\frac{\partial^2 \mathbf{A}}{\partial t^2} = 0$$

$$\nabla \cdot \mathbf{A} = 0 \tag{10-4}$$

It will be seen that this semiclassical description gives a correct account of the influence of an external radiation field on the particle (absorption and induced emission) but not of the influence of the particle on the field (spontaneous emission). The reason for the correct results for the former phenomena lies in the correspondence principle. When the radiation field is quantized it is regarded as a collection of quantized oscillators, with the nth excited state of the oscillator describing n photons in the electromagnetic field. For high values of n (many photons or intense beam), the correspondence principle allows a classical description of the field. Hence for intense external beams the semiclassical approximate treatment is expected to yield the correct results. But now we observe that (10-1) is linear in **A**. Thus the results that hold for the strong external beam case must also hold for the weak beam. That this is indeed so is related to the happy accident that the correspondence principle for a harmonic oscillator is already valid for low values of n.

These considerations do not hold for spontaneous emission. This emission occurs regardless whether an external field is initially present; that is, an accelerated charge radiates regardless whether or not it is in an external field. At least one quantum of radiation must be emitted; thus the effect is not linear in the field, and the correspondence principle cannot be extrapolated in a simple way to the emission of one quantum. For a completely satisfactory theory, we have to quantize the electromagnetic field, i.e., we need quantum field theory. However, we shall be able to obtain the correct probability of spontaneous emission from general conditions of equlibrium, and this result will be seen to be a plausible extrapolation of the classical theory of emission of radiation.

ABSORPTION AND INDUCED EMISSION

In order to obtain a description for the absorption of the external radiation, or for the (induced) emission of radiation in the presence of an external field, we must specify the external field, viz., we must obtain solutions of (10-4) for **A**. In practical applications we need concern ourselves only with two types of external fields: monochromatic plane waves and incoherent superpositions of plane waves with different frequencies.

The former case is of interest when we consider a transition from a definite initial state to a group of closely spaced or continuously distributed final states. Such a transition will proceed in the presence of monochromatic radiation of definite frequency ω so long as energy is conserved: $|E_n - E_f| = \hbar\omega$ (n and f denote the initial and and final states, respectively). Since the final state lies in the continuum, there will always exist a level E_f to which the transition can go, provided of course ω lies within suitable limits.

On the other hand if the final state is in the discrete spectrum and the radiation is monochromatic, energy cannot in general be conserved. In that case it is assumed that the external radiation covers a spread of frequencies, clustered around that frequency for which energy can be conserved in the transition.

We discuss both types of external fields, treating the simpler, monochromatic radiation first.

MONOCHROMATIC EXTERNAL FIELDS

The plane wave monochromatic solutions of (10-4) for **A** are of the form

$$2\mathbf{A}_0 \exp i(\mathbf{k}\cdot\mathbf{r} - \omega t) \tag{10-5}$$

where $2\mathbf{A}_0$ is a constant complex vector, describing both the intensity and the polarization, and **k** is the propagation vector. $\mathbf{A}_0$ is perpendicular to **k** and $kc = \omega$. Physical solutions correspond to the real part of (10-5). The vector potential, the electric and the magnetic fields are

$$\mathbf{A} = \mathrm{Re}\ 2\mathbf{A}_0 \exp i\,(\mathbf{k}\cdot\mathbf{r} - \omega t) \tag{10-6a}$$

$$\mathcal{E} = \mathrm{Re}\ 2ik\mathbf{A}_0 \exp i\,(\mathbf{k}\cdot\mathbf{r} - \omega t) \tag{10-6b}$$

$$\mathcal{H} = \mathrm{Re}\ 2i\mathbf{k} \times \mathbf{A}_0 \exp i\,(\mathbf{k}\cdot\mathbf{r} - \omega t) \tag{10-6c}$$

The Poynting vector $(c/4\pi)\,\mathcal{E} \times \mathcal{H}$ is in the direction $\mathbf{k}$. Averaged over a period $2\pi/\omega$ of the oscillation its magnitude is

$$\frac{\omega^2}{2\pi c}\,|\mathbf{A}_0|^2 \tag{10-7}$$

where $|\mathbf{A}_0|^2 = \mathbf{A}_0\cdot\mathbf{A}_0^*$. The quantity (10-7) is the intensity of the beam in ergs/cm^2 sec, which we shall denote by I. We can also introduce the number of quanta per unit area and time, $N = I/\hbar\omega$, so that (10-7) gives

$$|\mathbf{A}_0|^2 = \frac{2\pi\hbar c}{\omega}\,N \tag{10-8}$$

The polarization of the beam may be specified as follows. The complex vector $\mathbf{A}_0$ has a complex squared magnitude $\mathbf{A}_0\cdot\mathbf{A}_0$ which is a scalar and has a phase of 2θ, viz.,

$$\mathbf{A}_0\cdot\mathbf{A}_0 = |\mathbf{A}_0\cdot\mathbf{A}_0|\,e^{2i\theta} \tag{10-9a}$$

Defining a new complex vector $\mathbf{C}_0$ by

$$\mathbf{A}_0 = \mathbf{C}_0\,e^{i\theta} \tag{10-9b}$$

we note that $\mathbf{C}_0^2$ is real as it equals $|\mathbf{A}_0\cdot\mathbf{A}_0|$, while $|\mathbf{A}_0|^2 = \mathbf{A}_0\cdot\mathbf{A}_0^* = \mathbf{C}_0\cdot\mathbf{C}_0^* = |\mathbf{C}_0|^2$. Decomposing $\mathbf{C}_0$ into its real and imaginary parts

$$\mathbf{C}_0 = \mathbf{c}_1 + i\mathbf{c}_2 \tag{10-9c}$$

we see that $\mathbf{c}_1\cdot\mathbf{c}_2 = 0$ since $\mathbf{C}_0^2$ is real. We may choose $\mathbf{c}_1$ to be along the x axis, $\mathbf{c}_2$ along the $\pm$ y axis and $\mathbf{k}$, the propagation vector, along the z axis.

The vector potential is given by

$$\mathbf{A} = 2\ \mathrm{Re}\ \mathbf{C}_0 \exp i(\mathbf{k}\cdot\mathbf{r} - \omega t + \theta) \tag{10-10a}$$

$$\begin{aligned}\mathbf{A} = {} & 2\,|\mathbf{c}_1|\,\hat{\mathbf{i}}\cos(\mathbf{k}\cdot\mathbf{r} - \omega t + \theta) \\ & \pm 2\,|\mathbf{c}_2|\,\hat{\mathbf{j}}\sin(\mathbf{k}\cdot\mathbf{r} - \omega t + \theta)\end{aligned} \tag{10-10b}$$

Evidently

$$\frac{A_x^2}{4c_1^2} + \frac{A_y^2}{4c_2^2} = 1 \tag{10-11}$$

Equation (10-11) shows that when $|c_1| = |c_2|$, the radiation is circularly polarized; when either c_1 or c_2 is zero, the radiation is plane polarized; while if $0 \neq |c_1| \neq |c_2| \neq 0$, the beam is elliptically polarized. Since elliptically polarized light can be considered to be a superposition of plane and circular polarized light, we shall disregard the case $0 \neq |c_1| \neq |c_2| \neq 0$.

For plane or circularly polarized light $\mathbf{C}_0$ has the form

$$\mathbf{C}_0 = \sqrt{|\mathbf{C}_0|^2}\,\mathbf{P} = \sqrt{|\mathbf{A}_0|^2}\,\mathbf{P} = |\mathbf{A}_0|\,\mathbf{P} \tag{10-12a}$$

where $|\mathbf{A}_0|$ is defined as $\sqrt{|\mathbf{A}_0|^2}$ and $\mathbf{P}$ is a complex vector specifying the polarization,

$$\mathbf{P} = \frac{1}{\sqrt{2}}[\hat{\mathbf{i}} \pm i\hat{\mathbf{j}}] \quad \text{circular polarization} \tag{10-12b}$$

$$\mathbf{P} = \hat{\mathbf{i}} \text{ or } \pm i\hat{\mathbf{j}} \quad \text{plane polarization in the x or y direction} \tag{10-12c}$$

Having obtained a specific expression for the external field $\mathbf{A}$, we now calculate transitions of the quantum mechanical system in the presence of this external field.

We consider the term $(iq\hbar/mc)\mathbf{A}\cdot\nabla$ in (10-1) as a time-dependent perturbation with $\mathbf{A}$ given by (10-10a). If the system is initially in the state n, and the perturbation is turned on at time t = 0, the first-order amplitude is given by

$$a_f^{(1)}(t) = \frac{1}{i\hbar}\int_0^t H'_{fn}(t')\, e^{i\omega_{fn}t'}\, dt' \tag{1-13a}$$

with $\omega_{fn} = (E_f - E_n)/\hbar$. Then

$$a_f^{(1)}(t) = -\frac{T'_{fn}}{\hbar}\,\frac{e^{i(\omega_{fn} - \omega)t} - 1}{\omega_{fn} - \omega}\, e^{i\theta}$$

$$-\frac{T''_{fn}}{\hbar}\,\frac{e^{i(\omega_{fn}+\omega)t} - 1}{\omega_{fn} + \omega}\, e^{-i\theta} \tag{10-13a}$$

$$T'_{fn} = \frac{iq\hbar}{mc} |A_0| \int u_f^* e^{i\mathbf{k}\cdot\mathbf{r}} \mathbf{P}\cdot\nabla u_n \, d\tau \tag{10-13b}$$

$$T''_{fn} = \frac{iq\hbar}{mc} |A_0| \int u_f^* e^{-i\mathbf{k}\cdot\mathbf{r}} \mathbf{P}^*\cdot\nabla u_n \, d\tau \tag{10-13c}$$

The probability that a transition will occur is only appreciable for

$$\omega_{fn} = \pm\omega$$

that is,

$$\begin{aligned} E_f &= E_n + \hbar\omega \\ &= E_n - \hbar\omega \end{aligned} \tag{10-14}$$

The former condition corresponds to an absorption of one quantum from the radiation field, the latter to induced emission. It is quite remarkable that we obtain quantization of the energy emitted or absorbed without having assumed any quantization of the electromagnetic field initially. Energy conservation between particle and field is assured by (10-14). When $\omega_{fn} = \omega$, the probability of finding the system in a state f with higher energy is proportional to $|T'_{fn}|^2$. When $\omega_{fn} = -\omega$, the probability of finding the system in a state of lower energy is proportional to $|T''_{fn}|^2$.

We obtain the transition probability per unit time for transitions to a group of closely spaced or continuously distributed final states of the electron system. The transition will be either an absorption or a stimulated emission of a quantum. We can easily assume that the group of states covers an energy range small compared with $\hbar\omega$; then only one of the two relations $\omega = \pm\omega_{fn}$ is satisfied. In this case, the probability per unit time w for transition to this group of final states to occur is given by Fermi's golden rule,

$$w = \frac{2\pi}{\hbar} \rho(k_f) |A_{fn}|^2 \tag{1-15}$$

where $\rho(k_f)$ is the density of states in this region and A_{fn} is T'_{fn} or T''_{fn}, depending on whether an absorption or an emission is being considered. Here $\hbar k_f = \mathbf{p}$ is the momentum of the electron which has been ejected into the continuum or the group of closely spaced states. The final states approach plane waves at distances far from the neighborhood of the perturbation, so that

$$\rho(k_f) = \frac{V}{(2\pi\hbar)^3}\frac{p^2\,dp\,d\Omega}{dE} = \frac{V}{(2\pi\hbar)^3}\frac{p^2\,dp\,d\Omega}{v\,dp}$$

$$= \frac{V}{(2\pi\hbar)^3}\frac{p^2}{v}\,d\Omega \tag{10-15}$$

where $d\Omega$ is the solid angle into which the ejected electron with momentum $\mathbf{p}$ goes, and V is the quantization volume.

For absorption the transition probability becomes (we set $q = -e$, with $e > 0$ being the absolute magnitude of the electronic charge)

$$w = \frac{e^2}{(2\pi\hbar c)^2}\,v\,|\mathbf{A}_0|^2\,\left|\int u_f^* e^{i\mathbf{k}\cdot\mathbf{r}}\,\mathbf{P}\cdot\nabla u_n\,d\tau\right|^2\,V\,d\Omega \tag{10-16}$$

In (10-17) $\mathbf{P}$ is the direction of the polarization of $\mathbf{A}$ [see (10-12)]. The momentum $\mathbf{p}$ has been set equal to $m\mathbf{v}$. The final state u_f will asymptotically be a plane wave $V^{-1/2}e^{i\mathbf{k}_f\cdot\mathbf{r}}$, so that the dependence on the normalization volume in (10-16) cancels. The factor $|\mathbf{A}_0|^2$ can be expressed by (10-8) in terms of the number of incident quanta N per cm^2 and sec. The differential cross section for absorption of radiation is then

$$\frac{d\sigma(\theta,\varphi)}{d\Omega} = \frac{e^2}{2\pi\hbar c}\,\frac{v}{\omega}\,\left|\int u_{f1}^*\,e^{i\mathbf{k}\cdot\mathbf{r}}\,\mathbf{P}\cdot\nabla u_n\,d\tau\right|^2 \tag{10-17}$$

where u_{f1} is now normalized to unit amplitude at large distance from the atom. Equation (10-17) is the cross section for the photoelectric effect, where the photoelectron is ejected from the atom in the direction θ, φ relative to the incident beam. It will be further evaluated in Chapter 12.

A similar expression may be obtained for emission of radiation. However in most practical applications, emission accompanies a transition to a final state which is discrete. Accordingly, we must take the external radiation to be a superposition of monochromatic components.

NONMONOCHROMATIC EXTERNAL FIELDS

For a transition to a final state which lies in the discrete spectrum, we must demand that the incident radiation cover a frequency band of sufficient width to satisfy energy conservation. It is therefore assumed that the radiation consists of several *incoherent* frequency components i, closely spaced around an average frequency ω. Thus instead of (10-5) we have

$$2\,\mathrm{Re}\,\sum_i C_i \exp i(\mathbf{k}_i\cdot\mathbf{r} - \omega_i t + \theta_i) \tag{10-18}$$

The fields are

$$\mathbf{A} = 2\,\mathrm{Re}\sum_i \mathbf{A}_i \exp i(\theta_i - \omega_i t) \qquad \mathbf{A}_i = \mathbf{C}_i\, e^{i\mathbf{k}_i\cdot\mathbf{r}}$$

$$\boldsymbol{\mathcal{E}} = 2\,\mathrm{Re}\sum_i \boldsymbol{\mathcal{E}}_i \exp i(\theta_i - \omega_i t) \qquad \boldsymbol{\mathcal{E}}_i = ik_i\,\mathbf{C}_i\, e^{i\mathbf{k}_i\cdot\mathbf{r}}$$

$$\boldsymbol{\mathcal{H}} = 2\,\mathrm{Re}\sum_i \boldsymbol{\mathcal{H}}_i \exp i(\theta_i - \omega_i t) \qquad \boldsymbol{\mathcal{H}}_i = i\mathbf{k}_i \times \mathbf{C}_i\, e^{i\mathbf{k}_i\cdot\mathbf{r}} \tag{10-19a}$$

Since the various ω_i (hence the k_i) do not differ much among themselves, as they are clustered about ω, we may replace $\mathbf{C}_i$, $k_i\mathbf{C}_i$, and $\mathbf{k}_i \times \mathbf{C}_i$ in (10-19a) by average values $\mathbf{C}_0$, $k\mathbf{C}_0$, and $\mathbf{k} \times \mathbf{C}_0$ respectively. The factor $e^{i\mathbf{k}_i\cdot\mathbf{r}}$ can also be replaced by $e^{i\mathbf{k}\cdot\mathbf{r}}$, because we shall use only moderate values of $\mathbf{r}$. (on the other hand $e^{-i\omega_i t}$ cannot be set equal to $e^{-i\omega t}$ since we shall need to use large values of t.) Therefore, the fields are

$$\mathbf{A} = 2\,\mathrm{Re}\,\mathbf{A}_0\, R(t) \qquad \mathbf{A}_0 = \mathbf{C}_0\, e^{i\mathbf{k}\cdot\mathbf{r}}$$

$$\boldsymbol{\mathcal{E}} = 2\,\mathrm{Re}\,\boldsymbol{\mathcal{E}}_0\, R(t) \qquad \boldsymbol{\mathcal{E}}_0 = ik\,\mathbf{C}_0\, e^{i\mathbf{k}\cdot\mathbf{r}}$$

$$\boldsymbol{\mathcal{H}} = 2\,\mathrm{Re}\,\boldsymbol{\mathcal{H}}_0\, R(t) \qquad \boldsymbol{\mathcal{H}}_0 = i\mathbf{k} \times \mathbf{C}_0\, e^{i\mathbf{k}\cdot\mathbf{r}}$$

$$R(t) = \sum_i \exp i(\theta_i - \omega_i t) \tag{10-19b}$$

The Poynting vector is

$$\frac{c}{2\pi}\,\mathrm{Re}\left\{[\boldsymbol{\mathcal{E}}_0 \times \boldsymbol{\mathcal{H}}_0]\, R^2(t) + [\boldsymbol{\mathcal{E}}_0 \times \boldsymbol{\mathcal{H}}_0^*]\, |R(t)|^2\right\} \tag{10-20}$$

To obtain the intensity, we need the time average of this.

$$\langle R^2 \rangle = \sum_{ij} e^{i(\theta_i + \theta_j)} \langle e^{i(\omega_i + \omega_j)t} \rangle \tag{10-21a}$$

$$\langle |R|^2 \rangle = \sum_{ij} e^{i(\theta_i - \theta_j)} \langle e^{-i(\omega_i - \omega_j)t} \rangle \tag{10-21b}$$

Averaging over a sufficiently long time interval,

$$\langle e^{-i(\omega_i + \omega_j)t} \rangle = 0 \tag{10-21c}$$

$$\langle e^{-i(\omega_i - \omega_j)t} \rangle = \delta_{ij} \tag{10-21d}$$

Hence the magnitude of the time averaged Poynting vector, that is, the intensity, is simply

$$I = \frac{c}{2\pi} k^2 |C_0|^2 \sum_i 1 \tag{10-22a}$$

The sum Σ_i 1 is the number of components i contained in the beam. We write this as $n(\omega)\,\Delta\omega$, where $n(\omega)$ is the number of components per unit frequency, and $\Delta\omega$ is the spread of frequencies which occurs in the external beam. Similarly we write the intensity as $I(\omega)\,\Delta\omega$, where $I(\omega)$ is the intensity of the incident radiation per unit frequency. We have from (10-22a)

$$I(\omega) = \frac{\omega^2}{2\pi c} |C_0|^2 n(\omega) \tag{10-22b}$$

Introducing the number of incident quanta per cm² per sec per unit frequency

$$N(\omega) = \frac{I(\omega)}{\hbar\omega} \tag{10-22c}$$

we have finally the analog to (10-8)

$$|C_0|^2 n(\omega) = \frac{2\pi\hbar c}{\omega} N(\omega) \tag{10-22d}$$

We again assume plane or circular polarization. Thus $\mathbf{C}_0$ has the form (10-12).

We now repeat the perturbation calculation (10-13) using the expression for **A** given in (10-19b). We readily obtain

$$a_f^{(1)}(t) = -\frac{1}{\hbar} T'_{fn} \sum_i \frac{e^{i(\omega_{fn}-\omega_i)t} - 1}{\omega_{fn} - \omega_i} e^{i\theta_i}$$

$$-\frac{1}{\hbar} T''_{fn} \sum_i \frac{e^{i(\omega_{fn}+\omega_i)t} - 1}{\omega_{fn} + \omega_i} e^{-i\theta_i} \tag{10-23}$$

where the transition matrix elements T'_{fn} and T''_{fn} are as in (10-13) with $\mathbf{A}_0$ replaced by $\mathbf{C}_0$. Again the oscillating exponents allow a transition to occur only if $\omega_{fn} = \pm\omega_i$, i.e., only if ω_{fn} is included in the range of frequencies $\Delta\omega$ in the sum. The plus sign refers to absorption while the minus sign refers to emission.

To obtain the transition probability, we assume that the frequency range $\Delta\omega$ is sufficiently narrow so that only one of the two processes takes place. Taking the case of absorption we obtain

$$|a_f^{(1)}(t)|^2 = \frac{|T'_{fn}|^2}{\hbar^2} \sum_{ij} e^{i(\theta_i - \theta_j)} \frac{e^{i(\omega_{fn}-\omega_i)t} - 1}{\omega_{fn} - \omega_j}$$

$$\times \frac{e^{-i(\omega_{fn}-\omega_j)t} - 1}{\omega_{fn} - \omega_j} \tag{10-24a}$$

Here we make the incoherence assumption, viz., we average over the phases θ_i, θ_j. This is in general satisfied if the radiation was originally emitted by a large number of atoms, radiating independently —and this is the usual case. However, for laser light and similarly coherent radiation, it would not be correct. Taking the phase average, only the terms $i = j$ are retained and

$$|a_f^{(1)}(t)|^2 = \frac{|T'_{fn}|^2}{\hbar^2} \sum_i \frac{2}{(\omega_{fn} - \omega_i)^2} [1 - \cos(\omega_{fn} - \omega_i)t] \tag{10-24b}$$

The sum Σ_i may be replaced by $\int n(\omega_i)\, d\omega_i$ as $n(\omega)$ is the number of components per unit frequency. The sum therefore is

$$2 \int n(\omega_i)\, d\omega_i \left[\frac{1 - \cos(\omega_{fn} - \omega_i)t}{(\omega_{fn} - \omega_i)^2}\right] \tag{10-24c}$$

The quantity in brackets is sharply peaked at $\omega_i = \omega_{fn}$ and goes rapidly to zero away from that point. As we assume the point ω_{fn} lies in the region of integration we may evaluate $n(\omega_i)$ at that point, and get for (10-24c)

$$n(\omega_{fn}) \int d\omega \frac{2}{\omega^2}(1 - \cos \omega t)$$

$$\omega = \omega_i - \omega_{fn} \tag{10-24d}$$

As only the region $\omega \approx 0$, which lies in the range of integration, contributes significantly, we may assume that the integral extends from $-\infty$ to $+\infty$. It then can be readily evaluated by contour integration to give $2\pi t$. Therefore the transition probability per unit time for absorption is

$$w = \frac{(2\pi e)^2 \hbar}{m^2 c\, \omega_{fn}} N(\omega_{fn}) \left| \int u_f^* e^{i\mathbf{k}\cdot\mathbf{r}}\, \mathbf{p} \cdot \nabla u_n\, d\tau \right|^2 \tag{10-25}$$

where we have used (10-22d). The number of incident quanta can be replaced by the intensity per unit frequency, $N(\omega) = I(\omega)/\hbar\omega$; the resulting expression does not contain $\hbar$ and is thus quasi-classical.

The transition probability per unit time for emission of radiation is the same as (10-25) except that ω_{fn} is replaced by ω_{nf} and the integral is replaced by

$$\int u_f^* e^{-i\mathbf{k}\cdot\mathbf{r}} \mathbf{P}\cdot\nabla u_n \, d\tau \tag{10-26}$$

Here we may interchange the labels n and f; this has the advantage that then again f denotes the higher, n the lower energy state. We may then integrate by parts. Now $\mathbf{P}\cdot\nabla e^{-i\mathbf{k}\cdot\mathbf{r}} = 0$ because the polarization vector $\mathbf{P}$ is perpendicular to the propagation vector $\mathbf{k}$. This results in a transition probability per unit time given by

$$w = \frac{(2\pi e)^2 \hbar}{m^2 c\omega_{fn}} N(\omega_{fn}) \left| -\int u_f e^{-i\mathbf{k}\cdot\mathbf{r}} \mathbf{P}\cdot\nabla u_n^* \, d\tau \right|^2 \tag{10-27}$$

It is seen that (10-27) is the same as (10-25); the probabilities of reverse transitions between any pair of states under the influence of the same radiation field are the same. This is an example of the *principle of detailed balancing,* which is of fundamental importance for statistical mechanics.

MULTIPOLE TRANSITIONS

We expand the exponential in (10-25) and (10-27) and keep only the first term which leads to a nonvanishing integral. This is justified by observing that the ratio of two successive terms is O(ka), with a being a measure of the radius of the atom. Now for optical transitions,

$$ka = \frac{\omega}{c} a = \frac{a\Delta E}{\hbar c} \le \frac{a}{a_0} \frac{e^2}{2\hbar c} \approx \frac{1}{300} \tag{10-28}$$

Here we have assumed that the energy in the optical transition, $\Delta E < 1$ Rydberg $= e^2/2a_0$, and the radius of the atom $a \approx a_0$. Therefore, $ka \ll 1$. For X-rays ΔE is larger by a factor Z^2 and a is smaller by a factor Z. Hence $k \sim Z/300$, and for large Z the result $ka \ll 1$ is no longer true. Replacing $e^{i\mathbf{k}\cdot\mathbf{r}}$ by 1, the relevant integral becomes

$$\int u_f^* \nabla_{\mathbf{A}} u_n \, d\tau = \frac{i}{\hbar} \int u_f^* p_{\mathbf{A}} u_n \, d\tau$$

$$= \frac{i}{\hbar} (p_{\mathbf{A}})_{fn}$$

$$= \frac{mi}{\hbar} \frac{d}{dt} (r_A)_{fn}$$

$$= -\frac{m}{\hbar} \omega_{fn} (r_A)_{fn} \tag{10-29}$$

The subscript **A** indicates the component in the direction of the polarization of **A**; viz., $\mathbf{r}_A = \mathbf{P} \cdot \mathbf{r}$. For circular polarization, (10-29) is proportional to $(x \pm iy)_{fn}$, while for plane polarization along the z axis (10-29) is proportional to z_{fn}. The above approximation, is called the *dipole approximation.*

The probability per unit time, in the dipole approximation, is then

$$w = \frac{4\pi^2 e^2}{\hbar c} \omega_{fn} N(\omega_{fn}) |(r_A)_{fn}|^2 \tag{10-30a}$$

If we write $|\mathbf{r}_{fn}|^2 = \mathbf{r}_{fn} \cdot \mathbf{r}^*_{fn}$ we may define a transition probability which is an average of (10-30a) for the three types of polarization: right and left circular in the xy plane and plane along the z axis.

$$w_{av} = \frac{4\pi^2 e^2}{3\hbar c} \omega_{fn} N(\omega_{fn}) |\mathbf{r}_{fn}|^2 \tag{10-30b}$$

This has obviously the correct dimension: $e^2/\hbar c$ is the fine-structure constant, $\omega N(\omega)$ is a number per cm^2 per sec, and $|\mathbf{r}|^2$ is an area.

Transitions for which the probability is accurately given by the above approximation are called electric dipole transitions, since $e\mathbf{r}$ is the operator representing the electric dipole of the atom. If the dipole matrix element $(\mathbf{r})_{fn}$ is zero the transition is said to be forbidden. If the unapproximated integral in (10-25) or (10-27) vanishes, the transition is said to be strictly forbidden. In neither instance should it be concluded that no transition can occur. If the dipole transition is forbidden we must take further terms in the expansion of $e^{i\mathbf{k}\cdot\mathbf{r}}$. If the transition is strictly forbidden we must take higher orders of perturbation theory and must include the neglected term $e^2 A^2/2mc^2$; this then leads to the simultaneous emission of two photons.

SPONTANEOUS EMISSION

A classical description of the spontaneous emission of radiation by a current density **J**, oscillating with angular frequency ω, gives for the intensity of radiation in the radiation zone in the direction **k** the result

$$I = \frac{k^2}{2\pi r^2 c} \left| \int J_\perp(\mathbf{r}') \, e^{-i\mathbf{k}\cdot\mathbf{r}'} \, d\tau' \right|^2 \tag{10-31}$$

where $\mathbf{J(r)}$ is defined by assuming that the current density at $\mathbf{r}$ and t is given by the real part of

$$\mathbf{J}(\mathbf{r}, t) = \mathbf{J}(\mathbf{r})\, e^{-i\omega t}$$

Here $J_{\perp}$ is the component of $\mathbf{J}$ perpendicular to $\mathbf{k}$. In the dipole approximation (10-31) reduces to

$$\frac{k^2}{2\pi r^2 c} \left| \int J_{\perp}(\mathbf{r}')\, d\tau' \right|^2 \tag{10-32}$$

Writing

$$\mathbf{J}_0 = \int \mathbf{J}(\mathbf{r}')\, d\tau' \tag{10-33}$$

we conclude that the polarization is linear if $\mathbf{J}_0$ has only one component in the plane perpendicular to $\mathbf{k}$, circular if $\mathbf{J}_0$ has two equal perpendicular components in the plane perpendicular to $\mathbf{k}$ and 90° out of phase; and so forth. Thus the emission of radiation of *definite* polarization $\mathbf{A} \perp \mathbf{k}$ is obtained by substituting $\mathbf{P} \cdot \mathbf{J}$, the projection of $\mathbf{J}$ along the polarization vector $\mathbf{P}$, instead of $J_{\perp}$.

By writing $\mathbf{J}_0$ as $\mathbf{j}_1 + i\mathbf{j}_2$, equation (10-32) can be rewritten as

$$\frac{k^2}{2\pi r^2 c} \left(j_1^2 \sin^2 \theta_1 + j_2^2 \sin^2 \theta_2 \right) \tag{10-34}$$

where θ_i is the angle between $\mathbf{j}_i$ and $\mathbf{k}$. The total power radiated is the integral of (10-34) over a sphere of radius r. This gives

$$\frac{4k^2}{3c} \left| \mathbf{J}_0 \right|^2 \tag{10-35}$$

To convert this treatment to quantum mechanics we require a quantum mechanical operator corresponding to $\mathbf{J}$ and we interpret the radiated power as $\hbar\omega$ times the transition probability per unit time. In view of the usual identification of charge density with $-e\,|\psi|^2$, it is plausible to assume that the quantum mechanical operator corresponding to $\mathbf{J}$ is given by the current, familiar from the Schrödinger continuity equation:

$$\mathbf{J}(\mathbf{r}) = \frac{e\hbar}{2im} \left[u_n^* \nabla u_f - (\nabla u_n^*)\, u_f \right] \tag{10-36}$$

Here u_n (u_f) is the wave function of the initial (final) state.

The probability of emission of a quantum per unit time in a transition from state f to n is then given by (10-31) multiplied by $r^2\,d\Omega$ and divided by $\hbar\omega$, with (10-36) substituted for J, i.e.,

$$w = \frac{\omega}{2\pi\hbar c^3}\left(\frac{e\hbar}{m}\right)^2 \left| \int u_n^*(\mathbf{r}')\, e^{-i\mathbf{k}\cdot\mathbf{r}'}\, \nabla_\perp\, u_f(\mathbf{r}')\, d\tau' \right|^2 d\Omega \qquad (10\text{-}37)$$

Here $d\Omega$ is the element of solid angle subtended by $\mathbf{k}$, the propagation vector of the emitted radiation. An integration by parts of the second term occurring in (10-36) was used to obtain the form (10-37). For this integration it is important that only the component of ∇ perpendicular to $\mathbf{k}$ occurs; this ensures that the derivative of $e^{-i\mathbf{k}\cdot\mathbf{r}'}$ does not appear in (10-37).

Unlike induced emission, conservation of energy is not automatically obtained as a consequence of the theory; we must make the additional postulate $\omega_{fn} = \omega$. The emission of radiation of definite polarization $\mathbf{A} \perp \mathbf{k}$ is obtained by substituting $\mathbf{P}\cdot\nabla$ for $\nabla_\perp$ in (10-37).

In the dipole approximation, after averaging over angles, (10-37) reduces to the quantum analog of (10-35)

$$w_{av} = \frac{4e^2\omega_{fn}^3}{3\hbar c^3}\,|r_{fn}|^2 \qquad (10\text{-}38)$$

EINSTEIN TRANSITION PROBABILITIES

That the above transition from the classical description to the quantum mechanical description of spontaneous emission is consistent, can be shown by using an argument which Einstein employed for a different purpose.

We consider the thermal equilibrium between atoms and a radiation field which is achieved by the absorption and emission of photons of frequency $\omega_{fn} = (E_f - E_n)/\hbar > 0$. Two of the three available processes for attaining equilibrium, viz., stimulated emission and absorption, were found to be proportional to $\rho(\omega_{fn})$, the energy density of the external radiation field per unit frequency.

$$\rho(\omega_{fn}) = \frac{I(\omega_{fn})}{c} = \frac{\hbar\omega N(\omega_{fn})}{c} \qquad (10\text{-}39)$$

The third process, spontaneous emission, occurs even in the absence of external radiation, hence does not depend on $\rho(\omega_{fn})$. The rate at which atoms make transitions $n \to f$ (absorption) is

$$\frac{dN(n \to f)}{dt} = B_{nf}\, N(n)\,\rho(\omega_{fn}) \qquad (10\text{-}40a)$$

where N(n) is the number of atoms in state n. The rate at which the transition $f \to n$ (emission) proceeds is

$$\frac{dN(f \to n)}{dt} = B_{fn} N(f) \rho(\omega_{fn}) + A_{fn} N(f) \tag{10-40b}$$

A_{fn} and B_{fn} are called the Einstein spontaneous and induced transition probabilities, respectively. The requirement of equilibrium equates the two rates; the principle of detailed balance gives $B_{fn} = B_{nf}$. Thus

$$B_{fn} N(f) \rho(\omega_{fn}) + A_{fn} N(f) = B_{fn} N(n) \rho(\omega_{fn}) \tag{10-41a}$$

$$\rho(\omega_{fn}) = \frac{A_{fn}/B_{fn}}{[N(n)/N(f)] - 1} \tag{10-41b}$$

From statistical mechanics we have at thermal equilibrium at temperature T,

$$\frac{N(n)}{N(f)} = e^{\hbar\omega_{fn}/kT} \tag{10-42a}$$

$$\rho(\omega_{fn}) = \frac{\hbar\omega_{fn}^3}{\pi^2 c^3} \frac{1}{e^{(\hbar\omega_{fn}/kT)} - 1} \tag{10-42b}$$

which is the density of radiant energy per unit frequency at temperature T. Equation (10-42b) is, of course, the well-known formula due to Planck. (In fact this discussion is the method Einstein used to derive Planck's radiation formula. At that time there was no way of evaluating A_{fn}/B_{fn}.) For (10-41b) to agree with (10-42b) we must have

$$A_{fn} = \frac{\hbar\omega_{fn}^3}{\pi^2 c^3} B_{fn} \tag{10-43}$$

Comparing (10-38), (10-39) with (10-30b) we see that (10-43) is satisfied for dipole radiation. For the general case, we must compare (10-37) with (10-27), and use the fact that (10-37) includes two directions of polarization, which gives a factor of 2, and that it should be integrated over dΩ, which gives a factor 4π. If (10-27) is then averaged over the directions of **k** and **A**, and (10-39) is used, we again find (10-43) satisfied.

Thus we have justified (10-37) for the spontaneous emission, including the numerical factors, by invoking Einstein's argument on statistical equilibrium. The form (10-37) is also made plausible by noting that the integral has the same form as the induced emission

(10-27). The most satisfactory way to derive (10-37) is from field theory (e.g. Schiff[1], p. 531). Once (10-37), and hence (10-43), are justified, we may of course use Einstein's argument for his original purpose, viz., to deduce the Planck distribution (10-42b).

LINE BREADTH

Our discussion so far has led to the conclusion that spectral lines will be infinitely sharp, corresponding to the well-defined energies of the stationary states that are involved in the transition. This, of course, is an approximation as, due to the spontaneous emission, the energy states are not stable but decay. The probability of transition per unit time γ_n for this decay is given by (10-38) and is seen to be independent of time. It is well known from probability theory that such time-independent processes obey an exponential law of decay; i.e., the depletion of the occupation probability $|a_n|^2$ of a state n is given by $e^{-\gamma_n t}$, where $1/\gamma_n$ is called the lifetime. The amplitude a_n then decays as $e^{-1/2\gamma_n t}$. For optical transition $1/\gamma_n \approx 10^{-8}$ sec, and is very much larger than the characteristic period of electron motion (about 10^{-15} sec). Hence it is justified in first approximation to have considered the states stationary.

Weisskopf and Wigner analyzed the effects of this decay. They found, using the full quantum theory of the radiation field, that the spectrum of the emitted radiation is correctly given if we assume that the initial state n, as well as the final state m, have wave functions with the time dependence

$$\Psi_n = e^{-iE_n t/\hbar - \frac{1}{2}\gamma_n t}\, u_n \tag{10-44}$$

Then the "current" $\Psi_m^* \nabla \Psi_n$ has the time dependence

$$J(r, t) \sim e^{-i\omega_{nm} t - \frac{1}{2}(\gamma_n + \gamma_m)t} \tag{10-45}$$

If we make a Fourier analysis of this current, to determine the radiation of frequency ω, we find that its intensity (absolute square of amplitude) is proportional to

$$R(\omega) = \frac{1}{(\omega - \omega_{nm})^2 + \frac{1}{4}(\gamma_n + \gamma_m)^2} \tag{10-46}$$

This then gives the natural shape of the line (excluding Doppler effect, collision broadening, etc.). In emission, the intensity is distributed according to (10-46); in absorption, the absorption coefficient has that dependence.

The occurrence of $\gamma_n + \gamma_m$ in the width is, however, contrary to classical ideas, according to which one might expect a dependence on the initial state only. Alternatively we might expect a dependence on the Lorentz width, which is $\Delta\lambda_L = (2/3)(e^2/mc^2)$ in wavelength, or perhaps on $\Delta\lambda_L$ multiplied by the oscillator strength (see Chapter 11) of the line. Most physicists, before quantum mechanics, favored the latter idea, and thus expected weak spectral lines to be narrow. Equation (10-46), on the other hand, predicts lines to be broad if either the initial or the final state has a short lifetime, regardless of the strength of the line itself. Experiments have shown that the quantum prediction is correct. A simple example (although not directly verified experimentally) is the transition $3\,^1S - 2\,^1P$ of He: It is a weak transition because n and ℓ change in opposite directions (see Chapter 11), but the $2\,^1P$ state has a very short lifetime, owing to the strong transition $2\,^1P - 1\,^1S$; hence $3\,^1S - 2\,^1P$ should be broad.

It is clear that line width is connected with the uncertainty principle. The lifetime of a state, $1/\gamma$, measures the time a quantum system occupies that state. Hence the energy cannot be determined to greater accuracy than $\hbar\gamma$. If the energy is uncertain by that amount, the frequency will be broadened by γ.

PROBLEMS

1. The Schrödinger equation, in the presence of an electromagnetic field gauged by $\nabla \cdot \mathbf{A} = 0$, $\phi = 0$, is

$$i\hbar \frac{\partial\Psi}{\partial t} = \left[-\frac{\hbar^2}{2m}\nabla^2 + \frac{iq\hbar}{mc}\mathbf{A}\cdot\nabla + \frac{e^2}{2mc^2}A^2 + V\right]\Psi \qquad (1)$$

This is equation (10-1) where we have retained the term quadratic in A. Since electromagnetic potentials are arbitrary within gauge transformations, we may perform a gauge transformation on the potential:

$$\mathbf{A} \to \mathbf{A}' = \mathbf{A} + \nabla\chi$$

$$\phi \to \phi' = \phi - \frac{1}{c}\frac{\partial\chi}{\partial t} \qquad (2)$$

Show that observable quantities are not affected by such a gauge transformation. What is the probability density and current associated with equation (1)?

2. Derive (10-7).
3. Derive (10-13).
4. A spinless mass point (particle) of mass m and charge q moves in a static magnetic field $\mathcal{H}$, derivable from a vector potential $\mathbf{A} = \frac{1}{2}\mathcal{H} \times \mathbf{r}$. What is the Hamiltonian (including the quadratic

term)? What are the energy eigenvalues and eigenfunctions? Hint: Take $\mathcal{H}$ to be along the z axis; show that p_x and p_z commute with the Hamiltonian, hence the x and z dependence of the wave function is trivial. Finally, show that the y dependence of the wave function is related to the harmonic oscillator problem.

5. How are the energy eigenvalues of the previous problem modified if a uniform electric field is present, which is perpendicular to the magnetic field?

Chapter 11

INTENSITY OF RADIATION, SELECTION RULES

SUM RULES

In order to estimate the intensity of radiation and in other applications, it is useful to obtain closed expressions for sums of the form

$$\sum_k \omega_{kn}^p |x_{kn}|^2$$

$$\omega_{kn} = \frac{E_k - E_n}{\hbar} \qquad (11\text{-}1)$$

The summation is over the complete set of energy eigenstates; that is, a summation over the discrete states and an integration over the continuum states. The state n is a discrete energy eigenstate. The exponent p is a positive integer or zero. To derive the desired sum rules, we shall investigate a general sum rule of the form

$$S_p = \sum_k \omega_{kn}^p |A_{kn}|^2 = \sum_k \omega_{kn}^p A_{kn} A_{nk}^\dagger \qquad (11\text{-}2)$$

where A is any operator.

Consider the quantity

$$e^{itH/\hbar} A e^{-itH/\hbar} \qquad (11\text{-}3a)$$

which is equal to the series[1]

$$e^{itH/\hbar} A e^{-itH/\hbar} = \sum_{p=0}^{\infty} \left(\frac{it}{\hbar}\right)^p \frac{A_p}{p!} \qquad (11\text{-}3b)$$

---

[1]Problem 1 concerns itself with establishing (11-3).

where A_p is defined by the rule

$$A_0 = A$$

$$A_p = [H, A_{p-1}] \tag{11-3c}$$

Next consider

$$\langle n | A^\dagger e^{itH/\hbar} A e^{-itH/\hbar} | n \rangle \tag{11-4a}$$

Inserting a complete set of energy eigenstates, we obtain for (11-4a)

$$\langle n | A^\dagger e^{itH/\hbar} A e^{-itH/\hbar} | n \rangle$$

$$= \sum_k \langle n | A^\dagger e^{itH/\hbar} | k \rangle \langle k | A e^{-itH/\hbar} | n \rangle$$

$$= \sum_k e^{it\omega_{kn}} A_{kn} A^\dagger_{nk} \tag{11-4b}$$

Expanding the exponential in (11-4b) and interchanging the orders of summation yields

$$\langle n | A^\dagger e^{itH/\hbar} A e^{-itH/\hbar} | n \rangle$$

$$= \sum_p \frac{(it)^p}{p!} \sum_k \omega^p_{kn} |A_{kn}|^2 = \sum_p \frac{(it)^p}{p!} S_p \tag{11-4c}$$

On the other hand, we could use the formulas (11-3) to expand (11-4a). Upon use of (11-3b), (11-4a) can be written as

$$\langle n | A^\dagger e^{itH/\hbar} A e^{-itH/\hbar} | n \rangle$$

$$= \sum_p \frac{(it)^p}{p!} \frac{1}{\hbar^p} \langle n | A^\dagger A_p | n \rangle \tag{11-4d}$$

Comparing (11-4c) with (11-4d) gives a preliminary result

$$S_p = \frac{1}{\hbar^p} \langle n | A^\dagger A_p | n \rangle \tag{11-5}$$

Equation (11-5), in principle, gives a closed expression for S_p. One needs to commute the operator A with H p times to get A_p and then evaluate $\langle n | A^\dagger A_p | n \rangle$.

However (11-5) can be simplified by making use of the relation

$$\langle n | A^\dagger A_p | n \rangle = \langle n | A^\dagger_m A_{p-m} | n \rangle$$

$$p \geq m \tag{11-6a}$$

To prove (11-6a) we note that

$$\langle n | A^\dagger A_p | n \rangle = \langle n | A^\dagger H A_{p-1} - A^\dagger A_{p-1} H | n \rangle \tag{11-6b}$$

Since $| n \rangle$ is an energy eigenstate,

$$\langle n | A^\dagger A_{p-1} H | n \rangle = E_n \langle n | A^\dagger A_{p-1} | n \rangle$$

$$= \langle n | H A^\dagger A_{p-1} | n \rangle \tag{11-6c}$$

Hence (11-6b) becomes

$$\langle n | [A^\dagger, H] A_{p-1} | n \rangle = \langle n | [H, A]^\dagger A_{p-1} | n \rangle$$

$$= \langle n | A_1^\dagger A_{p-1} | n \rangle \tag{11-6d}$$

Evidently, this procedure may be continued m times to give (13-6a). Therefore (11-5) may be rewritten as

$$S_{2m} = \frac{1}{\hbar^{2m}} \langle n | A^\dagger_m A_m | n \rangle \tag{11-7a}$$

$$S_{2m+1} = \frac{1}{\hbar^{2m+1}} \langle n | A^\dagger_m A_{m+1} | n \rangle \tag{11-7b}$$

$$= \frac{1}{\hbar^{2m+1}} \langle n | A^\dagger_{m+1} A_m | n \rangle \tag{11-7c}$$

$$= \frac{1}{2\hbar^{2m+1}} \langle n | A^\dagger_m A_{m+1} + A^\dagger_{m+1} A_m | n \rangle \tag{11-7d}$$

Of the three equivalent expressions for S_{2m+1}, the last is usually the most convenient.

We now return to the problem at hand of evaluating (11-1). Here $A = A^\dagger = x$ and $H = (p^2/2m) + V$. We shall be interested in values of the exponent p up to 4, hence we need A_0, A_1, and A_2. These are easy to calculate

$$A_0 = A_0^\dagger = x$$

$$A_1 = [H, A_0] = \frac{\hbar}{im} p^x = -A_1^\dagger$$

$$A_2 = [H, A_1] = \frac{\hbar^2}{m} \frac{\partial V}{\partial x} = A_2^\dagger \qquad (11\text{-}8)$$

Therefore the various sum rules are as follows.

1. *Dipole Moment*

$$\sum_k |x_{kn}|^2 = \langle n | A_0^\dagger A_0 | n \rangle = \langle n | x^2 | n \rangle \qquad (11\text{-}9a)$$

If the state n is isotropic

$$\sum_k |x_{kn}|^2 = \tfrac{1}{3} \langle n | r^2 | n \rangle \qquad (11\text{-}9b)$$

If $|\mathbf{r}_{kn}|^2$ is defined as $\mathbf{r}_{kn} \cdot \mathbf{r}_{kn}^*$

$$\sum_k |\mathbf{r}_{kn}|^2 = \langle n | r^2 | n \rangle \qquad (11\text{-}9c)$$

2. *Oscillator Strength*

The oscillator strength f_{kn} is defined as

$$f_{kn}^x = \frac{2m\,\omega_{kn}}{\hbar} |x_{kn}|^2 \qquad (11\text{-}10a)$$

$$f_{kn} = f_{kn}^x + f_{kn}^y + f_{kn}^z \qquad (11\text{-}10b)$$

The oscillator strength sum rule, first discovered by Thomas, Reiche, and Kuhn, is

$$\sum_k f_{kn}^x = \frac{2m}{\hbar} \sum_k \omega_{kn} |x_{kn}|^2$$

$$= \frac{2m}{\hbar} \cdot \frac{1}{2\hbar} \langle n | A_0^\dagger A_1 + A_1^\dagger A_0 | n \rangle$$

$$= \frac{1}{i\hbar} \langle n | [x, p^x] | n \rangle = 1 \qquad (11\text{-}10c)$$

Therefore

$$\sum_k f^x_{kn} = \sum_k f^y_{kn} = \sum_k f^z_{kn} = 1 \tag{11-10d}$$

$$\sum_k f_{kn} = 3 \tag{11-10e}$$

Because of the importance and simplicity of this result, we now present an alternate, simpler derivation of the oscillator strength sum rule. Recalling that [see (10-29)]

$$\langle k | p^x | n \rangle = im\,\omega_{kn} \langle k | x | n \rangle \tag{11-10f}$$

we have

$$f^x_{kn} = -\frac{i}{\hbar}(x_{nk} p^x_{kn} - p^x_{nk} x_{kn})$$

$$\sum_k f^x_{kn} = -\frac{i}{\hbar} \langle n | [x, p^x] | n \rangle = 1 \tag{11-10g}$$

The Thomas-Reiche-Kuhn sum rule played a role in the invention of quantum mechanics.

3. *Momentum*

$$\sum_k \omega^2_{kn} |x_{kn}|^2 = \frac{1}{\hbar^2} \langle n | A^\dagger_1 A_1 | n \rangle = \frac{1}{m^2} \langle n | (p^x)^2 | n \rangle$$

$$= \frac{1}{m^2} \sum_k |p^x_{kn}|^2 \tag{11-11a}$$

If the state n is spherically symmetric, we have

$$\sum_k \omega^2_{kn} |x_{kn}|^2 = \frac{1}{3m^2} \langle n | p^2 | n \rangle$$

$$= \frac{2}{3m} \langle n | H - V | n \rangle$$

$$= \frac{2}{3m} (E_n - \langle n | V | n \rangle) \tag{11-11b}$$

Also

$$\sum_k \omega^2_{kn} |\mathbf{r}_{kn}|^2 = \frac{2}{m} (E_n - \langle n | V | n \rangle) \tag{11-11c}$$

4. *Force Times Momentum*

$$\sum_k \omega_{kn}^3 |x_{kn}|^2 = \frac{1}{2\hbar^3} \langle n | A_1^\dagger A_2 + A_2^\dagger A_1 | n \rangle$$

$$= \frac{1}{2m^2 i} \left\langle n \middle| \left[\frac{\partial V}{\partial x}, p^x \right] \middle| n \right\rangle$$

$$= \frac{\hbar}{2m^2} \left\langle n \middle| \frac{\partial^2 V}{\partial x^2} \middle| n \right\rangle \tag{11-12a}$$

If the state is spherically symmetric

$$\sum_k \omega_{kn}^3 |x_{kn}|^2 = \frac{\hbar}{6m^2} \langle n | \nabla^2 V | n \rangle$$

$$= \frac{2\pi}{3} \frac{\hbar e^2}{m^2} \langle n | \rho(\mathbf{r}) | n \rangle \tag{11-12b}$$

where e $\rho(\mathbf{r})$ is the density of positive charge which generates the potential energy V. For a bare nucleus, $\rho(\mathbf{r}) = Z\,\delta(\mathbf{r})$, and $\langle n | \rho(\mathbf{r}) | n \rangle = Z\,|u_n(0)|^2$. Also

$$\sum_k \omega_{kn}^3 |\mathbf{r}_{kn}|^2 = \frac{2\pi \hbar e^2}{m^2} \langle n | \rho(\mathbf{r}) | n \rangle \tag{11-12c}$$

We note that because of the relations

$$p_{kn}^x = im\,\omega_{kn} x_{kn} \tag{11-13a}$$

$$\frac{d}{dt} p_{kn}^x = i\,\omega_{kn} p_{kn}^x = -\left(\frac{\partial V}{\partial x}\right)_{kn} \tag{11-13b}$$

the quantity $\omega_{kn}^3 |x_{kn}|^2$ is equal

$$\frac{1}{m^2} \omega_{kn} |p_{kn}^x|^2 = \frac{i}{m^2} p_{nk}^x \left(\frac{\partial V}{\partial x}\right)_{kn} \tag{11-13c}$$

Hence this sum rule measures force times momentum.

5. *Force Squared*

$$\sum_k \omega_{kn}^4 |x_{kn}|^2 = \frac{1}{\hbar^4} \langle n | A_2^\dagger A_2 | n \rangle$$

$$= \frac{1}{m^2} \left\langle n \left| \left(\frac{\partial V}{\partial x} \right)^2 \right| n \right\rangle$$

$$= \frac{1}{m^2} \sum_k \left| \left\langle k \left| \frac{\partial V}{\partial x} \right| n \right\rangle \right|^2 \tag{11-14a}$$

If the state n is spherically symmetric

$$\sum_k \omega_{kn}^4 |x_{kn}|^2 = \frac{1}{3m^2} \langle n | (\nabla V)^2 | n \rangle \tag{11-14b}$$

Also

$$\sum_k \omega_{kn}^4 |\mathbf{r}_{kn}|^2 = \frac{1}{m^2} \langle n | (\nabla V)^2 | n \rangle \tag{11-14c}$$

SUMMARY

The above results for $\Sigma_k \, \omega_{kn}^p \, |x_{kn}|^2$ are summarized in Table 11-1. In the table it is assumed for all p, except p = 1, that the state n is spherically symmetric. The values of $\Sigma_k \, \omega_{kn}^p \, |\mathbf{r}_{kn}|^2$ can be obtained by multiplying the sums by 3. Clearly, in this case no assumptions about spherical symmetry are made.

Table 11-1
Sum Rules

p	Sum
0	$\frac{1}{3}\,(r^2)_{nn}$
1	$\frac{\hbar}{2m}$
2	$\frac{2}{3m}\,[E_n - (V)_{nn}]$
3	$\frac{1}{6}\,\frac{\hbar}{m^2}\,(\nabla^2 V)_{nn}$
4	$\frac{1}{3m^2}\,((\nabla V)^2)_{nn}$

MANY ELECTRONS

The above sum rules are for transitions of one electron. For Z electrons, the integral relevant for the transition amplitude is

$$\int d\tau \; u_k^* \sum_{j=1}^{Z} e^{-i\mathbf{k}\cdot\mathbf{r}_j} \frac{\partial}{\partial x_j} u_n \tag{11-15}$$

which in the dipole approximation is proportional to

$$\int u_k^* \left(\sum_{j=1}^{Z} x_j \right) u_n \, d\tau = X_{kn} \tag{11-16}$$

The sum rules of interest are

$$\sum_k \omega_{kn}^p |X_{kn}|^2 \tag{11-17}$$

Only for p = 0 and p = 1 are the results simple. Applying the scheme we developed above for one electron, we take

$$A = \sum_{j=1}^{Z} x_j \qquad H = \sum_{j=1} \frac{p_j^2}{2m} + V$$

$$A_1 = [A, H] = \frac{\hbar}{im} \sum_j p_j^x = -A_1^\dagger \tag{11-18}$$

$$\sum_k |X_{kn}|^2 = \langle n | A^2 | n \rangle$$

$$= \int d\tau \, |u_n|^2 \left(\sum_{j=1}^{Z} x_j^2 + 2 \sum_{i>j} x_i x_j \right) \tag{11-19}$$

For many electrons the oscillator strength is defined by

$$f_{kn}^x = \frac{2m\,\omega_{kn}}{\hbar} \, |\langle k | \sum_j x_j \, | n \rangle |^2 \tag{11-20}$$

The oscillator strength sum rule is

$$\sum_k f^x_{kn} = \frac{2m}{\hbar} \sum_k \omega_{kn} |\langle k | \sum_j x_j | n \rangle|^2$$

$$= \frac{2m}{\hbar} \frac{1}{2\hbar} \langle n | A_0^\dagger A_1 + A_1^\dagger A_0 | n \rangle$$

$$= \frac{1}{i\hbar} \left\langle n \left| \left[\sum_i x_i , \sum_j p_j^x \right] \right| n \right\rangle$$

$$= \sum_{ij} \delta_{ij} = Z \tag{11-21}$$

HYDROGEN

As an example of the application of sum rules we give results for hydrogen, the electron in the lowest (isotropic) state (principal quantum number = 1).

$$\sum_k \omega^p_{k1} |\mathbf{r}_{k1}|^2 = 3 \sum_k \omega^p_{k1} |x_{k1}|^2 \qquad \text{for } \ell = 0$$

p	0	1	2	3
sum	$3a_0^2$	$3a_0^2$ Ry	$4a_0^2$ (Ry)^2	$16a_0^2$ (Ry)^3

where Ry = $e^2/2a_0\hbar$ is the Rydberg frequency. From this table, we deduce that the average frequency of transitions from the ground state is just Ry, the root-mean-square frequency is $\sqrt{4/3}$ Ry = 1.14 Ry, the cube root of the mean cube frequency is $(16/3)^{1/3}$ Ry = 1.75 Ry.

For the case p = 3, $(\rho(\mathbf{r}))_{nn} \propto (\delta(\mathbf{r}))_{nn} = |u_n(0)|^2$. Since $u_{n\ell} \sim r^\ell$, we have $\Sigma_k \omega^3_{kn} |\mathbf{r}_{kn}|^2 = 0$ for $\ell \neq 0$.

For p = 4, $(\nabla V)^2 \propto 1/r^4$, $((\nabla V)^2)_{nn} \propto \int |u_{n\ell}|^2 (1/r^2)\, dr$. For $\ell = 0$ this diverges, for $\ell \neq 0$ it is finite.

From these sum rule results, we can now draw some approximate conclusions on the behavior of the matrix elements $|x_{kn}|^2$ for large energy E(k), without explicitly calculating them. While explicit calculation is possible for hydrogen, it is laborious for other elements, while calculation of the sum rule expressions is usually simple.

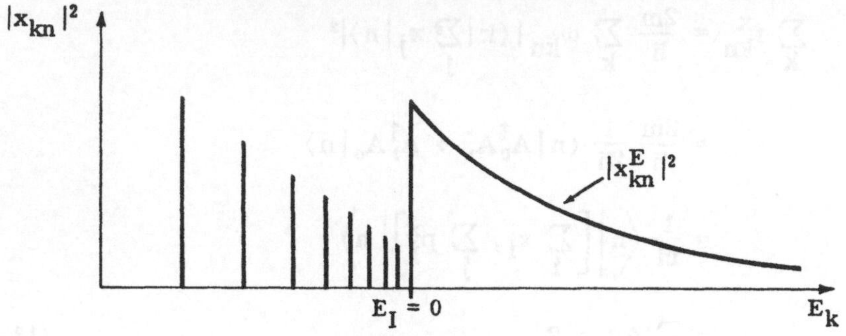

Figure 11-1

Figure 11-1 gives (schematically) the transition matrix element $|x_{kn}|^2$ for fixed n in hydrogen, as a function of the energy E_k. The point $E_I = 0$ marks the beginning of the continuum; here the usual energy scale has been used. The first discrete line, k = 2, lies at $E_k = -\frac{1}{4}\,\text{Ry}$. The $|x_{kn}|^2$ for discrete levels k are not directly comparable to the $|x_{kn}|^2$ for transitions to the continuum, and Fig. 11-1 has been drawn accordingly.[2] We have normalized the continuum wave functions per unit energy

$$\int u^*_{E_n \ell m}(\mathbf{r})\, u_{E'_n \ell' m'}(\mathbf{r})\, d\tau = \delta_{\ell\ell'}\, \delta_{mm'}\, \delta(E_n - E'_n) \qquad (11\text{-}22)$$

These continuum wave functions have the dimension $(\text{energy})^{-1/2}$ $(\text{volume})^{-1/2}$ as compared to the discrete wave functions which have the dimension $(\text{volume})^{-1/2}$. Thus the matrix element $|x_{kn}|^2$ has the dimension $(\text{energy})^{-1}$ $(\text{distance})^2$ when k is a continuum state normalized according to (11-22); and $(\text{distance})^2$ when k is a discrete state. We indicate the former case by $|x^E_{kn}|^2$.

Explicitly the sums which occur in the sum rules are

$$\sum_{\ell'=0}^{\infty} \sum_{m'=-\ell'}^{\ell'} \left\{ \sum_{k=\ell'+1}^{\infty} \left[\frac{E_{k\ell'} - E_{n\ell}}{\hbar}\right]^p |\langle k\ell' m' | x | n\ell m\rangle|^2 \right.$$

$$\left. + \int_0^\infty dE_{k\ell'} \left[\frac{E_{k\ell'} - E_{n\ell}}{\hbar}\right]^p |\langle k\ell' m' | x^E | n\ell m\rangle|^2 \right\}$$

[2] The two kinds of $|x_{kn}|^2$ can be made comparable if we divide the discrete ones by the spacing between neighboring energy levels which is about $2\,\text{Ry}/k^3$ near the discrete energy level of principal quantum number k. When thus normalized, discrete and continuous matrix elements join smoothly at $E_I = 0$.

The sum rules for p = 3 and 4 emphasize the high energies. If we assume that in this limit,

$$\sum_{\ell'm'} |\langle k\ell'm' \,|\, x^E \,|\, n\ell m\rangle|^2 = C\,\omega_{kn}^{-r} \tag{11-23a}$$

then for $\ell = 0$, the sum rules for p = 3 and 4 state that

$$\int^{\infty} \omega_{kn}^{3-r}\, d\omega_{kn} \quad \text{converges}$$

$$\int^{\infty} \omega_{kn}^{4-r}\, d\omega_{kn} \quad \text{diverges} \tag{11-23b}$$

Hence $4 < r \leq 5$. In Chapter 12, we shall find that actually $r = 4.5$.

SELECTION RULES AND MATRIX ELEMENTS

Since the operator $\mathbf{r} = \Sigma_i \mathbf{r}_i$, which we saw in Chapter 10 determines the transition probability, has odd parity it follows that $\mathbf{r}$ has no matrix elements between states of the same parity. Hence transitions between states of the same parity are forbidden. This is known as *Laporte's rule*.

Since $\mathbf{r}$ is a one-electron operator, it will have matrix elements between states that are described by determinantal wave functions that differ at most by one orbital. According to (4–13) the matrix element reduces to $\langle i' \,|\, \mathbf{r} \,|\, i\rangle$, where i is the different orbital. We want, therefore,

$$\langle n'\ell' m_\ell' m_s' \,|\, x_\mu \,|\, n\ell m_\ell m_s\rangle$$

$$= \delta(m_s', m_s) \int R_{n'\ell'} R_{n\ell}\, x_\mu\, Y^*_{\ell'm'} Y_{\ell m}\, d\tau \tag{11-24}$$

where the various x_μ are

$$z = r\cos\theta = arY_{10}$$

$$2^{-1/2}(x + iy) = 2^{-1/2}\, r\sin\theta\, e^{i\varphi} = -arY_{11}$$

$$2^{-1/2}(x - iy) = 2^{-1/2}\, r\sin\theta\, e^{-i\varphi} = arY_{1-1}$$

$$a = \left(\frac{4\pi}{3}\right)^{1/2}$$

Equation (11-24) then reduces to

$$\pm a \int \mathcal{R}_{n'\ell'}\mathcal{R}_{n\ell}\, r\, dr \int Y^*_{\ell'm'} Y_{\ell m} Y_{1t}\, d\Omega \tag{11-25}$$

$t = 0, 1, -1$. Considering the angular integration first we obtain according to (4-48) and (4-49), the nonvanishing result

$$\int Y^*_{\ell'm'} Y_{\ell m} Y_{1t}\, d\Omega = \left(\frac{3}{4\pi}\right)^{1/2} c^1(\ell'm', \ell m) \tag{11-26a}$$

only when

$$t = m' - m$$

$$|\ell' - \ell| \le 1 \le \ell' + \ell = \text{odd integer} \tag{11-26b}$$

Set $\ell' = \ell + \Delta\ell$. Equation (11-26) then implies $|\Delta\ell| \le 1$, $\Delta\ell$ odd; therefore $\Delta\ell = \pm 1$. We conclude therefore that the only possible transitions are

$$z\colon m' = m$$

$$x + iy\colon m' = m + 1$$

$$x - iy\colon m' = m - 1$$

and, in any case,

$$\Delta\ell = \pm 1$$
$$m'_s = m_s \tag{11-27}$$

The selection rules for m (which become relevant for Zeeman splitting) are associated (in the classical limit) with the angular momentum carried away by the radiation. Emitted light propagating in the z direction and circularly polarized, $\mathbf{A} = A\mathbf{i} + iA\mathbf{j}$, carries off ±1 unit of angular momentum about the z axis (see Panofsky and Phillips,[3] p. 269). Hence $\Delta m = \pm 1$. The classical analogy is not easy to draw for $\Delta m = 0$, which leads to light polarized linearly in the z direction, and hence propagating in the xy plane. If we wanted to discuss its angular momentum, we would have to consider the angular momentum of the electron about the direction of propagation of the light; this cannot be done easily since we have chosen L_z as quantized,

---

[3]W. K. H. Panofsky and M. Phillips, *Classical Electricity and Magnetism*, 2nd ed., Addison-Wesley, Reading, MA., 1962.

and L_z fails to commute with L_x and L_y. However, the total orbital angular momentum changes by ±1 in quantum mechanics, thus allowing for the unit angular momentum of the emitted light.

If we are dealing with one-electron spectra such as hydrogen or the alkali atoms, the selection rules (11-27) entirely determine the spectrum. In an atom with many electrons, if the states are described by a single configuration, then because $\Sigma\, \mathbf{r}_i$ is a one-electron operator, the configuration change is restricted to just the $n\ell$ of one electron, the change in ℓ being ±1. These are called one-electron jumps. Transitions are observed in which two $n\ell$'s appear to change (two-electron jumps). This is interpreted as a breakdown in the characterization of energy levels by configuration assignment.

The c's in (11-26) can be evaluated and are found to be (we have suppressed m_s and replaced m_ℓ by m)

$$\langle n'\ell + 1\, m\,|z|\,n\ell m\rangle$$

$$= \sqrt{\frac{(\ell+1)^2 - m^2}{(2\ell+3)(2\ell+1)}}\ \langle n'\ell + 1|r|n\ell\rangle$$

$$\langle n'\ell - 1\, m\,|z|\,n\ell m\rangle$$

$$= \sqrt{\frac{\ell^2 - m^2}{(2\ell+1)(2\ell-1)}}\ \langle n'\ell - 1|r|n\ell\rangle$$

$$\langle n'\ell + 1\, m \pm 1|x \pm iy|n\ell m\rangle$$

$$= \pm\sqrt{\frac{(\ell \pm m + 2)(\ell \pm m + 1)}{(2\ell+3)(2\ell+1)}}\ \langle n'\ell + 1|r|n\ell\rangle$$

$$\langle n'\ell - 1\, m \pm 1|x \pm iy|n\ell m\rangle$$

$$= \pm\sqrt{\frac{(\ell \mp m)(\ell \mp m - 1)}{(2\ell+1)(2\ell-1)}}\ \langle n'\ell - 1|r|n\ell\rangle$$

$$\langle n\ell|r|n'\ell'\rangle$$

$$= \int_0^\infty r\, \mathcal{R}_{n\ell}\mathcal{R}_{n'\ell'}\, dr \tag{11-28}$$

When ± signs are given, the upper signs go together and so do the lower signs.

Equations (11-28) have the following consequences. Consider the following sums which measure the lifetime of the state $n\ell m$

$$\sum_{m'} |\langle n'\ell + 1\ m' | \mathbf{r} | n\ell m\rangle|^2 = \frac{\ell + 1}{2\ell + 1} \langle n'\ell + 1 | r | n\ell\rangle^2$$

$$\sum_{m'} |\langle n'\ell - 1\ m' | \mathbf{r} | n\ell m\rangle|^2 = \frac{\ell}{2\ell + 1} \langle n'\ell - 1 | r | n\ell\rangle^2 \qquad (11\text{-}29)$$

We have added the intensities of transition regardless of polarization from a certain state $n\ell m$ to all levels m' of $n'\ell'$; i.e., $m' = m - 1$, m, $m + 1$. We conclude that the lifetime of a state depends only on n', and ℓ. Next consider the sums

$$\sum_{m=-\ell}^{\ell} |\langle n'\ell - 1\ m | z | n\ell m\rangle|^2 = \tfrac{1}{3}\ell \langle n'\ell - 1 | r | n\ell\rangle^2$$

$$\sum_{m=-\ell}^{\ell} |\langle n'\ell - 1\ m + 1 | x | n\ell m\rangle|^2 + |\langle n'\ell - 1\ m - 1 | x | n\ell m\rangle|^2$$

$$= \tfrac{1}{3}\ell \langle n'\ell - 1 | r | n\ell\rangle^2 \qquad (11\text{-}30)$$

In (11-30) we have summed the intensities of all the Zeeman components of a spectral line that have the same polarization and concluded that the total intensity is the same for each of the three components of a Lorentz triplet in the normal Zeeman effect. Both (11-29) and (11-30) are consequences of the isotropy of space.

The following two "partial oscillator strength sum rules" can now be proved (see Bethe and Salpeter,[4] pp. 260–261). [See also (11-10a).]

$$\sum_{n'} \langle n'\ell + 1 | f^x | n\ell\rangle = \frac{1}{3}\ \frac{(\ell + 1)(2\ell + 3)}{2\ell + 1}$$

$$\sum_{n'} \langle n'\ell - 1 | f^x | n\ell\rangle = -\frac{1}{3}\ \frac{\ell(2\ell - 1)}{2\ell + 1}$$

---

[4]H. A. Bethe and E. E. Salpeter, *Quantum Mechanics of One- and Two-Electron Atoms*, Plenum, New York, 1977.

where

$$\langle n'\ell' | f^x | n\ell \rangle$$

$$= \frac{1}{2\ell + 1} \sum_{m'=-\ell'}^{\ell'} \sum_{m=-\ell}^{\ell} f^x_{n'\ell'm', n\ell m} \qquad (11\text{-}31)$$

If the two sums in (11-31) are added, we obtain 1 as expected from (11-10c). Since the first sum is positive we conclude that among the transitions $\ell \to \ell + 1$ absorption ($\omega_{n'\ell+1,n\ell} > 0$) predominates. Since the second sum is negative, in $\ell \to \ell - 1$ transitions, emission ($\omega_{n'\ell-1,n\ell} < 0$) predominates. Since energy increases with increasing principal quantum number, (11-31) shows that a change of principal and orbital quantum numbers in the same sense is more probable than a jump in the opposite sense. Similarly, the two last equations of (11-28) show that $|m|$ is likely to jump in the same direction as ℓ. Both results have classical analogues, derivable from a study of the motion of an ellipse.

SELECTION RULES FOR MANY ELECTRONS

Returning to the problem of determining the selection rules for a complex atom i.e. an atom not described in general by determinantal wavefunctions, we first consider the LS coupling scheme. Since **r** commutes with **S**, **r** cannot connect states with different S or M_S. Hence transitions between levels of different multiplicity are forbidden, and $\Delta M_S = 0$.

Since **r** is a vector of type **A** with respect to **L** and **J**, i.e., of the type considered in (8-17), we have $\Delta L = 0, \pm 1$; $\Delta J = 0, \pm 1$. As for the magnetic quantum number, the matrix element of z is nonvanishing only for $M_L' = M_L$; of $x \pm iy$ only for $M_L' = M_L \pm 1$. The polarization of the light follows the same rules as for one-electron spectra.

For any atom, J = 0 to J = 0 is strictly forbidden. For consider

$$\int u_f^* \sum_i e^{i\mathbf{k}\cdot\mathbf{r}_i} \frac{\partial}{\partial x_i} u_n \, d\tau$$

Since $J_f = J_n = 0$, the wave functions will not change if we rotate the system in any manner. If in particular we rotate about **k** as an axis, then $e^{i\mathbf{k}\cdot\mathbf{r}_i}$ also does not change. For instance, we can rotate the system about **k** through 180° for each i. Then $x_i \to -x_i$ and the integral changes sign, without a change in value. Hence it is 0.

We summarize the selection rules for LS coupling:

1. Parity changes.
2. Configuration must change by $\Delta\Sigma\ell_i = \pm 1$.
3. Multiplicity does not change, $\Delta S = 0$.
4. $\Delta M_S = 0$.
5. $\Delta L = 0, \pm 1$.
6. $\Delta M_L = 0, \pm 1$.
7. $\Delta J = 0, \pm 1$.
8. $J = 0 \rightarrow J = 0$ strictly forbidden.
9. $\Delta M = 0, \pm 1$.

It is seen that these are quite similar to those previously discussed for one-electron spectra. The first, seventh, eighth, and ninth rules hold for arbitrary coupling. The second holds as long as the configuration description is valid. The others holds for LS coupling only. Whenever the description is in terms of orbitals, we recall that parity is + (−) when $\Sigma_j \ell_j$ is even (odd). If the system cannot be approximated by orbitals, we still know the parity from the fact that the admixtures to the orbital eigenfunctions must be of the same parity, i.e., as far as parity is concerned, the orbitals are always a good approximation.

It is important to note that rule 5 permits $\Delta L = 0$, which is not permitted for one-electron spectra. The latter prohibition is a consequence of the parity selection rule: for one electron, $L = \ell$ directly determines the parity, so $\Delta\ell = 0$ would mean no parity change. For a many-electron atom, on the other hand, L has no direct relation to the parity, but the sum of the ℓ's of the individual orbitals does. We have seen, e.g., that for two equivalent electrons of a given ℓ all triplet states have odd L, but of course all these states have even parity. Thus $\Delta L = 0$ is compatible with change of parity.

It should further be noticed that although rule 9 follows from 4 and 6, rule 7 does not follow from 3 and 5. For consider $J_1 = L + \frac{1}{2}$, $J_2 = L - \frac{1}{2}$; set $L' = L - 1$, then $J_1' = L - \frac{1}{2}$, $J_2' = L - \frac{3}{2}$. Rules 3 and 5 would allow all four transitions, however rule 7 prohibits $J_1 \rightarrow J_2'$. Rules 4, 6, and 9 are useful only when an external field has removed the magnetic degeneracy.

The following intensity sum rule can be proved for LS coupling. In a transition array going from one multiplet LS to another, L'S, the sum of the strengths of the lines having a given initial level J is proportional to 2J + 1. The sum of the strengths having a given final J' is proportional to 2J' + 1 (Condon and Shortley,[5] p. 238).

---

[5]E. U. Condon and G. H. Shortley, *The Theory of Atomic Spectra*, Cambridge University Press, Cambridge, 1959.

In intermediate, or jj, coupling, only rules 1, 2, 7, 8, and 9 apply. The intensity sum rule is no longer valid as formulated above, but other sum rules exist (Condon and Shortley,[5] pp. 278–281).

HIGHER MOMENTS

If a transition is forbidden, i.e., if the dipole matrix element is zero, higher terms in the expansion of $e^{i\mathbf{k}\cdot\mathbf{r}}$ must be taken. Taking the second term $\mathbf{k}\cdot\mathbf{r}$, assuming it equal to $k_z z$, we wish to discuss the physical significance of this expression. If the polarization is in the x direction, the operator appearing in the matrix element of (10-25) is proportional to $\Sigma_i z_i (\partial/\partial x_i)$, which we may conveniently write as

$$\sum_i \left[\left(z_i \frac{\partial}{\partial x_i} - x_i \frac{\partial}{\partial z_i}\right) + \left(x_i \frac{\partial}{\partial z_i} + z_i \frac{\partial}{\partial x_i}\right)\right] \tag{11-32}$$

The first half of this expression is proportional to the total orbital angular momentum operator. Angular momentum is in turn proportional to the magnetic (dipole) moment of the atom (see (8-4)). If we make the usual extension to include spin angular momentum, we conclude that the second term in the multipole expansion leads to magnetic dipole radiation with the matrix elements proportional to

$$\langle f | \mathbf{L} + 2\mathbf{S} | i \rangle = \langle f | \mathbf{J} + \mathbf{S} | i \rangle \tag{11-33}$$

Since $[\mathbf{J}, H] = 0$, (11-33) reduces to $\langle f | \mathbf{S} | i \rangle$, which would be zero for exact LS coupling. That is, if there were no spin-orbit interaction to break exact LS coupling, there would no no magnetic dipole radiation. Since the spin-orbit interaction breaks LS coupling by mixing states with different L, S but in general of the same configuration, magnetic dipole radiation occurs, but it is quite weak, since the energy difference of two such states belonging to the same configuration is rather small.

The selection rules for magnetic dipole radiation are

1. Parity must remain unchanged.
2. $\Delta J = 0, \pm 1$.
3. $\Delta M = 0, \pm 1$.
4. $J = 0 \rightarrow J = 0$ strictly forbidden.
5. $\Delta \ell_i = \Delta n_i = 0$ (i runs over all orbitals).

(For nuclei the magnetic dipole radiation is quite prominent. For one thing, nuclear coupling is jj rather than LS. However, even if it were LS, the relevant matrix element would be proportional to

$$\langle f | \mathbf{L}_p + \mu_p \mathbf{S}_p + \mu_n \mathbf{S}_n | i \rangle \tag{11-34}$$

where p and n refer to proton and neutron, respectively, and $\mu_p = 2.8$, $\mu_n = -1.9$. The quantity that is conserved is $\mathbf{L}_p + \mathbf{S}_p + \mathbf{S}_n$ and not $\mathbf{L}_p + \mu_p \mathbf{S}_p + \mu_n \mathbf{S}_n$, hence magnetic dipole transitions are strong.)

The remaining part of (11-32) when operating on u_n satisfies the following identity:

$$2 \sum_i \left(x_i \frac{\partial}{\partial z_i} + z_i \frac{\partial}{\partial x_i} \right) u_n$$

$$= \sum_i \nabla_i^2 (z_i x_i u_n) - z_i x_i \nabla_i^2 u_n \qquad (11\text{-}35)$$

Therefore the matrix element of the left-hand side between states f and n is proportional to

$$\left[\left(\sum_i z_i x_i \right) H - H \left(\sum_i z_i x_i \right) \right]_{fn}$$

$$= \left[\left(\sum_i z_i x_i \right), H \right]_{fn} \propto \left(\frac{d}{dt} \sum_i z_i x_i \right)_{fn}$$

$$\propto \omega_{fn} \left(\sum_i z_i x_i \right)_{fn} \qquad (11\text{-}36)$$

We recognize this to be the electric quadrupole moment of the atom and therefore this term leads to electric quadrupole radiation.

The selection rules for quadrupole radiation of one-electron atoms can be obtained as follows. From matrix multiplication

$$(xy)_{fn} = \sum_{n'} x_{fn'} y_{n'n}$$

Applying the dipole selection rules to the matrix elements of x and y, we obtain $\Delta \ell = 0, \pm 2$; $\Delta m_\ell = 0, \pm 1, \pm 2$; $\Delta S = 0$; $\Delta M_S = 0$; $\Delta J = 0, \pm 1, \pm 2$; ℓ: $0 \to 0$ forbidden, J: $0 \to 0$ forbidden. It is also seen that parity must remain the same. For many-electron atoms the selection rules can be obtained by general arguments (Condon and Shortley,[5] p. 93). With the substitution of L for ℓ and the modification $\Delta L = 0, \pm 1, \pm 2$, the results are the same as the above one electron rules. The transition L: $0 \to 0$ is still forbidden in the LS coupling. It can also be shown that L: $0 \to 1$ or $1 \to 0$, and J: $0 \to 0$, and $\frac{1}{2} \to \frac{1}{2}$ is forbidden.

We saw that in the expansion of $e^{i\mathbf{k}\cdot\mathbf{r}}$, the ratio of the successive integrals is $0(ka) \sim 1/300$ if a is taken to be a_0 and $\hbar\omega_{fn} \sim$ Ry. The quadrupole radiation is further reduced by the fact that the angular average of zx is small ($zx \propto \sin\theta\cos\theta\sin\varphi$). It is found that the ratio of transition probabilities of quadrupole to dipole radiation is usually less than 10^{-6}, making the lifetime of states which can decay *only* by quadrupole radiation greater than 10^{-2} seconds.

As an example we consider oxygen. It has a p^4 configuration leading to states ^3P, ^1D, ^1S. The ground state is the ^3P and the successive separation of the other two states is about 1 eV. Dipole transitions between these states are forbidden since the parity is the same for all of them, as the configuration does not change. The quadrupole transitions $^1S \to {}^1D$ is allowed and has a long life. $^1D \to {}^3P$ is doubly forbidden since S changes. Since spin-orbit coupling breaks LS coupling, $^1D \to {}^3P_{2,1}$ can actually occur, via a magnetic dipole transition, but has a very long lifetime. In a discharge tube filled with oxygen under reasonable pressure, atoms in the ^1D state decay by collision. But under low pressure, e.g., in the ionosphere, the atom has time to radiate before making a collision. This is the origin of the famous red line of the aurora. The oxygen is excited by incoming protons and then decays. Forbidden transitions are also seen frequently in the sun's corona and in some nebulae. It was first thought that these lines corresponded to a new element, nebulium. But it was eventually realized that they are forbidden transitions in highly ionized familiar atoms.

ABSOLUTE TRANSITION PROBABILITIES

The total transition probability for spontaneous emission from state k to n is given by

$$A_{kn} = \frac{4}{3}\frac{e^2\omega^3_{kn}}{\hbar c^3}\,|(\mathbf{r})_{kn}|^2 \tag{10-38}$$

In terms of the oscillator strength this becomes

$$A_{kn} = \frac{2}{3}\frac{e^2}{mc^3}\,\omega^2_{kn}\,f_{kn} \tag{11-37}$$

If one sums (10-38) or (11-37) over all states n which have energy less than that of the initial state k, one arrives at the total probability per unit time that the state k is vacated through spontaneous emission.

$$\beta_k = \sum_{E_n < E_k} A_{kn} \tag{11-38}$$

The reciprocal of this quantity is called the lifetime of the state k; see also the end of Chapter 10.

$$T_k = \frac{1}{\beta_k} \tag{11-39}$$

Bethe and Salpeter[4] (pp. 262–269) list the squares of the dipole moments, the oscillator strengths, and the transition probabilities for hydrogen for transition $n\ell \to n'\ell \pm 1$. We discuss the qualitative behavior of their results.

The oscillator strength decreases rapidly with increasing n'. To see this we recall that the oscillator strength involves the radial integral

$$\left| \int r \mathcal{R}_{n'\ell'} \mathcal{R}_{n\ell} \, dr \right|^2$$

If $n \ll n'$ then $\mathcal{R}_{n\ell}$ is large only for small r. But for small r, we found that $\mathcal{R}_{n'\ell'} \propto n'^{-3/2}$; see (3-31). Therefore,

$$f_{n\ell, n'\ell'} \propto \frac{1}{n'^3} \tag{11-40}$$

for large n'.

We found from sum rules that if $n' > n$, ℓ' tends to be greater than ℓ. This rule is verified. For example, see Table 11-2.

For small orbital quantum numbers (eccentric orbits in the Bohr theory) transitions into the continuum are more frequent than for large quantum numbers (circular orbits). For example, the oscillator strengths for transitions from $n\ell$ to the continuum are given in Table 11-3.

Table 11-2

Transitions between Various States

Initial state	Final state	Average oscillator strength
2p	3s	0.014
	3d	0.7
4f	5d	0.009
	5g	1.35

The emission probability is largest if the final state is the ground state. For example, the emission probabilities from the 6p state to various states are: 1s, 0.20×10^8 sec^{-1}; 2s, 0.03×10^8 sec^{-1}; 5s, 0.002×10^8 sec^{-1}.

Table 11-3
Transitions from State $n\ell$ to Continuum

$n\ell$	Total oscillator strength of the continuum
1s	0.43
4s	0.25
4f	0.056

Table 11-4
Lifetimes of Various States $n\ell$

$n\ell$	Lifetime
2p	0.16×10^{-8} sec
6p	4.1×10^{-8} sec
6h	61×10^{-8} sec

The lifetimes of highly excited states are longer than those of moderately excited states. For example, the lifetimes of various $n\ell$ states are given in Table 11-4.

For alkalis, where the valence electron is subject to a non-Coulombic potential, the first (so-called resonance) line, ns → np, has a much larger oscillator strength than the lines from the ground state ns to higher states. As an example, we list in Table 11-5 the oscillator strengths for Na for transitions from the 3s level, and compare them with the corresponding ones in H.

Finally we observe that in hydrogen the transition 2s → $2p_{1/2}$ has a lifetime ~2 days for dipole radiation. A faster way for the 2s to decay is by emitting two quanta and going to the 1s level. The lifetime for this process is $\frac{1}{7}$ second. The energies of the individual quanta can be arbitrary as long as the sum equals the energy difference of the levels.

Table 11-5
Oscillator Strengths for Transitions from the 3s Level

	Na	H
3p	0.98	0
4p	0.014	0.48
6p	0.001	0.05

PROBLEMS

1. Show that

$$e^{itH/\hbar} A e^{-itH/\hbar} = \sum_{p=0}^{\infty} \left(\frac{it}{\hbar}\right)^p \frac{A_p}{p!}$$

$$A_0 = A$$

$$A_p = [H, A_{p-1}]$$

This establishes (11-3). For further discussion and application of sum rules, see R. Jackiw, Phys. Rev. 157, 1220 (1967).

2. Obtain a sum rule for

$$\sum_k \omega_{kn}^s \, |x_{kn}|^2$$

With the assumption that for high energies E_k

$$\sum_{\ell' m'} |\langle k\ell' m' \,|\, x^E \,|\, n\ell m\rangle|^2 = C\omega_{kn}^{-r}$$

show that for hydrogen $5 < r \leq 6$ when $\ell = 1$.

3. Verify all the entries in Table 11-1 on p. 215 using the ground state wave function of the hydrogen atom.
4. Calculate the relative intensities of the 6 lines in the transition array going from a multiplet LS to another L − 1 S with S = 1. Use the sum rule in the last paragraph on p. 224 plus the following information on the ratio of two of the intensities

$$\frac{P(L, J = L - 1 \to L - 1, J = L - 1)}{P(L, J = L - 1 \to L - 1, J = L)} = (2L + 1)(2L - 1)$$

From the result, show that J is most likely to change in the same direction as L.

5. Verify (11-35) and (11-36).

Chapter 12

PHOTOELECTRIC EFFECT

In Chapter 10 we found the differential cross section for the absorption of radiation of frequency ω by an atom and removal of an electron into the continuum. This is the experimental condition for the *photoelectric effect,* which we now study in detail. In Chapter 10, the differential cross section was found to be

$$\frac{d\sigma(\theta, \varphi)}{d\Omega} = \frac{e^2}{2\pi mc} \frac{k_f}{\omega} \left| \int u_f^* e^{i\mathbf{k}\cdot\mathbf{r}} \mathbf{P}\cdot\nabla u_n \, d\tau \right|^2 \qquad (10\text{-}18)$$

where the subscript f indicates the final state. The direction of the ejected electron is specified by (θ, φ). For definiteness we take the propagation vector to be along the z axis and assume the vector potential A to be plane polarized along the x axis.

We now discuss three approaches to evaluating this integral assuming that the atom is hydrogen and the initial state is the ground state.

BORN APPROXIMATION

We replace the final state by $e^{i\mathbf{k}_f\cdot\mathbf{r}}$. This is reasonably good if $p_f \gg p_{1s}$, the momentum of the ground state. We have

$$\frac{\hbar^2 k_f^2}{2m} = \hbar\omega - \frac{e^2}{2a_0} > 0$$

and require

$$\hbar k_f \gg \sqrt{\frac{e^2 m}{a_0}} \qquad \text{or} \qquad a_0^2 k_f^2 \gg 1 \qquad (12\text{-}1)$$

The integral becomes

$$\int e^{-i\mathbf{q}\cdot\mathbf{r}} \frac{\partial u_{10}}{\partial x} \, d\tau \tag{12-2}$$

with $\mathbf{q} = \mathbf{k}_f - \mathbf{k}$. We have used the assumption that the polarization vector is along the x axis and thus $\mathbf{k}$ has no x component. We integrate by parts, substitute the ground state wave function for hydrogen and obtain, with the aid of some algebra for the differential cross section,

$$\frac{d\sigma(\theta, \varphi)}{d\Omega} = \frac{32e^2}{mc\omega} \frac{(k_f a_0)^3 \cos^2 \alpha}{(1 + q^2 a_0^2)^4}$$

$$\cos \alpha = \frac{\mathbf{k}_f \cdot \mathbf{x}}{k_f x} = \sin\theta \cos\varphi \tag{12-3}$$

Equation (12-3) depends on angle in two ways. First, $\cos^2 \alpha$ indicates that the electron preferentially comes out along the direction of the electric field of the incident light wave (polarization vector). If the initial state had not been an isotropic state, then an expression of the form $A + B\cos^2 \alpha$ would replace the $\cos^2 \alpha$ term.

The second dependence on angle arises from the q^2 dependence in the denominator.

$$q^2 = k_f^2\left(1 - 2\frac{k}{k_f}\cos\theta + \left(\frac{k}{k_f}\right)^2\right)$$

$$\cos\theta = \frac{\mathbf{k}\cdot\mathbf{k}_f}{kk_f} \tag{12-4}$$

From (12-1) we conclude that

$$\frac{k}{k_f} = \frac{\omega}{ck_f} \approx \frac{\hbar k_f}{2mc} = \frac{v}{2c} \tag{12-5}$$

where v is the velocity of the ejected electron. Since we are dealing with a nonrelativistic theory we can write

$$q^2 = k_f^2\left(1 - \frac{v}{c}\cos\theta\right) \tag{12-6}$$

Further $q^2 a_0^2 \sim k_f^2 a_0^2 >> 1$. Therefore we can replace $1 + q^2 a_0^2$ by $k_f^2 a_0^2(1 - (v/c)\cos\theta)$.

$$\frac{d\sigma(\theta,\varphi)}{d\Omega} = \frac{32e^2}{mc\omega(k_f a_0)^5} \frac{\sin^2\theta\cos^2\varphi}{\left(1 - \frac{v}{c}\cos\theta\right)^4} \tag{12-7}$$

The angular dependence of (12-7) is easily understood. If we neglect the term with v/c, which is reasonable in the nonrelativistic case, the angular dependence is as $\cos^2\alpha = \sin^2\theta \cos^2\varphi$, having a maximum in the direction of the electric field of the incident light wave. The result means that the electron goes from the initial s state into a final state, whose m = 0 with respect to the x axis (field direction). This corresponds to the selection rules derived in Chapter 11. With respect to the direction of propagation, the distribution has a maximum in the equatorial plane, $\theta = \pi/2$. If we now take the retardation term with v/c into account, this maximum is shifted forward to approximately

$$\theta_{max} \approx \frac{\pi}{2} - 2\frac{v}{c} \tag{12-8}$$

There is no special significance to the factor 2; it is different for other initial states.

If we integrate (12-7) over angles, neglecting terms of relative order $(v/c)^2$, we obtain

$$\sigma = 2^8 \frac{\pi}{3} \frac{e^2}{\hbar c} a_0^2 \left(\frac{Ry}{\hbar\omega}\right)^{7/2} \tag{12-9}$$

DIPOLE APPROXIMATION

The second method of calculating the photoelectric effect is to ignore the retardation, viz., the exponent in (10-18) but use correct wave functions for the continuum. This approximation is valid when $ka_0 \ll 1$. In particular, near threshhold,

$$\hbar\omega = \hbar ck \approx \frac{e^2}{2a_0} \qquad ka_0 \approx \frac{e^2}{2\hbar c} \approx \frac{1}{300} \ll 1$$

Thus we shall be able to obtain the behavior of the cross section near threshold where the Born approximation certainly fails. Moreover in the region of k, $1 >> ka_0 >> 1/300$, we have $k_f^2 a_0^2 >> 1$, i.e., the Born approximation is also valid. Hence we shall be able to extrapolate the result to be derived below and compare it to the Born approximation.

In the dipole approximation, (10-18) is

$$\frac{d\sigma(\theta,\varphi)}{d\Omega} = \frac{e^2}{\omega} \frac{k_f}{2\pi mc} \frac{1}{\hbar^2} |(p_x)_{fn}|^2 = \frac{e^2}{\omega} \frac{k_f m}{2\pi c} \frac{\omega_{fn}^2}{\hbar^2} |(x)_{fn}|^2 \tag{12-10}$$

Here $|x_{fn}|^2$ is evaluated for a continuum state u_f normalized to unit amplitude, $e^{i\mathbf{k}_f\cdot\mathbf{r}}$. We wish to express this in terms of $|x_{fn}^E|^2$, i.e., u_f normalized on the energy scale.

The connection between $|x_{fn}^E|^2$ and $|x_{fn}|^2$ is given by[1]

$$|x_{fn}^E|^2 = \frac{p_f^2}{(2\pi\hbar)^3 v_f}|x_{fn}|^2 \tag{12-11}$$

The proportionality constant is simply the number of states per unit volume, per energy, per solid angle [see (10-16)]. Thus we frequently speak of $|x_{fn}^E|^2$ as the dipole moment strength per energy and write $d|x_{fn}|^2/dE$ for $|x_{fn}^E|^2$.

Returning now to (12-10) we have, with the help of (12-11),

$$\frac{d\sigma(\theta,\varphi)}{d\Omega} = \frac{4\pi^2 e^2 \omega_{fn}}{c}|x_{fn}^E|^2 \tag{12-12}$$

where we have used $mv_f = p_f$ which is justified in the dipole approximation.

Before proceeding with the explicit evaluation of (12-12), we note that (12-12) implies that $d\sigma/d\Omega$ and therefore $\sigma \propto \omega|x_{fn}^E|^2$. Comparing with (12-9), we find that the dipole matrix element in the energy normalization behaves for large ω as

$$\left|x_{fn}^E\right|^2 \propto \omega^{-9/2} \tag{12-13}$$

This agrees with our previous conclusion from sum rules, p. 219.

We now evaluate (12-12). The final state $|f\rangle$ is a continuum Coulomb wave function with momentum k_f, which satisfies the usual scattering boundary conditions; viz., it is proportional at large distances to $e^{i(\mathbf{k}\cdot\mathbf{r} + \varphi(\mathbf{r}))}$ where $\varphi(\mathbf{r})$ is the logarithmic phase characteristic of the Coulomb wave functions. This state may be expanded in terms of the radial continuum wave functions in a fashion similar to the expression of the plane wave in terms of spherical Bessel functions.

$$|f\rangle = \sum_{\ell m} R_{E_f \ell}(r) Y^*_{\ell m}(\Omega_k) Y_{\ell m}(\Omega_r) \tag{12-14}$$

Here $R_{E_f\ell}$ is a radial continuum wave function with energy normalization. Since n, the initial state, is the ground state $R_{10}(r)\,Y_{00}(\Omega_r)$, the Ω_r integration in (12-12) leaves only the $\ell = 1$, $m = \pm 1$ term in the sum (12-14).

$$\langle f | x^E | n \rangle = \sum_{\ell m} Y_{\ell m}(\Omega_k) \int d\tau\, R_{E_f \ell}(r)\, Y^*_{\ell m}(\Omega_r) \times R_{10}(r)\, Y_{00}(\Omega_r)$$

$$= -[Y_{11}(\Omega_k) - Y_{1\,-1}(\Omega_k)] \sqrt{\tfrac{1}{8}} \int_0^\infty r^3\, dr\, R_{E_f 1}(r)\, R_{10}(r)$$

$$= \frac{1}{\sqrt{4\pi}} \sin\theta \cos\varphi \int_0^\infty r^3\, dr\, R_{E_f 1}(r)\, R_{10}(r) \qquad (12\text{-}15)$$

The differential cross section then is

$$\frac{d\sigma(\theta, \varphi)}{d\Omega} = \frac{\pi e^2 \omega_{fn}}{c} \sin^2\theta \cos^2\varphi\, J(k_f) \qquad (12\text{-}16a)$$

and the total cross section is

$$\sigma = \frac{4\pi^2}{3} \frac{e^2 \omega_{fn}}{c} J(k_f) \qquad (12\text{-}16b)$$

where $J(k_f)$ is given by

$$J(k_f) = \left| \int_0^\infty r^3\, dr\, R_{E_f 1}(r)\, R_{10}(r) \right|^2 \qquad (12\text{-}17)$$

The calculation of $J(k_f)$ is described, e.g., in Bethe and Salpeter,[2] pp. 303-304. The result is

$$\sigma = \frac{2^9 \pi^2}{3} a_0^2 \frac{e^2}{\hbar c} \left(\frac{Ry}{\hbar\omega} \right)^4 f(n')$$

$$f(n') = \frac{e^{-4n' \operatorname{arccot} n'}}{1 - e^{-2\pi n'}}$$

$$\hbar\omega = \left(1 + \frac{1}{n'^2} \right) Ry \qquad (12\text{-}18)$$

---

[2]H. A. Bethe and E. E. Salpeter, *Quantum Mechanics of One- and Two-Electron Atoms*, Plenum, New York, 1977.

For large $\hbar\omega$, n′ becomes small, and

$$f(n') \approx \frac{1}{2\pi n'} \approx \frac{1}{2\pi}\left(\frac{\hbar\omega}{Ry}\right)^{1/2} \qquad (12\text{-}19a)$$

Then

$$\sigma = \frac{2^8\pi}{3}\, a_0^2 \frac{e^2}{\hbar c}\left(\frac{Ry}{\hbar\omega}\right)^{7/2} \qquad (12\text{-}19b)$$

which is identical with the Born approximation. This is to be expected because for $\hbar\omega \gg Ry$ the Born approximation should be valid, and for $v \ll c$ the retardation is negligible, and has in fact been neglected in (12-9).

Near threshold, $n' \to \infty$, thus

$$f(n') \approx e^{-4}\left(1 + \frac{4}{3n'^2}\right) \approx e^{-4}\left(1 + \frac{1}{n'^2}\right)^{4/3}$$

$$f(n') \approx e^{-4}\left(\frac{\hbar\omega}{Ry}\right)^{4/3} \qquad (12\text{-}20a)$$

Then

$$\sigma = \frac{2^9\pi^2 e^{-4}}{3}\, a_0^2 \frac{e^2}{\hbar c}\left(\frac{Ry}{\hbar\omega}\right)^{8/3} \qquad (12\text{-}20b)$$

Since also $\hbar\omega \approx Ry$, this can be written

$$\sigma = 31\, \frac{a_0^2 e^2}{\hbar c} \qquad (12\text{-}20c)$$

The factors a_0^2 and $e^2/\hbar c$ could have been expected from dimensional arguments; the numerical factor 31 can only be obtained from direct calculation. Since this factor is large, there is strong absorption of X-rays above threshold.

ROUGH ESTIMATE

The third method of calculating the photoelectric effect cross section is a very rough approximation and is useful when little is known, e.g., for calculations for atoms other than hydrogen.

We return to the dipole approximation for σ (12-16b) which of course is valid for any one-electron atom. In general, we cannot calculate $J(k_f)$. We recall however that $J(k_f)$ arose from the matrix elements of x, viz.,

$$|\langle n'\ \ell' = 1\ m' = 1\,|\,x\,|\,1\ 0\ 0\rangle|^2$$

$$+\ |\langle n'\ \ell' = 1\ m' = -1\,|\,x\,|\,1\ 0\ 0\rangle|^2 = \tfrac{1}{3}\,J(k_f)$$

These matrix elements may be related to the partial oscillator strengths defined in (11-31). We have

$$\tfrac{1}{3}\,J(k_f) = \frac{\hbar}{2m\omega}\,\langle n'\ \ell' = 1\,|\,f^{xE}\,|\,1\ 0\rangle \tag{12-21}$$

Thus

$$\sigma = 2\pi^2 e^2\,\frac{\hbar}{mc}\,\langle n'\ \ell' = 1\,|\,f^{xE}\,|\,1\ 0\rangle \tag{12-22}$$

An approximate expression for the oscillator strength may be inferred from empirical data. For hydrogen, for example, we assume

$$\frac{df}{dE} \equiv \langle n'\ \ell' = 1\,|\,f^{xE}\,|\,1\ 0\rangle = 0.86\,\frac{(Ry)^2}{(\hbar\omega)^3}$$

This is chosen so that

$$\int_{Ry}^{\infty} d(\hbar\omega)\,\frac{df}{dE} = 0.43$$

which we know to be the total oscillator strength from the 1s state to the continuum (see Table 11-3). The dependence on ω is chosen in accord with empirical data on X-ray absorption, and is an average between the $\omega^{-7/2}$ of (12-9) and the $\omega^{-8/3}$ of (12-20b). At $\hbar\omega = Ry$, (12-22) gives

$$\sigma = 34a_0^2\,\frac{e^2}{\hbar c} \tag{12-23}$$

instead of (12-20c), an error of 10 percent.

PROBLEMS

1. Show that

$$|x_{fn}^E|^2 = \frac{p_f^2}{(2\pi\hbar)^3 v_f}\,|x_{fn}|^2$$

Here the state n is a discrete state; the state f is a continuum state. x_{fn}^E is the matrix element of x between states f and n, where the state f is normalized on the energy scale as in (11-22); while x_{fn} is calculated with the state f normalized to unit amplitude, $e^{i\mathbf{k}_f \cdot \mathbf{r}}$.

2. Derive equation (12-3).
3. Verify that the small and large n′ limits of f(n′), defined in (12-18), are as given in (12-19a) and (12-20a).
4. Obtain the maximum of (12-7) as a function of θ, and verify that the estimate (12-8) is correct.
5. Integrate (12-7) over angles and verify that the estimate (12-9) is correct.

Part III

ATOMIC COLLISIONS

IN this part, we concentrate entirely on applications. We assume the general quantum theory of collisions to be known to the reader. For most of this part, the collision theory as given in Schiff's[1] book or Gottfried's[2] is sufficient. Only for Chapter 18 the more elaborate, general theory is required; see, e.g., Messiah.[3] Two excellent textbooks on collision theory are available, that of Goldberger and Watson[4] emphasizing the fundamental theory, and that of Mott and Massey[5] treating both general theory and applications. We shall frequently refer to the last two works.

---

[1]L. I. Schiff, *Quantum Mechanics,* 3rd ed., McGraw-Hill, New York, 1968.

[2]K. Gottfried, *Quantum Mechanics,* Vol. I, W. A. Benjamin, New York, 1966.

[3]A. Messiah, *Quantum Mechanics,* Vols. I and II, Wiley, New York, 1961, 1962.

[4]M. L. Goldberger and K. M. Watson, *Collision Theory*, Krieger, Melbourne FL., 1975.

[5]N. F. Mott and H. S. Massey, *The Theory of Atomic Collisions*, 3rd ed., Oxford University Press, Oxford, 1965.

Chapter 13

ELASTIC SCATTERING AT HIGH ENERGIES

We shall study the elastic collision of a particle of charge ze ($e > 0$) and mass m with an atom of atomic number Z. An exact formulation of this problem requires the use of a many-body Hamiltonian which describes all the particles of the system, viz., the nucleus, the atomic electrons, and the incident particle which interacts with the nucleus and each atomic electron individually. For the moment, however, we shall make the assumption that for purposes of studying the elastic scattering of the incident particle, the complicated interaction of the incident particle with the constituents of the atom can be summarized by an effective electrostatic potential V in which the incident particle travels. In later chapters, we shall examine to what extent this approximation is valid.

It is physically reasonable that the electrostatic potential in which the incident particle travels is well approximated by

$$V(\mathbf{r}) = ze \left[\frac{Ze}{r} - \int \frac{\rho(\mathbf{r}')\, e}{|\mathbf{r} - \mathbf{r}'|}\, d\tau'\right] \tag{13-1}$$

Here $\mathbf{r}$ is the position vector of the incident particle. The above potential satisfies Poisson's equation.

$$\nabla^2 V = -4\pi z e^2 [Z\, \delta(\mathbf{r}) - \rho(\mathbf{r})] \tag{13-2}$$

The first term is due to the field of the nucleus, which is taken to be a charged point particle at the origin of the coordinate system. The second term is the potential of the atomic electrons, described in terms of an effective electron density ρ. Evidently, we are ignoring all effects of symmetry and spin for the time being. It is seen that (13-1) is exactly the Hartree approximation (see Chapter 4), but now applied to an external particle.

If the atom is neutral, then the density satisfies

$$\int \rho(\mathbf{r})\, d\tau = Z \tag{13-3}$$

We define $F(\mathbf{q})$ to be the Fourier transform of ρ

$$F(\mathbf{q}) = \int \rho(\mathbf{r})\, e^{i\mathbf{q}\cdot\mathbf{r}}\, d\tau \tag{13-4a}$$

$$|F(\mathbf{q})| \leq \int \rho(\mathbf{r})\, d\tau = Z \tag{13-4b}$$

$F(\mathbf{q})$ is called the *form factor* of the atom and, as we shall see, is closely related to the cross section for high energy scattering off the atom.

We shall later show (Chapter 17) that the above, intuitive potential can be derived from first principles, and that the electron density is given by the usual expression

$$\rho(\mathbf{r}) = \sum_{i=1}^{Z} \int |\phi_0(\mathbf{r}_1, \ldots, \mathbf{r}_{i-1}, \mathbf{r}, \mathbf{r}_{i+1}, \ldots, \mathbf{r}_Z)|^2 \prod_{j\neq i} d\tau_j$$

$$= \sum_{i=1}^{Z} \int |\phi_0(\mathbf{r}_1, \ldots, \mathbf{r}_Z)|^2\, \delta(\mathbf{r} - \mathbf{r}_i) \prod_j d\tau_j \tag{13-5a}$$

where ϕ_0 is the ground state wave function for a Z-electron atom. If the Hartree assumption is made, viz., if ϕ_0 is taken to be a product of individual orbitals $\phi_0 = \Pi_i u_i(\mathbf{r}_i)$, then (13-5a) becomes

$$\rho(\mathbf{r}) = \sum_{i=1}^{Z} |u_i(\mathbf{r})|^2 \tag{13-5b}$$

Since the ground state wave function has definite parity, we have

$$\rho(\mathbf{r}) = \rho(-\mathbf{r}) \tag{13-5c}$$

even if the general expression (13-5a) is used. Therefore the form factor $F(\mathbf{q})$, (13-4a), is real.

ELASTIC SCATTERING IN THE BORN APPROXIMATION

When the incident particle carries sufficiently high energy, the scattering amplitudes can be easily evaluated by the Born approximation. The two criteria for the validity of the Born approximation are[1] (Schiff, p. 326)

[1] L. I. Schiff, *Quantum Mechanics*, 3rd ed., McGraw-Hill, New York, 1968.

Table 13-1
Energy Threshold for Validity of Born Formula According to (13-7)

	Z	$15\ Z^2$ eV
Hydrogen	1	15 eV
Helium	2	60 eV
Neon	10	1500 eV
Argon	18	5 keV
Krypton	36	19 keV
Xenon	54	45 keV

$$\text{(I)}\quad Va^2 \ll \frac{\hbar^2}{m} \qquad \text{(condition for validity at all energies)}$$

or

$$\text{(II)}\quad Va \ll \hbar v \qquad \text{(condition for validity at high energy)} \tag{13-6}$$

V is the strength of the potential, a is the range, and v the velocity of the incident particle. In our case we take a to be the radius of the atom and $V \sim Ze^2/a$. The two criteria become

$$\text{(I)}\quad Za \ll \frac{\hbar^2}{me^2} \qquad \text{(Bohr radius if the scattered particle is an electron, less for other particles)}$$

or

$$\text{(II)}\quad Z \ll \frac{\hbar v}{e^2} = \frac{\hbar c}{e^2}\frac{v}{c} = 137\frac{v}{c} \tag{13-7}$$

It is seen that the first criterion is never satisfied in our problem. The second condition, however, will be satisfied for any given Z if only v is sufficiently large. In this chapter, we shall assume, therefore, that the incident particles are fast enough for the Born approximation to hold, but nonrelativistic. (We shall nevertheless write many of the subsequent formulas in a fashion which makes them relativistically correct.) Thus for the present we assume the initial particle velocity to be between $(Z/137)c$ and $\frac{1}{2}c$, which for incident electrons corresponds to a kinetic energy between $15\ Z^2$ eV and 60 keV.

Table 13-1 lists the energy at which the criterion (13-7) indicates that the Born formula becomes valid for electron scattering in various atoms.

The elastic scattering amplitude is given in the Born approximation by (Schiff[1], p. 324)

$$f(\Omega) = \frac{-m}{2\pi\hbar^2} \int e^{i\mathbf{q}\cdot\mathbf{r}}\, V(\mathbf{r})\, d\tau \tag{13-8a}$$

$$\mathbf{q} = \mathbf{k}_0 - \mathbf{k} \qquad q = 2k \sin \tfrac{1}{2}\theta \qquad \cos\theta = \frac{\mathbf{k}_0 \cdot \mathbf{k}}{k^2} \tag{13-8b}$$

Here we have given the amplitude for scattering in the direction of the final momentum $\hbar\mathbf{k}$. The initial momentum $\hbar\mathbf{k}_0$ is along the polar axis. Since the scattering is elastic, $|\mathbf{k}| = |\mathbf{k}_0| \equiv k$. When the potential is spherically symmetric, the angular integration can be performed to give

$$f(\theta) = -\frac{2m}{\hbar^2} \int_0^\infty \frac{\sin qr}{qr}\, V(r)\, r^2\, dr \tag{13-9}$$

The following transformation facilitates the evaluation of the Fourier transform of the potential which is proportional to the scattering amplitude.

$$\int V(\mathbf{r})\, e^{i\mathbf{q}\cdot\mathbf{r}}\, d\tau = -\frac{1}{q^2} \int V(\mathbf{r})\, \nabla^2 e^{i\mathbf{q}\cdot\mathbf{r}}\, d\tau$$

$$= -\frac{1}{q^2} \int e^{i\mathbf{q}\cdot\mathbf{r}}\, \nabla^2 V(\mathbf{r})\, d\tau$$

$$= \frac{4\pi z e^2}{q^2} [Z - F(\mathbf{q})] \tag{13-10}$$

where we have integrated by parts, and used (13-2), and the definition (13-4a) of the form factor. The integrated terms from the partial integration go to zero as we assume the potential $V(\mathbf{r})$ to fall off rapidly at large distances. Then

$$f(\Omega) = -\frac{2mze^2}{\hbar^2} \frac{1}{q^2} [Z - F(\mathbf{q})] \tag{13-11}$$

The differential scattering cross section for elastic processes thus becomes

$$\frac{d\sigma}{d\Omega} = \left(2z\,\frac{me^2}{\hbar^2}\,\frac{1}{q^2}\right)^2 [Z - F(\mathbf{q})]^2 \tag{13-12}$$

For incident electrons, we set $z = -1$ and

$$\frac{\hbar^2}{me^2} = a_0 \qquad \text{(Bohr radius)} \tag{13-13}$$

and (13-12) reduces to

$$\frac{d\sigma}{d\Omega} = \left(\frac{2}{a_0 q^2}\right)^2 [Z - F(\mathbf{q})]^2 \tag{13-14}$$

The cross section (13-12) is seen to be proportional to m^2. This is characteristic of other collision processes as well, for example the scattering of slow neutrons by nuclei, and has the consequence that the scattering of neutrons off a bound proton is much larger than that off a free proton, owing to the reduced mass effect.[2]

For large q, $F(\mathbf{q}) \to 0$ since $\rho(\mathbf{r})$ is sufficiently spread out in space, thus for $qa \gg 1$

$$\frac{d\sigma}{d\Omega} = \left(\frac{2me^2 Zz}{\hbar^2 k^2 4 \sin^2 \frac{1}{2}\theta}\right)^2 = \left(\frac{e^2 Zz}{2pv \sin^2 \frac{1}{2}\theta}\right)^2 \tag{13-15}$$

where we have set $\hbar k = p$, the momentum of the incident particle. In its relativistic generalization (13-15) acquires an additional factor of $(1 - v^2/c^2 \sin^2 \frac{1}{2}\theta)$, but the overall factor $1/p^2v^2$ remains; see Chapter 23. In the non-relativistic region pv is twice the kinetic energy and the differential cross section depends only on the kinetic energy of the incident particle. In the extreme relativistic case $v \to c$, pv approaches the total energy. Thus we can obtain a measurement of the energy of a fast particle by determining the differential cross section for elastic scattering. In practice, this is done by measuring the multiple scattering of a particle track in emulsion, cloud or bubble chamber; the multiple scattering can be derived from the differential cross section by the theory of E. J. Williams.[3] The result (13-15) is also identical with that derived from classical mechanics by Rutherford in 1910; in particular, the angular dependence $\csc^4 \frac{1}{2}\theta$ is familiar.

For small q, we can expand $Z - F(\mathbf{q})$ in powers of q. The zero-order term is 0 from (13-4a) and (13-3). The first-order term is

---

[2] H. A. Bethe and P. Morrison, *Elementary Nuclear Theory*, 2nd ed., Wiley, New York, 1965, p. 62.

[3] N. F. Mott and H. S. W. Massey, *Theory of Atomic Collisions*, 3rd. ed., Oxford University Press, Oxford, 1965, p. 467.

proportional to $\int \rho(\mathbf{r})\mathbf{r}\, d\tau$, which is the mean value of the dipole moment of the atom and vanishes identically [equation (13-5c)]. Similarly all odd orders vanish also. The first nonvanishing term is quadratic in q.

$$Z - F(q) = \frac{1}{2} \int (\mathbf{r}\cdot\mathbf{q})^2 \rho(\mathbf{r})\, d\tau + O(q^4) \qquad (13\text{-}16a)$$

For spherically symmetric atoms,

$$\frac{1}{2} \int (\mathbf{r}\cdot\mathbf{q})^2 \rho(\mathbf{r})\, d\tau = \frac{1}{6} q^2 \langle r^2 \rangle Z \qquad (13\text{-}16b)$$

Inserting (13-16b) into (13-14), we see that the differential cross section for small angle scattering is independent of angle and depends simply on the mean square distance of the electrons from the nucleus. In particular, the singularity at $q = 0$ which occurs in the Rutherford formula disappears.

The diamagnetic susceptibility of atoms is proportional to $\langle r^2 \rangle$ (see p. 171). Hence an unsuspected relationship exists between this quantity and small-angle elastic scattering (but only in the Born approximation).

The measurement of form factors of charged systems determines, as is seen from (13-16), the rms radius. Specifically,

$$\left.\frac{\partial F}{\partial q^2}\right|_{q^2=0} = -\frac{\langle r^2 \rangle}{6} Z$$

For a proton, experiment gives $\sqrt{\langle r^2 \rangle} = 0.8$ fermis.

The total elastic cross section for spherically symmetric atoms is

$$\sigma = 2\pi \left(\frac{2m\, e^2 z}{\hbar^2 k}\right)^2 \int_0^{2k} \frac{dq}{q^3} (Z - F(q))^2 \qquad (13\text{-}17)$$

where the angular integral has been reexpressed by the change of variable from θ to q [see (13-8b)]. Thus the total cross section as a function of energy is proportional to $(1/E) \int_0^{2k} [(Z - F)^2/q^3]\, dq$. Since the integral $\int_0^{\infty} [(Z - F)^2/q^3]\, dq$ exists, $\sigma(E)$ varies as $1/E$ for large energies.

CALCULATION OF FORM FACTORS

For hydrogen and helium we can calculate the electron density and so the form factor from the available wave functions. For hydrogen we use the exact ground state function; for helium the simplest variational function equation (3-40). This gives for the density in these two cases

$$\rho(r) = Z\left(\frac{\alpha^3}{\pi a_0^3}\right) e^{-2\alpha r/a_0} \tag{13-18}$$

α is the effective charge, and is 1 for hydrogen and 1.69 for helium. The form factor becomes

$$F(q) = \frac{Z}{\left(1 + \left(\frac{a_0}{\alpha}\right)^2 \frac{q^2}{4}\right)^2} \tag{13-19}$$

For complex atoms, the charge density can be determined by the statistical method of Thomas-Fermi which is valid for heavy atoms; or for any atom, and more accurately, by the self-consistent Hartree theory. Further improvement is obtained for each of the two methods when exchange is included by the use of the Thomas-Fermi-Dirac theory, or the Hartree-Fock theory, respectively.

In the Thomas-Fermi method, all atoms are identical in shape, and the scale of length is proportional to $Z^{-1/3}$. Thus, the form factor will be a universal function for all (heavy) atoms. Specifically, the electron density is given by (see Chapter 5)

$$\rho(r) = \frac{Z^2}{a_0^3} f\left(\frac{rZ^{1/3}}{b}\right) \tag{13-20a}$$

$$f(x) = \frac{32}{9\pi^3}\left(\frac{\Phi(x)}{x}\right)^{3/2} \tag{13-20b}$$

$$b = 0.885\, a_0$$

where $\Phi(x)$ is the universal Thomas-Fermi function. The form factor is

$$F(q) = ZG(bqZ^{-1/3}) \tag{13-21a}$$

$$G(Q) = \frac{4\pi}{Q}\frac{b^3}{a_0^3}\int_0^\infty x \sin Qx\, f(x)\, dx \tag{13-21b}$$

or

$$G(Q) = \frac{1}{Q}\int_0^\infty [\Phi(x)]^{3/2} x^{-1/2} \sin Qx\, dx \tag{13-21c}$$

The universal function G(x) is tabulated.[4]

---

[4] L. D. Landau and E. Lifshitz, *Quantum Mechanics*, 3rd ed., Pergamon, Oxford, 1977, p. 578; Mott and Massey,[3] p. 462. The latter authors also give, on p. 460, a table of the differential cross section for various atoms, from Z = 3 to 80, on the basis of the Hartree-Fock distribution.

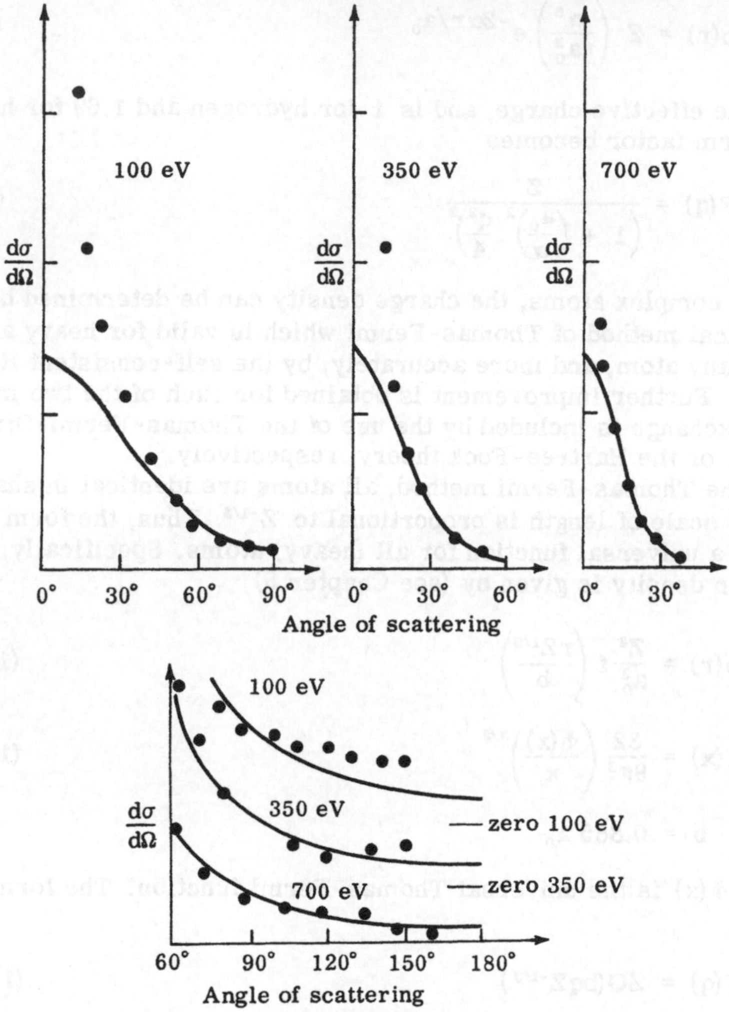

Figure 13-1

Theoretical and experimental differential cross section for electrons off helium (Mott and Massey[3], p. 458); ● experimental points; — theoretical curves.

COMPARISON WITH EXPERIMENT

Comparison between the theoretical and experimental results for elastic scattering of electrons off helium is afforded by Figure 13-1. The theoretical curves are calculated by the Born approximation, (13-14) and (13-19). The experimental results have been normalized to agree with theory at 700 eV, and scattering angles greater than about 10°. It is seen that in the energy range considered, 100 to 700

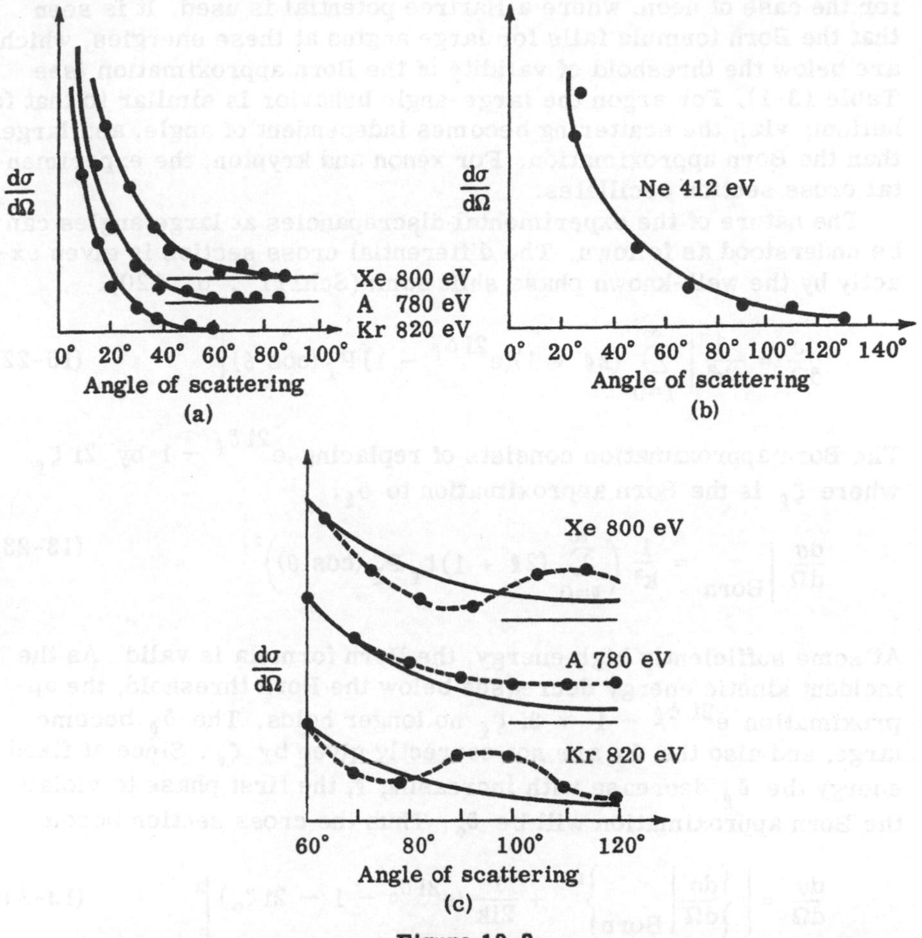

Figure 13–2

Differential cross section for electrons off noble gases (Mott and Massey [3], p. 467). Solid line represents theoretical curves, dashed line, experimental curves; points are experimental points for (a) and (b) small angle scattering; (c) large angle scattering.

eV, the qualitative agreement is good. This is to be expected from the validity criterion for the Born approximation in helium: energy $> 15Z^2$ eV = 60 eV. At small and large angles the Born formula deviates from experiment. At large angles, for electron energies below 500 volts, the experimental differential cross section becomes independent of angle. At small angles, the observed cross section is larger than the calculated one.

Figure 13-2 compares the Born formula with experiment for elastic electron scattering off various heavier atoms. The form factors are calculated with the help of the Thomas-Fermi potential, except for the case of neon, where a Hartree potential is used. It is seen that the Born formula fails for large angles at these energies, which are below the threshold of validity of the Born approximation (see Table 13-1). For argon the large-angle behavior is similar to that for helium; viz., the scattering becomes independent of angle, and larger than the Born approximation. For xenon and krypton, the experimental cross section oscillates.

The nature of the experimental discrepancies at large angles can be understood as follows. The differential cross section is given exactly by the well-known phase shift sum: (Schiff[1], p. 120).

$$\frac{d\sigma}{d\Omega} = \frac{1}{4k^2}\left|\sum_{\ell=0}^{\infty}(2\ell+1)(e^{2i\delta_\ell}-1)P_\ell(\cos\theta)\right|^2 \tag{13-22}$$

The Born approximation consists of replacing $e^{2i\delta_\ell}-1$ by $2i\zeta_\ell$ where ζ_ℓ is the Born approximation to δ_ℓ.

$$\left.\frac{d\sigma}{d\Omega}\right|_{\text{Born}} = \frac{1}{k^2}\left(\sum_{\ell=0}^{\infty}(2\ell+1)\zeta_\ell P_\ell(\cos\theta)\right)^2 \tag{13-23}$$

At some sufficiently high energy, the Born formula is valid. As the incident kinetic energy decreases below the Born threshold, the approximation $e^{2i\delta_\ell}-1 \approx 2i\zeta_\ell$ no longer holds. The δ_ℓ become large, and also the δ_ℓ are not correctly given by ζ_ℓ. Since at fixed energy the δ_ℓ decrease with increasing ℓ, the first phase to violate the Born approximation will be δ_0. Thus the cross section becomes

$$\frac{d\sigma}{d\Omega} = \left|\left\{\left.\frac{d\sigma}{d\Omega}\right|_{\text{Born}}\right\}^{1/2} + \frac{1}{2ik}(e^{2i\delta_0}-1-2i\zeta_0)\right|^2 \tag{13-24}$$

Since the Born term goes to zero at large angles, the correction which is isotropic dominates, and the differential cross section becomes independent of angle. As the energy continues to decrease below threshold, δ_1 becomes too large for the Born approximation. The cross section becomes

$$\frac{d\sigma}{d\Omega} = \left|\left\{\left.\frac{d\sigma}{d\Omega}\right|_{\text{Born}}\right\}^{1/2} + \frac{1}{2ik}\{(e^{2i\delta_0}-1-2i\zeta_0) + 3(e^{2i\delta_1}-1-2i\zeta_1)\cos\theta\}\right|^2 \tag{13-25}$$

At large angles, the dominating correction now will exhibit an oscillatory behavior due to $\cos\theta$. At still lower energies, the large angle behavior of the cross section will oscillate in a complicated fashion due to the superposition of various Legendre polynomials. This is exactly the behavior observed.

The deviation at small angles will be explained in Chapter 15.

PROBLEMS

1. Supply the missing steps in the derivation of (13-17).
2. Supply the missing steps in the derivation of (13-21).
3. Consider N static, spherically symmetric scatters placed on a straight line, such that the nth scatterer is at the point $(n-1)\mathbf{a}$. A particle, with incident momentum $\hbar\mathbf{k}_0$, $\mathbf{k}_0 \cdot \mathbf{a} = 0$, is scattered off this system. The interaction between the particle and the nth center is of the form $V(|\mathbf{r} - \mathbf{r}_n|)$, and is sufficiently weak for the Born approximation to be valid. Show that the elastic differential cross section in this approximation is of the form

   $$\frac{d\sigma}{d\Omega} = \frac{d\sigma_0}{d\Omega} F^2(\theta)$$

   where $d\sigma_0/d\Omega$ is the Born approximation to the differential cross section for the scattering by a single scatterer. Find the form factor $F(\theta)$.
4. Verify the entries in Table 13-1.
5. Exhibit the explicit connection between the diamagnetic susceptibility of atoms and the Born approximation to the small angle, elastic cross section.

Chapter 14

ELASTIC SCATTERING AT LOW ENERGIES

As we have seen in the previous chapter the Born approximation becomes invalid at energies below the Born threshold. To solve the scattering problem at these low energies we must calculate the phase shifts and use the exact expression for the scattering cross section (13-22). The potential is that of (13-1) and the electron density is calculated by whatever method is practical: Hartree-Fock, Hartree, Thomas-Fermi-Dirac or Thomas-Fermi.

The Thomas-Fermi gives the simplest result; the potential energy is a universal function (see Chapter 5).

$$V(r) = \frac{zZe^2}{r}\,\Phi\left(\frac{rZ^{1/3}}{b}\right)$$

$$b = 0.885a_0 \qquad (14\text{-}1)$$

This is valid for heavy atoms (large Z). Since the charge density of the Thomas-Fermi atom decreases as $1/r^6$ rather than exponentially as it should, the above tends to overestimate the charge density and the potential at large distances. Thus the energy of the incoming particle must be large compared to the potential energy in those regions in which the Thomas-Fermi potential is unphysical. (The Thomas-Fermi potential becomes invalid for $r > a_0 Z^{1/6}$, since $a_0 Z^{1/6}$ is roughly the radius of a heavy atom. Therefore, $E \gg zZ^{5/6}(e^2/a_0)\Phi(a_0 Z^{1/2}/b)$. Taking $Z = 50$, $z = 1$, we find that this will be satisfied for electrons with energies of 100 eV or more.)

Given a suitable potential, present-day computing machines can easily obtain the phase shifts. Nevertheless, we will present here a semi-analytic method which can be used to obtain the cross section.

We use the semiclassical WKB approximation (see Chapter 1).

Recall that the phase shift within this approximation is given by

$$\delta_\ell^{WKB} = \lim_{r\to\infty} \left[\int_a^r \Phi^{1/2} dr' - kr + \left(\ell + \frac{1}{2}\right)\frac{\pi}{2}\right] \qquad (1\text{-}25b)$$

where

$$\Phi(r) = k^2 - \frac{2m}{\hbar^2} V(r) - \frac{(\ell+1/2)^2}{r^2}$$

and a is the largest zero of Φ. The integral can be evaluated numerically once $V(r)$ is given.

For the Thomas-Fermi potential, the phase shift is a universal function satisfying the equation

$$\delta_{[\ell+\frac{1}{2}]}(E, Z) = \frac{1}{\beta}\, \delta_{[\beta(\ell+\frac{1}{2})]}(\beta^4 E, \beta^3 Z)$$

$$\beta \equiv \sqrt[3]{\frac{Z'}{Z}} \qquad \delta_{[\ell+\frac{1}{2}]} \equiv \delta_\ell \qquad (14\text{-}6)$$

It is therefore known for any Z' once it has been calculated for one value of Z.

The accuracy of this approach is exhibited in Table 14-1, which lists the values of the phase shifts calculated for 54 eV electrons off krypton. The accurate solution was obtained by numerical integration

Table 14-1

Phase Shifts in Radians Calculated for 54 eV Electrons off Kr. (After Mott and Massey,[1] p. 467)

ℓ \ δ_ℓ	Accurate solution	WKB	Born approximation
0	9.70	9.60	–
1	7.45	7.54	–
2	4.47	4.50	–
3	1.24	1.36	0.78
4	0.44	0.54	0.41
5	0.143	0.174	0.144

[1]N. F. Mott and H. S. W. Massey, *The Theory of Atomic Collisions*, 3rd ed., Oxford University Press, Oxford, 1965.

of Schrödinger's equation using a Hartree potential. The WKB method as well as the Born approximation used the Thomas-Fermi potential.

On the whole, the agreement between the WKB and the accurate solution is rather good, especially so for small ℓ. One can expect from the WKB approximation an accuracy of about 0.1 radian. The Born approximation is seen to be exceedingly good for $\ell = 5$, and quite good for $\ell = 4$. When the Born phase shift reaches $\pi/4$, however, it is no longer reliable.

The results of a WKB, Thomas-Fermi calculation of the phase shifts as a function of energy in mercury are given in Figure 14-1 and Table 14-2.

The first three phase shifts at zero energy are undoubtedly exaggerated by the Thomas-Fermi potential being too large at large distances. It is seen that the phase shifts for $\ell \leq 3$ are finite at zero energy while those for $\ell \geq 4$ vanish. This can be understood on the basis of the calculation of the lowest Z for which bound atomic states with a given ℓ first appear (see p. 91). Thallium (Z = 81) presents an atomic configuration similar to that of mercury (Z = 80) + 1 electron. For thallium, which binds s, p, d, f electrons, the effective radial potential

$$V(r) + \frac{\hbar^2}{2m}\,\frac{(\ell + \frac{1}{2})^2}{r^2}$$

is attractive (for some r) for $\ell \leq 3$; and repulsive for all r for $\ell \geq 4$. Therefore incident electrons with vanishing energy and $\ell \leq 3$ can still penetrate deep into the atom and into a region of strong potential, while those with $\ell \geq 4$ are shielded by the centrifugal barrier. It is only at high energies that the electrons of $\ell \geq 4$ can overcome the centrifugal repulsion.

In Figure 14-2 we exhibit the differential cross section obtained, by summing sufficient terms in the phase shift sum.

The WKB theory gives the right positions and well approximates the height of the various maxima. (It is seen that the Born approximation fails to predict the cross section at 812 eV.) The absolute maximum for forward scattering arises from the fact that all the Legendre polynomials attain their maxima and contribute the same amount, viz., $P_\ell(1) = 1$. For back scattering we observe a relative maximum; for 812 eV and 480 eV it is the second largest maximum. This is understood by noting that $P_\ell(-1) = (-1)^\ell$. At other angles,

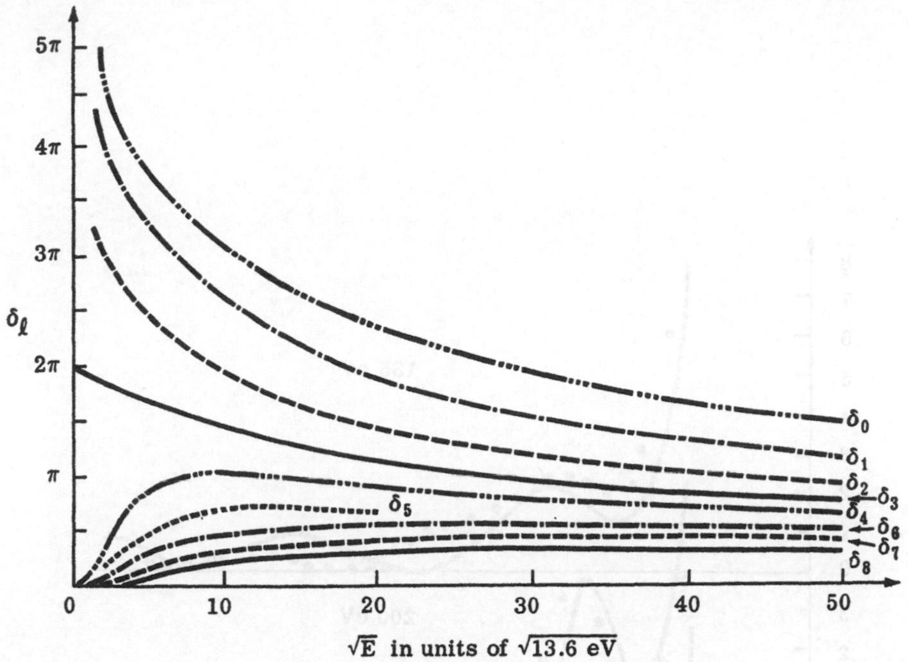

Figure 14-1
Phase shifts for electrons off Hg

E \ ℓ	0	1	2	3	4	5
0 eV	14	11	7	2	0	0
60 eV	5	4	3	1.8	0.5	0.2

Table 14-2
Phase Shifts in Units of π for Electrons off Hg

P_ℓ is smaller in magnitude and varies more smoothly with ℓ. This leads to very effective destructive interference. On the other hand, at $\theta = \pi$, the large magnitude and the abrupt alternation in sign of P_ℓ lead to a result for f which is of the order of magnitude of one term ℓ in the series, and this is in general larger than f for intermediate angles.

The angular position of the relative maxima (those not at 0 or π) decreases with increasing energy. This is familiar from optical diffraction where shorter wavelengths are diffracted through a smaller angle.

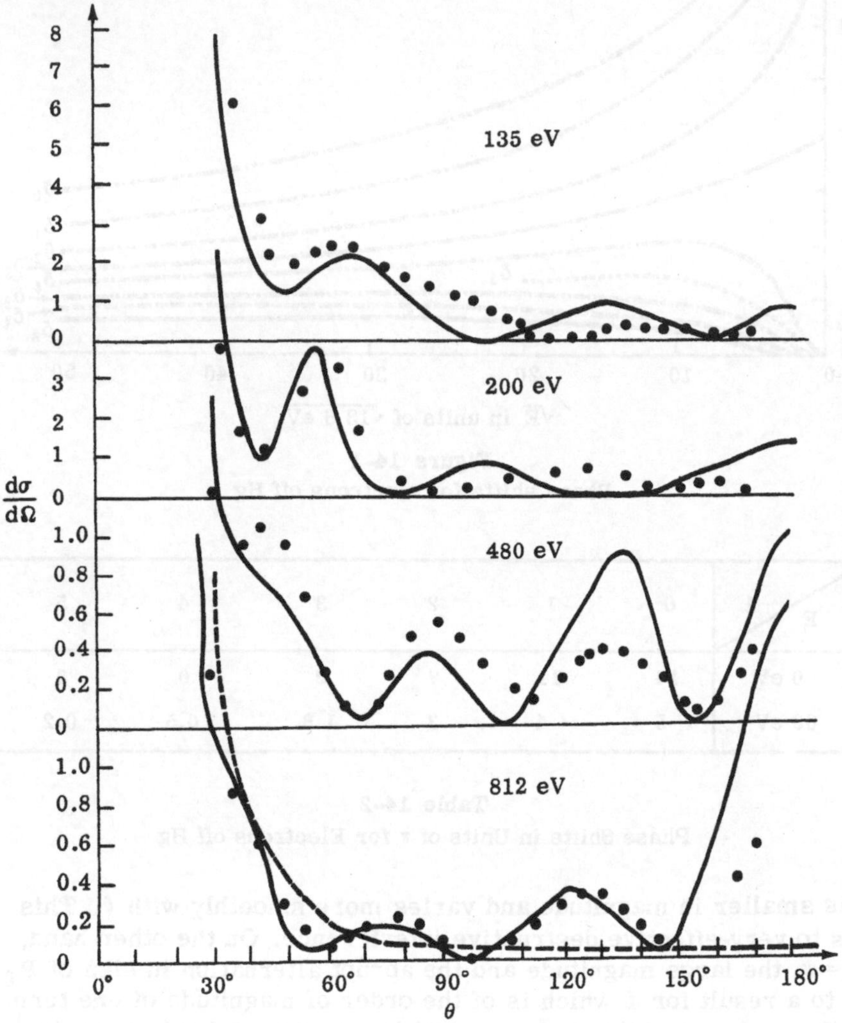

Figure 14-2

Differential elastic cross section for electrons off Hg are various energies; • Experimental points. —Theoretical WKB curve. --- Theoretical Born curve.

RAMSAUER-TOWNSEND EFFECT

In Figures 14-3 to 14-5 we exhibit the observed variation of the total cross section $\sigma = \int d\Omega \, d\sigma/d\Omega$ with energy.

At low energy, σ is seen to vary between wide limits; the low energy magnitude of σ for mercury being over 10 times larger than that for noble gases. There appears a familial similarity in the curves for elements of the same column in the periodical table. The heavier noble gases, argon, krypton, and xenon have a minimum below 0.7 eV. At larger energies the variation is not so marked.

To obtain a qualitative understanding for this behavior of σ, recall the phase shift formula for total cross section

$$\sigma(k^2) = \frac{4\pi}{k^2} \sum_{\ell=0}^{\infty} (2\ell + 1) \sin^2 \delta_\ell \tag{14-7}$$

At higher energies, many ℓ's contribute, and since it is unlikely that all the $\sin^2 \delta_\ell$ are small (we are below the Born approximation threshold), we expect a statistical variation from atom to atom following the same general pattern. It is only at small energies when only few ℓ's contribute that individual differences appear. For any given k, the phase shift will be small for all $\ell > ka$, where a is the radius of the atom, because these ℓ correspond to classical particles passing by the atom at distances (impact parameters) greater than a. For these large ℓ, $\sin \delta_\ell$ is very small. The atomic radius a will of course be large for alkalis, small for noble gases and intermediate for other atoms.

In particular, when $k < 1/a$, then δ_ℓ will be small for all $\ell \neq 0$, so that only s waves contributes to the scattering,

$$\sigma = \frac{4\pi}{k^2} \sin^2 \delta_0 \qquad k < 1/a \tag{14-8}$$

It is then possible, if the potential has just the correct and sufficient strength, that δ_0 attains the value $n\pi$ and $\sigma = 0$. This is known as the Ramsauer-Townsend effect. For small atoms, pure s wave scattering can occur at larger k than for large atoms ($k \sim 1/a$). Therefore, it is more likely that the Ramsauer-Townsend effect occurs with small atoms, because a wider interval of k^2 is available to fulfill the condition $\sin \delta_0 = 0$.

To put these considerations on a quantitative basis we take for a, the radius of the outer shell, which is $1.3a_0$ for argon, $1.7a_0$ for krypton, and $6.1a_0$ for potassium. (It would be better to define a as the distance beyond which there is

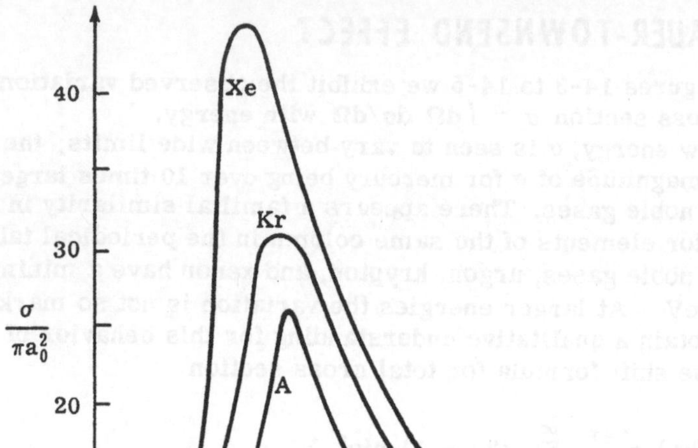

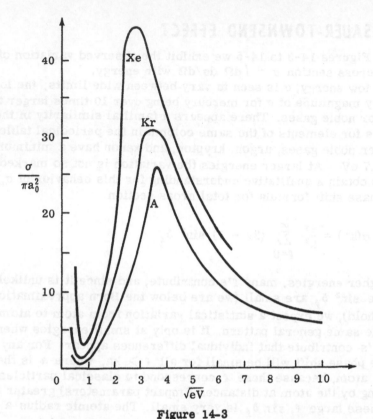

Figure 14-3

Observed total cross section for electrons off various atoms. (After Massey and Burhop[2], pp. 10 and 11.)

exactly one electron.) Pure s wave scattering can occur at energies below 8 eV for argon, 4.6 eV for krypton, 0.4 eV for potassium. It is very unlikely that δ_0 attains $n\pi$ within 0.4 eV, hence the Ramsauer-Townsend effect will not occur for potassium. (Figure 14-5 shows that it does not occur for zinc, cadmium, and silver.) But it is of course not necessary[3] that it will occur for the "small" atoms argon and krypton. To determine whether it does, we list in Table 14-3 the phase shifts in argon and krypton at low energies. These are calculated from a Hartree potential with a polarization correction (see next subsection).

---

[2]H. S. W. Massey and E. H. S. Burhop, *Electronic and Ionic Impact Phenomena*, Oxford University Press, Oxford, 1952.

[3]The discussion in many textbooks is misleading on this point.

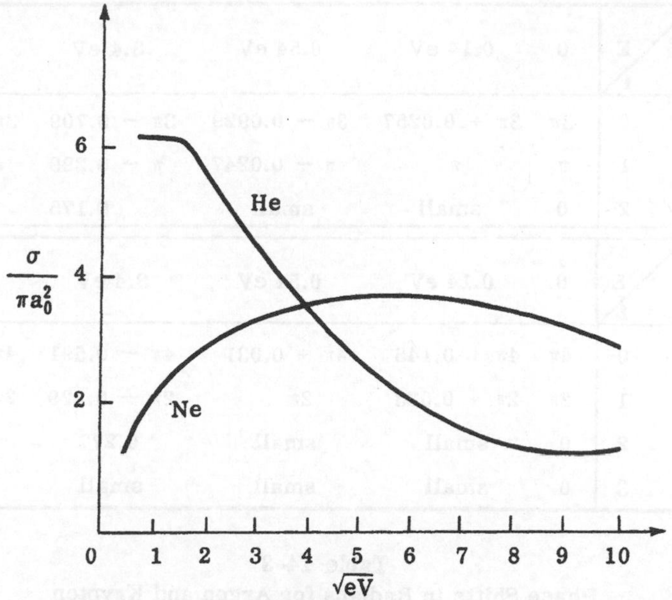

Figure 14–4

Observed total cross section for electrons off various atoms. (After Massey and Burhop[2], pp. 10 and 11.)

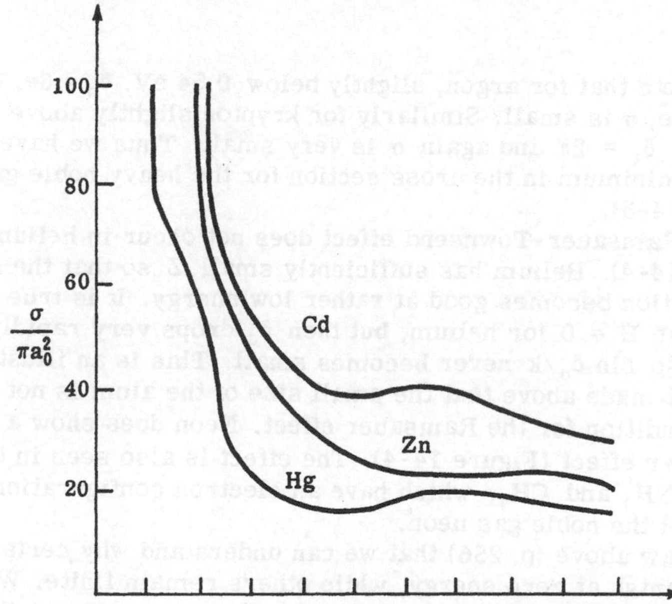

Figure 14–5

Observed total cross section for electrons off various atoms. (After Massey and Burhop[2], pp. 10 and 11.)

Ar — E / ℓ	0	0.14 eV	0.54 eV	3.4 eV	13.6 eV
0	3π	$3\pi + 0.0257$	$3\pi - 0.0929$	$3\pi - 0.709$	$3\pi - 1.084$
1	π	π	$\pi - 0.0247$	$\pi - 0.299$	$\pi - 0.888$
2	0	small	small	0.175	1.232

Kr — E / ℓ	0	0.14 eV	0.54 eV	3.4 eV	13.6 eV
0	4π	$4\pi + 0.143$	$4\pi + 0.031$	$4\pi - 0.591$	$4\pi - 1.57$
1	2π	$2\pi + 0.023$	2π	$2\pi - 0.229$	$2\pi - 0.936$
2	0	small	small	0.277	1.226
3	0	small	small	small	0.226

Table 14-3

Phase Shifts in Radians for Argon and Krypton

We note that for argon, slightly below 0.54 eV, $\delta_0 = 3\pi$, $\delta_1 = \pi$, and therefore, σ is small. Similarly for krypton slightly above 0.54 eV, $\delta_0 = 4\pi$, $\delta_1 = 2\pi$ and again σ is very small. Thus we have accounted for the minimum in the cross section for the heavy noble gases (see Figure 14-3).

The Ramsauer-Townsend effect does not occur in helium (see Figure 14-4). Helium has sufficiently small Z so that the Born approximation becomes good at rather low energy. It is true that $\delta_0 = \pi$ at E = 0 for helium, but then δ_0 drops very rapidly with energy so $\sin \delta_0/k$ never becomes small. This is an illustration of the point made above that the small size of the atom is not a sufficient condition for the Ramsauer effect. Neon does show a weak Ramsauer effect (Figure 14-4). The effect is also seen in the compounds NH_3 and CH_4, which have an electron configuration similar to that of the noble gas neon.

We saw above (p. 256) that we can understand why certain phase shifts vanish at zero energy, while others remain finite. We can also understand the specific multiple of π occurring in the nonzero phase shift: For Kr, $E = 0, \ell = 0$, the wave function of the incident electron has 4 nodes inside the atom. Furthermore, an incident electron of vanishingly small energy moving

E	0.14 eV	0.54 eV	3.4 eV	13.6 eV
Calculated	$28a_0^2$	$2.5a_0^2$	$42.5a_0^2$	$101a_0^2$
Observed	$23a_0^2$	$2.5a_0^2$	$36.5a_0^2$	$95a_0^2$

Table 14-4
Total Elastic Cross Section for Electrons off Krypton

in the field of the krypton atom (Z = 36), is physically similar to the valence electron of the rubidium atom (Z = 37); both "see" closed electron shells containing 36 electrons. In rubidium the valence electron is in a 5s state which also has 4 nodes. We further note that the above argument fails for the p wave, since the rubidium 5p state has 3 nodes, yet $\delta_1(0) = 2\pi$ for krypton. This is merely a reflection that the two situations, viz., krypton + electron and valence electron of rubidium, are not entirely equivalent.[4]

Finally in Table 14-4, we list the observed and calculated total cross section for krypton. The agreement is good.

POLARIZATION OF ATOM BY INCIDENT PARTICLE

For sufficiently slow incident particles the electronic configuration has time to become polarized under the influence of the field of the external electron. That is, the energy of the atom is lowered due to a second-order Stark effect. This lowering in energy can be considered as a potential which acts on the external electron.

To obtain this effective potential we note that the potential energy of the incident particle in the field of the nucleus and the atomic electrons is

[4]The field in which the Rb valence electron moves is stronger (by one net charge) than that acting on the free electron interacting with Kr. This makes it qualitatively understandable that the Rb p-electron has one extra node. We may further remember (Chapter 4, p. 76) that the $\ell = 2$ shell in Hartree calculations of transition elements is very sensitive to the nuclear charge, because of the near compensation between the attractive Coulomb and the repulsive centrifugal potential. It is reasonable that some such effect, although smaller, exists also for $\ell = 1$ but not for $\ell = 0$. We may thus understand that for $\ell = 1$ the Rb wave function has one more node than that in Kr, while for $\ell = 0$ the number of nodes is equal.

$$ze\left[\frac{Ze}{r} - \sum_{i=1}^{Z} \frac{e}{|\mathbf{r} - \mathbf{r}_i|}\right] \tag{14-9}$$

which for large distances is approximately

$$\frac{ze^2}{r^2} \sum_{i=1}^{Z} \hat{\mathbf{r}} \cdot \mathbf{r}_i \tag{14-10}$$

Since the effect is second order, the energy of the atom is lowered by

$$\Delta E = -\frac{Ce^2 z^2}{r^4} \tag{14-11}$$

where the constant C is proportional to the polarizability of the atom. The singularity at the origin is spurious since the polarization effect is unimportant at small distances. Hence the polarization potential is given by (14-11) suitably cut off at small distances.

There are some difficulties in justifying the above. The argument is of course modeled after the treatment of molecules (Chapter 9). But in that case, we could expand in m/M, the ratio of the mass of an atomic electron to that of a nucleus; and it was therefore reasonable to treat the motion of the atomic electrons for fixed position of the nuclei. In our case, M is the mass of the external electron and hence $m/M = 1$. Nevertheless, the use of a polarization potential can be justified. Without the polarization potential, the agreement is not good; the improvement is not very sensitive to the exact choice of the polarization potential. The importance of polarization in the scattering of electrons by hydrogen atoms is especially striking. For $E = 0$, δ_0 goes to 0 without, and to π with polarization. For $E = 1$ to 15 eV, δ_0 stays practically constant without polarization ($\delta_0 \approx 1.0$ radian), while it decreases from about 2 to 1 radian with polarization. For $E > 15$ eV, the influence of polarization is small (~0.1 radian) as might be expected since the polarization potential is nowhere very large.[5]

[5]M. L. Goldberger and K. M. Watson, *Collision Theory*, Krieger, Melbourne FL, 1975, pp. 850, ff.

PROBLEMS

1. Using the WKB expression for the phase shift (14-5), and the Thomas-Fermi potential (14-1), derive the formulas (14-6) which give the connection between the phase shifts of different Z.
2. Verify (14-1).
3. Consider a particle scattering of a potential of the form $V(r) = \alpha/r^2$. Compare the phase shifts in this potential as determined by the exact Schrödinger equation with those determined by the WKB method, (14-5).
4. Show that the WKB phase shifts for a Coulomb potential diverge logarithmically. Verify that this same divergence is present in the exact solution of the Schrödinger equation. Show also that when this divergent term is removed from both the WKB and the exact Coulomb phase shifts, the remaining finite parts agree asymptotically at low energy. (Consider only the s wave, for simplicity.)
5. Dicuss in detail the derivation of (14-11).

Chapter 15
FURTHER CORRECTIONS TO ELASTIC SCATTERING FORMULAS

ELECTRON EXCHANGE

A possibility that we have not considered yet which may occur when the incident particle is an electron, is that the incident electron is captured in the atom and an atomic electron is emitted; or more correctly, we have not taken into account the fact that the electrons are identical. This effect is important at small energies. In the analysis of this effect, we may proceed in a fashion similar to the derivation of the Hartree-Fock theory (see Chapter 4). In order to obtain the potential energy for the incident electron, we form Slater determinant trial functions and minimize the expectation value of the Hamiltonian. We may make the simplifying assumption that the incident electron does not affect the equations for the atomic orbitals. This may be justified by noting that the wave function of the incident electron may be normalized in a large box of volume Ω; then its probability density inside the atom will be proportional to Ω^{-1} and so will be the potential it exerts on the atomic electrons in the Hartree-Fock approximation. Of course, physically the external electron *will* affect the atomic ones; this is described by the polarization effect discussed in the last section of the previous chapter, and we can see now that this polarization will be modified by exchange. Obviously, this problem goes beyond the Hartree-Fock approximation and is quite complicated; it is discussed by Goldberger and Watson,[1] p. 855.

We shall remain strictly within the Hartree-Fock approximation. Then the atomic orbitals are unchanged, and the equation for the incident electron wave function ψ becomes

---

[1]M. L. Goldberger and K. M. Watson, *Collision Theory*, Krieger, Melbourne FL, 1975.

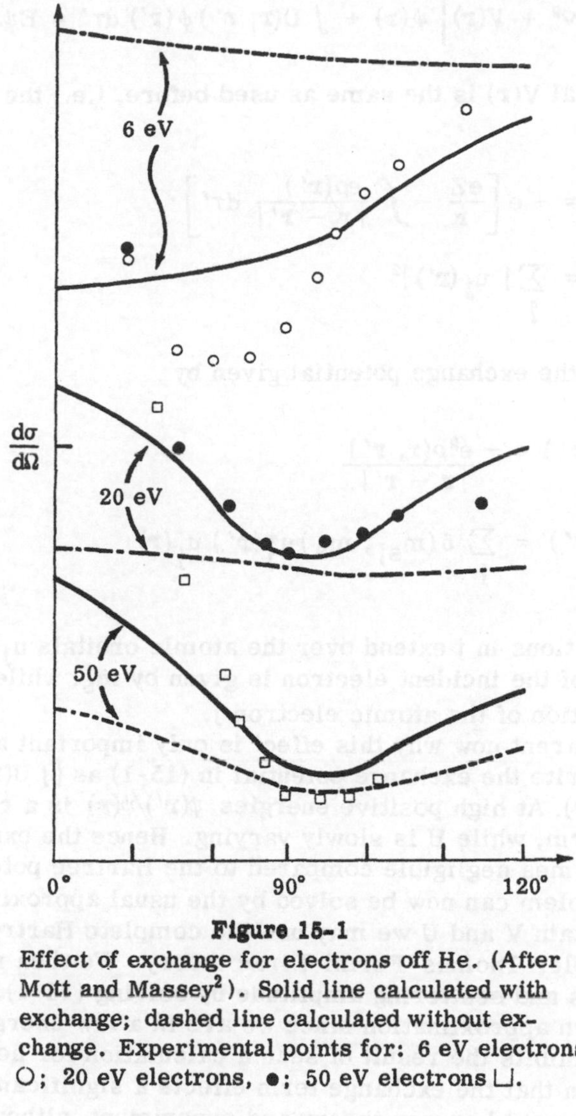

Figure 15-1

Effect of exchange for electrons off He. (After Mott and Massey[2]). Solid line calculated with exchange; dashed line calculated without exchange. Experimental points for: 6 eV electrons, ○; 20 eV electrons, ●; 50 eV electrons □.

---

[2]N. F. Mott and H. S. W. Massey, *The Theory of Atomic Collisions*, 3rd ed., Oxford University Press, Oxford, 1965.

$$\left[-\frac{\hbar^2}{2m}\nabla^2 + V(\mathbf{r})\right]\psi(\mathbf{r}) + \int U(\mathbf{r}, \mathbf{r}')\psi(\mathbf{r}')\,d\tau' = E\psi(\mathbf{r}) \qquad (15\text{-}1)$$

The potential $V(\mathbf{r})$ is the same as used before, i.e., the usual Hartree expression

$$V(\mathbf{r}) = -e\left[\frac{eZ}{r} - \int \frac{e\rho(\mathbf{r}')}{|\mathbf{r} - \mathbf{r}'|}\,d\tau'\right]$$

$$\rho(\mathbf{r}') = \sum_j |u_j(\mathbf{r}')|^2 \qquad (15\text{-}2)$$

$U(\mathbf{r}, \mathbf{r}')$ is the exchange potential given by

$$U(\mathbf{r}, \mathbf{r}') = -\frac{e^2\rho(\mathbf{r}, \mathbf{r}')}{|\mathbf{r} - \mathbf{r}'|}$$

$$\rho(\mathbf{r}, \mathbf{r}') = \sum_j \delta(m_{sj}, m_s)u_j^*(\mathbf{r}')u_j(\mathbf{r}) \qquad (15\text{-}3)$$

The summations in j extend over the atomic orbitals $u_j(\mathbf{r})$. The spin projection of the incident electron is given by m_s, while m_{sj} is the spin projection of the atomic electron j.

It is apparent now why this effect is only important at low energies. We write the exchange potential in (15-1) as $\{\int U(\mathbf{r}, \mathbf{r}')[\psi(\mathbf{r}')/\psi(\mathbf{r})]d\tau'\}\psi(\mathbf{r})$. At high positive energies $\psi(\mathbf{r}')/\psi(\mathbf{r})$ is a rapidly oscillating term, while U is slowly varying. Hence the exchange potential becomes negligible compared to the Hartree potential.

The problem can now be solved by the usual approximation methods. To obtain V and U we may use the complete Hartree-Fock theory or the simpler Thomas-Fermi-Dirac theory. We then must obtain the phase shifts and scattering amplitude by solving (15-1). (We cannot use the Born approximation since we are in a low energy region.) Figure 15-1 exhibits the result of such a calculation for helium.

It is seen that the exchange term effects a significant improvement in the agreement between theory and experiment, although a discrepancy at small angles still remains, especially at the higher energies, 20 and 50 eV (see the next section on shadow scattering). The differential cross section without exchange is almost isotropic, which is to be expected at these small energies. The result with exchange is no longer isotropic. The presence of the exchange potential permits higher angular momenta to scatter. To understand this, we note that the usual Hartree term is quadratic in $u_j(\mathbf{r}')$ whereas the exchange potential is linear in $u_j(\mathbf{r}')$. Thus the range of the exchange potential is twice that of the Hartree potential. The greater the range of a potential the larger the phase shifts. Hence high angular momentum phase shifts, which are negligible for a Hartree potential, may contribute significantly with an exchange potential.

We can understand the back-scattering maximum at 6 eV by assuming that only s and p wave scattering contribute. Then

$$f \sim \sin\delta_0 \, e^{i\delta_0} + 3\sin\delta_1 \, e^{i\delta_1} \cos\theta \tag{15-4}$$

At this low energy we may assume that $\delta_0 \sim \pi - \epsilon$, $\epsilon >> \delta_1 > 0$. Then

$$f \sim -\epsilon + 3\delta_1 \cos\theta$$

$$|f|^2 \sim \epsilon^2 - 6\epsilon\,\delta_1 \cos\theta \tag{15-5}$$

which attains a maximum at $\theta = 180°$.

For the case of hydrogen, the atom is in a definite spin state say ↑. For incident ↑ electrons there will be an exchange potential and a definite scattering amplitude f_x. For incident ↓ electrons, the exchange term disappears and we obtain a different scattering amplitude f_n. If the incident beam is unpolarized and we do not measure spins of the scattered electrons, the differential cross section is

$$\frac{d\sigma}{d\Omega} = \tfrac{1}{2}(|f_x|^2 + |f_n|^2) \tag{15-6}$$

Expressions similar to (15-6) may be derived for other atoms with nonvanishing total spin.

An interesting consequence of exchange scattering by atoms of nonvanishing spin is that, at any given angle θ, the scattered beam will be polarized even if the incident beam is not, since $f_x \neq f_n$ in general. Thus we have the possibility of producing polarized electron beams by exchange scattering. This method however is available only in principle rather than practice since strong exchange occurs only at small energies.

SHADOW SCATTERING

We recall that in the region of validity of the Born approximation, the experimental data showed a marked discrepancy from the Born formula at small angles; viz., the experimental points were much higher. A similar effect was observed when exchange was included. This is due to inelastic processes, which diminish the intensity of the outgoing elastic waves, and thus affect the elastic scattering. This phenomenon is known as *shadow scattering*. (Sometimes this is called a polarization effect, but it must not be confused with what we called polarization in Chapter 14.) A direct but extremely complicated treatment proceeds by explicitly evaluating the second-order contribution from inelastic processes, and summing over all the inelastic intermediate states. Historically, this was the way the problem was first done. We shall give here a simpler account. We confine ourselves to the case when the scattered particle is an electron carrying initial momentum $\hbar k$.

In the presence of inelastic scattering, the elastically scattered amplitude is given by [3]

$$f = \frac{i}{2k} \sum_{\ell} (2\ell + 1)(1 - \eta_{\ell}) P_{\ell}(\cos\theta) \tag{15-7}$$

where η_{ℓ} is the amplitude of the outgoing wave with angular momentum ℓ. We may write this

$$\eta_{\ell} = a_{\ell} e^{2i\delta_{\ell}} \tag{15-8}$$

where δ_{ℓ} and a_{ℓ} are real. In the absence of inelastic scattering, the real factor a_{ℓ} is equal to one; in the presence of such scattering $a_{\ell} < 1$. The contribution of the partial wave ℓ to the inelastic cross section is [3]

$$\sigma_{\text{inel},\ell} = \frac{\pi}{k^2}(2\ell + 1)(1 - a_{\ell}^2) \tag{15-9}$$

Now it is well known that electrons may cause excitation and ionization of atoms even if they pass at large impact parameters $b = \ell/k$, namely at $b \gg a$, where a is the atomic radius. The reason for this is, of course, the very long range of the Coulomb interaction between external and atomic electron. Therefore $\sigma_{\text{inel},\ell}$ will remain appreciable for very large ℓ, for which δ_{ℓ} is negligible. Therefore, for

$$\ell = bk \gg ak \tag{15-10}$$

we have essentially $\delta_{\ell} = 0$ and

[3] L. D. Landau and E. Lifshits, *Quantum Mechanics*, 3rd ed., Pergamon, Oxford, 1977, p. 592.

$$\eta_\ell \approx a_\ell \tag{15-11}$$

This is the typical condition for shadow scattering. Moreover, since $a_\ell < 1$ up to very large ℓ, this shadow scattering is concentrated at very small angles.

A good estimate for $\sigma_{inel,\ell}$ may be obtained by Williams' classical method, which we discuss in detail in Chapter 19. This estimate will give a value for a_ℓ, and thus from (15-11) for η_ℓ (at least for $\ell >> ak$). Once we have obtained η_ℓ we can calculate the elastic scattering amplitude from (15-7).

Since the impact parameter is large, the incident electron may be treated classically and its trajectory as a straight line. Its interaction with the atom is

$$V = -\frac{Ze^2}{r_0} + \sum_{i=1}^{Z} \frac{e^2}{|\mathbf{r}_0 - \mathbf{r}_i|} \approx e^2 \frac{\mathbf{r}_0}{r^3} \cdot \sum_i \mathbf{r}_i \tag{15-12}$$

where $\mathbf{r}_0$ is the position of the incident electron and i labels the atomic electrons. We write

$$\mathbf{r}_0 = \mathbf{b} + \mathbf{v}t \tag{15-13}$$

where $\mathbf{b}$ is the impact parameter, $\mathbf{b} \cdot \mathbf{v} = 0$. Only the $\mathbf{b}$ component of $\mathbf{r}_0$ in the numerator of (15-12) contributes appreciably to the transition amplitude (see Chapter 19); choosing $\mathbf{b}$ in the x direction, we obtain from (15-12)

$$V = \frac{e^2 b}{(b^2 + v^2 t^2)^{3/2}} D^x$$

$$D^x = \sum_i x_i \tag{15-14}$$

In Chapter 19 we show that

$$\sigma_{inel,\ell} = \frac{\pi}{k^2}(2\ell + 1)\left(\frac{2e^2}{\hbar v b}\right)^2 \sum_n |\langle n | D^x | 0 \rangle|^2 \tag{15-15a}$$

that is,

$$1 - a_\ell^2 = 4\left(\frac{e^2}{\hbar v b}\right)^2 \sum_n |\langle n | D^x | 0 \rangle|^2 \tag{15-15b}$$

In the above, the sum in n is over a complete set of states, and the state $|0\rangle$ is the ground state.

We now proceed to obtain an explicit expression for $1 - a_\ell$. To evaluate the sum, we first use closure:

$$\sum_n |\langle n|D^x|0\rangle|^2 = \langle 0|D^{x2}|0\rangle$$

$$= \sum_i \langle 0|x_i^2|0\rangle + 2\sum_{i<j} \langle 0|x_i x_j|0\rangle \qquad (15\text{-}16a)$$

In the Hartree approximation, there is no correlation between electrons so that the last term vanishes. (This is a good approximation also for exact wave functions.) For the first term we have, averaging over direction,

$$\langle 0|x_i^2|0\rangle = \tfrac{1}{3}\langle 0|r_i^2|0\rangle \qquad (15\text{-}16b)$$

Now we group the atomic electrons occurring the sum over in (15-16a) into shells α. With z_α electrons in the shell and $\langle 0|r_i^2|0\rangle = r_\alpha^2$ for the shell, (15-15b) becomes

$$1 - a_\ell^2 \approx \frac{4}{3}\left(\frac{e^2}{\hbar vb}\right)^2 \sum_\alpha z_\alpha r^2_\alpha \qquad (15\text{-}16c)$$

The atomic radius may be taken to be equal to the largest r_α which we shall call r_1, thus

$$a = r_1 \equiv (r_\alpha)_{max} \qquad (15\text{-}17)$$

[The expression $\Sigma_\alpha z_\alpha r_\alpha^2$ is proportional to the diamagnetic susceptibility of the atom (see p. 171)]. It is clear that for the important values of $b\,(\geq a)$, (15-16c) will be very small, viz., $a_\ell \approx 1$, so that we may write

$$1 - a_\ell^2 = (1 - a_\ell)(1 + a_\ell) \approx 2(1 - a_\ell)$$

$$1 - a_\ell \approx \tfrac{1}{2}(1 - a_\ell^2) = \frac{2}{3}\left(\frac{e^2}{\hbar vb}\right)^2 \sum_\alpha z_\alpha r^2_\alpha \qquad (15\text{-}18)$$

or, in terms of $\ell = kb$:

$$1 - a_\ell = \frac{2}{3}\sum_\alpha \frac{z_\alpha r^2_\alpha}{a_0^2}\,\frac{1}{\ell^2} = \frac{C}{\ell^2} \qquad (15\text{-}19)$$

where $a_0 = \hbar^2/me^2$ is the Bohr radius. This is a very simple result; it is remarkable that it is independent of v (except of course for the implicit dependence of ℓ on v). It is valid, for the contribution from shell α, if (see the next paragraph below)

$$kr_\alpha < \ell = kb < \frac{kv}{\omega_\alpha} = \frac{2E}{\hbar\omega_\alpha} \equiv \ell_1 \qquad (15\text{-}20)$$

where $\hbar\omega_\alpha$ is the average excitation energy of the excited states reached by the matrix element $|\langle n | D^x | 0\rangle|^2$. It is reasonable to take

$$\hbar\omega_\alpha = \frac{\hbar^2}{2mr_\alpha^2} \qquad (15\text{-}21)$$

This relation is connected with the oscillator strength sum rule. Then the upper limit in (15-20) may be written

$$\ell_1 = 2k^2 r_\alpha^2 \leq 2\, k^2 a^2 \qquad (15\text{-}22)$$

where $a = r_1$ is the atomic radius.

The upper limit on ℓ in (15-20) is derived in Chapter 19 where it is shown that the collision time, b/v, must be less than $1/\omega_\alpha$. The lower limit on ℓ in (15-20) comes from the fact that for $b < r_\alpha \leq a$, the expansion of $|\mathbf{r}_0 - \mathbf{r}_i|^{-1}$ in (15-12) is no longer justified. The transition probability $1 - a_\ell^2$ is then smaller than (15-19); probably, a good estimate is to substitute kr_α instead of ℓ in (15-19).

Fortunately, the inelastic effect for $b < a$ is relatively unimportant because there the phase shift is expected to be larger than $1 - a_\ell$, as can be seen from the following argument. The semi-classical approximation for the phase shift δ_ℓ gives

$$\delta_\ell = (\hbar v)^{-1} \int_{-\infty}^{\infty} V[(b^2 + z^2)^{1/2}]\, dz \qquad (15\text{-}23)$$

where $b = \ell/k$ and V(r) is the potential. Roughly, (15-23) gives

$$\delta_\ell \approx \frac{bV(b)}{\hbar v} = Z_p(b)\frac{e^2}{\hbar v} = \frac{Z_p(b)}{ka_0} \qquad (15\text{-}24)$$

where $Z_p(r)$ is the "effective nuclear charge for the potential" used by Hartree, see p. 75. If the "atomic radius" is defined as in (15-17), i.e., as the radius of the outermost shell, we expect approximately

$$Z_p(a) = \tfrac{1}{2}z_1 \qquad (15\text{-}25)$$

where z_1 is the number of electrons in this shell. Then

$$\delta_{\ell = ka} \approx \frac{z_1}{2ka_0} \tag{15-26}$$

On the other hand, if we take only the term $\alpha = 1$ in (15-19) (i.e., only the outermost shell which gives the largest contribution), then

$$1 - a_{\ell = ka} \approx \frac{2}{3} \frac{z_1}{(ka_0)^2} \ll \delta_{\ell = ka} \tag{15-27}$$

since we assume

$$ka_0 \gg 1 \tag{15-28}$$

Thus for $b < a$, the phase shift dominates, while for $b > a$, it goes rapidly to zero.

We may write the scattering amplitude for elastic scattering as

$$f = \frac{1}{k} \sum_{\ell} (2\ell + 1) \sin \delta_{\ell} \, e^{i\delta_{\ell}} P_{\ell} (\cos \theta)$$

$$+ \frac{i}{2k} \sum_{\ell} (2\ell + 1) e^{2i\delta_{\ell}} (1 - a_{\ell}) P_{\ell} (\cos \theta)$$

This expression is identical with (15-7). We have written η_{ℓ} as $a_{\ell} e^{2i\delta_{\ell}}$; and the total amplitude has been separated into two terms, the first being of the same form as in pure elastic scattering; and the second term representing the effect of inelastic scattering on the elastic amplitude. We shall call this the *absorptive part* of the elastic amplitude. The first term we shall continue to call the ***elastic part.*** According to our discussion, the sum in the elastic part may be terminated at $\ell = ka$, as the phase shifts δ_{ℓ} are negligible beyond this value of ℓ. The sum in the absorptive part extends to $\ell_1 \gg ka$. If we divide the absorptive sum into two terms $0 \le \ell \le ka$ and $ka \le \ell \le \ell_1$, we may set δ_{ℓ} to zero in the second term. Thus we may write (15-7) in the form

$$f = f_1 + f_2 + f_3 \tag{15-29}$$

$$f_1 = \frac{1}{k} \sum_{0}^{ka} (2\ell + 1) \sin \delta_{\ell} \, e^{i\delta_{\ell}} P_{\ell} (\cos \theta) \tag{15-30}$$

$$f_2 = \frac{i}{2k} \sum_{ka}^{\ell_1} (1 - a_{\ell})(2\ell + 1) P_{\ell} (\cos \theta) \tag{15-31}$$

$$f_3 = \frac{i}{2k} \sum_{0}^{ka} (1 - a_\ell)(2\ell + 1)\, e^{2i\delta_\ell}\, P_\ell(\cos\theta) \tag{15-32}$$

(15-30) is the usual elastic scattering without absorption; (15-31) is the absorptive effect on the total elastic amplitude due to inelastic processes arising from distant collisions; and (15-32) is the absorptive effect due to close collisions. We shall show below that f_3 is unimportant.

We now examine f_2. In the expression for f_2, (15-31), we may insert (15-19) for $1 - a_\ell$. Since ka $>>$ 1, we may neglect 1 compared with ℓ and replace the sum by an integral, thus

$$f_2 = \frac{iC}{k} \int_{ka}^{\ell_1} \frac{\ell\, d\ell}{\ell^2}\, P_\ell(\cos\theta) \tag{15-33}$$

Since all ℓ's are large, this integral will be appreciable only for small θ. It is therefore a good approximation to write

$$P_\ell(\cos\theta) \approx J_0(\ell\theta) \tag{15-34}$$

so that[4]

$$f_2 = \frac{iC}{k} \int_{x_1}^{x_2} \frac{dx}{x}\, J_0(x) \tag{15-35a}$$

$$x_1 = ka\theta \tag{15-35b}$$

$$x_2 = \ell_1\theta = 2k^2a^2\theta \tag{15-35c}$$

Since the energy is high, equation (15-28), we have also ka $>>$ 1 and

$$x_2 >> x_1 \tag{15-35d}$$

We have then 3 regions to consider:

(a) $x_1 >> 1$. In this case, the integral (15-35) is small. In other words, the absorptive scattering is small outside the diffraction region, i.e., for

$$\theta > \frac{1}{ka} \tag{15-36}$$

Thus we have a typical case of shadow scattering.

---

[4]We replace r_α in ℓ_1 by a for simplicity.

(b) $x_1 \ll 1 \ll x_2$. In this case, we may replace x_2 by ∞ in (15-35), the error being of order $x_2^{-3/2}$. Then

$$\int_{x_1}^{\infty} \frac{dx}{x} J_0(x) = -\ln ka\theta + 0.11593 + \int_0^{x_1} \frac{dx}{x}(1 - J_0(x)) \tag{15-37}$$

The number 0.11593 represents $(\ln 2) - \epsilon$, where ϵ is Euler's constant. The last integral is very small.

(c) $x_2 < 1$. In this case, we may replace $J_0(x)$ by 1, i.e., we have the constant, forward cross section,

$$f_2(0) = \frac{iC}{k} \ln \frac{x_2}{x_1} = \frac{iC}{k} \ln 2ka \tag{15-38}$$

In f_3, we assume, as previously discussed

$$1 - a_\ell = \frac{C}{(ka)^2} \tag{15-39}$$

An upper limit will be obtained if we replace δ_ℓ in (15-32) by 0. In cases (b) and (c), i.e., if $x_1 \ll 1$, we may replace P_ℓ by 1. Then

$$-if_3 < \frac{C}{2k} \tag{15-40}$$

which amounts to adding $\frac{1}{2}$ to (15-37) and to the log in (15-38). Thus f_3 is indeed not very important. However, if (15-26) is small (Born approximation good), then the right-hand side of (15-40) is a good estimate of $-if_3$.

An alternative way to derive the effect of the inelastic processes on the *forward* elastic scattering amplitude is from the optical theorem, which states that

$$\mathrm{Im}\, f(0) = \frac{k}{4\pi} \sigma_{\mathrm{tot}} \tag{15-41}$$

σ_{tot} is the *total* scattering cross section; viz, it is the sum of the total elastic cross section σ_{el}, and the total inelastic cross section σ_{inel}. The total elastic cross section includes, of course, the shadow effects of inelastic scattering; while the total inelastic cross section measures the inelastic processes themselves. We may therefore use (15-41) to calculate the contribution to $\mathrm{Im}\, f(0)$ of σ_{inel}, which we now proceed to do.

The total inelastic scattering cross section is from (15-9), (15-19), (15-20), and (15-39).

$$\sigma_{\text{inel}} = \frac{\pi}{k^2}\, 2C \left[\sum_{ka}^{\ell_1} \frac{(2\ell + 1)}{\ell^2} + \sum_{0}^{ka} \frac{(2\ell + 1)}{(ka)^2} \right]$$

$$\approx \frac{4\pi}{k^2}\, C\left(\ln \frac{\ell_1}{ka} + \frac{1}{2} \right)$$

$$= \frac{4\pi}{k^2}\, C(\ln 2ka + \tfrac{1}{2}) \qquad (15\text{-}42)$$

Inserting C from (15-19), and replacing a inside the logarithm by r_α, as we should according to (15-20), (15-22), we get

$$\sigma_{\text{inel}} = \frac{8\pi}{3k^2 a_0^2} \sum_\alpha z_\alpha r_\alpha^2 (\ln 2kr_\alpha + \tfrac{1}{2}) \qquad (15\text{-}43)$$

Calling the contribution to Im f(0) due to the inelastic total cross section $[\text{Im } f(0)]_{\text{inel}}$, we have according to (15-41) and (15-42)

$$[\text{Im } f(0)]_{\text{inel}} = \frac{k}{4\pi}\, \sigma_{\text{inel}}$$

$$= \frac{C}{k}(\ln 2ka + \tfrac{1}{2}) \qquad (15\text{-}44)$$

This agrees with what our previous calculation of $f_2(0) + f_3(0)$, equations (15-38) and (15-40), would have given

We shall now compare our forward shadow scattering, (15-44), with other contributions to the forward scattering. There are two of these, (a) the real part of the forward elastic amplitude, and (b) the contribution of the elastic cross section to the imaginary part, as given by the optical theorem (15-41). We first calculate the latter, for which we need the total elastic cross section. In the Born approximation, we have the formula (13-17)

$$\sigma_{\text{el}} = \frac{8\pi}{k^2 a_0^2} \int_0^{2k} \frac{dq}{q^3} (Z - F(q))^2 \qquad (15\text{-}45)$$

where the form factor is given by

$$F(q) = \int \rho(r) \frac{\sin qr}{qr}\, d\tau \qquad (15\text{-}46a)$$

With our assumption of separate electron shells α, we have approximately

$$\rho(r) = \sum_{\alpha} \frac{z_\alpha}{4\pi r^2}\,\delta(r - r_\alpha) \tag{15-46b}$$

$$Z - F(q) = \sum_{\alpha} z_\alpha \left(1 - \frac{\sin qr_\alpha}{qr_\alpha}\right) \tag{15-46c}$$

For any given q, usually one of the shells α dominates, viz., that for which $qr_\alpha \approx 1$ to 3. Accordingly, we evaluate (15-45) by adding the squares of the terms in (15-46c), leaving out the mixed terms; this underestimates (15-45), but probably not greatly. Then the integral in (15-45) can be evaluated and we get

$$\sigma_{el} = \frac{8\pi}{3k^2 a_0^2} \sum_{\alpha} z_\alpha^2\, r_\alpha^2 \left(\ln 2 + \tfrac{1}{12}\right) \tag{15-47}$$

The outermost electrons ($\alpha = 1$, $r_\alpha = a$) give the main contribution, both here and in (15-43).

For any given ℓ, or given impact parameter $b = \ell/k$, the Born approximation is fairly good as long as the phase shift is less than one. According to (15-24), this means for a given b

$$Z_p(b) \equiv \frac{bV(b)}{e^2} < ka_0 \tag{15-48a}$$

For (15-47) to be a good approximation, it is necessary that (15-48a) be fulfilled for $b = r_1 = a$; then the contribution of the outermost shell, $\alpha = 1$, to (15-47) is correctly given, and those of the inner shells are small by comparison. Therefore, we must have

$$Z_p(a) < ka_0 \tag{15-48b}$$

[If (15-48b) is not fulfilled, a rough approximation is obtained by the assumption that all phase shifts up to $\ell = ka$ are large; then $\sin^2 \delta$ is on the average $\frac{1}{2}$ for $\ell < ka$, and we obtain, e.g., from (15-30),

$$\sigma_{el} \approx 2\pi a^2 \tag{15-49}$$

This includes the shadow scattering due to the elastic collisions. An alternative criterion for the Born approximation is that (15-47) should be less than (15-49) which requires

$$ka_0 > z_1 \tag{15-50}$$

a condition very similar to (15-48b).]

Taking just the shell of largest radius, $\alpha = 1$, in (15-47) and (15-43), we get

$$\frac{\sigma_{\text{inel}}}{\sigma_{\text{el}}} = \frac{1.29}{z_1} (\ln 2ka + \tfrac{1}{2}) \qquad (15\text{-}51)$$

The ratio of inelastic to elastic scattering increases slowly with energy. However, it is never very large unless the number of electrons in the outermost shell, z_1, is very small. Therefore the shadow scattering discussed here should be most important for H and He. Indeed it is for He that Massey and Mohr[3] noticed the large forward scattering, and interpreted it as shadow scattering.

Equation (15-51) gives only the ratio

$$\frac{[\text{Im } f(0)]_{\text{inel}}}{[\text{Im } f(0)]_{\text{el}}} = \frac{\sigma_{\text{inel}}}{\sigma_{\text{el}}} \qquad (15\text{-}52)$$

Here, $[\text{Im } f(0)]_{\text{el}}$ means of course the imaginary part of the forward amplitude of *elastic* scattering which is due to the elastic cross section. We must still compare this with the real part of f(0). The latter is given by the Born approximation formula (13-11), with $q \to 0$. Using (15-46c) for $Z - F(q)$, and using in (13-11) $z = -1$, we have

$$\text{Re } f(0) = \frac{2}{a_0} \lim_{q \to 0} \frac{1}{q^2} (Z - F) = \frac{2}{6a_0} \sum_{\alpha} z_{\alpha} r_{\alpha}^2$$

$$\approx \frac{z_1}{3} \frac{a^2}{a_0} \qquad (15\text{-}53)$$

Comparing with (15-44), (15-19)

$$\frac{[\text{Im } f(0)]_{\text{inel}}}{\text{Re } f(0)} = \frac{2}{ka_0} (\ln 2ka + \tfrac{1}{2}) \qquad (15\text{-}54)$$

independent of the z_{α} and r_{α}. This has a maximum for rather low energy, $ka \approx 1$ where, however, the theory is not valid. In the range of validity, the ratio (15-54) decreases with increasing energy; for $ka = 3$, $a = a_0$ (about 120 eV), it is 1.5. Thus the shadow scattering, i.e., $[\text{Im } f(0)]_{\text{inel}}$ *can* be larger than the Born approximation, in agreement with experiment, Figure 13-1 for 100 eV. The same figure shows also the relative decrease of shadow scattering from helium with increasing energy, according to experiment.

Our theory is particularly simple for hydrogen and helium. These atoms have only a single electron shell, so that the estimate (15-47) for the elastic scattering, and (15-51) for the ratio, are both good (better expressions could easily be obtained for both, using explicit wave functions for the atomic electrons). Moreover, the Born approximation is valid already for very low electron energy. Then the contribution f_1 in (15-30) is purely real (ordinary eleastic scattering), while f_2 and f_3 are purely imaginary (absorptive scattering). The differential elastic cross section is simply

$$\frac{d\sigma}{d\Omega} = f_1^2 + |f_2 + f_3|^2 \tag{15-55}$$

The forward peak due to $f_2 + f_3$ appears in its purest form. For the quantitative results, see Massey and Mohr.[3]

For alkalis and similar atoms, the contributions of the *two* outer shells must be taken into account in (15-43) and (15-47), because the next-to-outer shell has many more electrons (8 or 18) than the outermost one. For noble gases other than helium, z_1 is large, 8 or 18, so $\sigma_{inel} < \sigma_{el}$ in the experimental region. Thus the imaginary forward scattering is *mostly* due to the elastic total cross section, cf. (15-51). Nevertheless, the narrow peak, of angular width 1/ka, which represents the "shadow" of the inelastic scattering, should still be noticeable over the smoother background of the "purely elastic" scattering $|f_1|^2$, when experiments are extended to really small angles. For the experiments at angles $> 15°$ shown in Figure 13-2, the theory without shadow scattering seems sufficient.

A similar effect should exist in the scattering of very fast (>100 MeV) electrons by nuclei. The excitation of the giant resonance is a dipole interaction which can therefore occur for relatively distant collisions of the electron. This also should give rise to a sharply forward-peaked shadow. However, because of the strong direct Coulomb scattering, this is probably difficult to observe.

PROBLEMS

1. Obtain the scattering amplitude for a particle moving in a separable, nonlocal potential (see Chapter 4, Problem 4) of the form $V(\mathbf{r}, \mathbf{r}') = \lambda u(\mathbf{r}) u^*(\mathbf{r}')$
2. Verify that, in the Hartree approximation, correlation terms of the form $\Sigma_{i<j} \langle 0 | x_i x_j | 0 \rangle$, occurring in (15-16a), vanish. What is the contribution of such terms in the Hartree-Fock approximation?
3. Verify the approximations leading to equation (15-37).
4. Verify (15-42).

Chapter 16

ELASTIC SCATTERING OF SPIN $\frac{1}{2}$ PARTICLES

In this Chapter we discuss the nonrelativistic theory of the elastic scattering of spin $\frac{1}{2}$ particles of mass μ in a spin-dependent potential which contains an $\boldsymbol{\ell}\cdot\mathbf{s}$ term. The Schrödinger equation for the particle is

$$\left(-\frac{\hbar^2}{2\mu}\nabla^2 + V(r) + 2\boldsymbol{\ell}\cdot\mathbf{s}\,U(r) - E\right)\Psi = 0 \qquad (16\text{-}1)$$

For simplicity V and U are taken to be spherically symmetric. Ψ, which describes the scattered particle, depends on space and spin coordinates, and the spin is $\frac{1}{2}$. Although orbital angular momentum $\boldsymbol{\ell}$ and spin $\mathbf{s}$ are no longer conserved, $\mathbf{j}$ as well as ℓ^2 and s^2 remain constants of the motion. Equation (16-1) separates by setting $\Psi = \eta_{\ell j}(r)/r\,|jm\rangle$ where $|jm\rangle$ is an eigenfunction of j^2 and j_z, as well as of ℓ^2 and s^2. The radial equation is

$$\left\{\frac{d^2}{dr^2} + k^2 - \frac{\ell(\ell+1)}{r^2} - \frac{2\mu}{\hbar^2}\left[V(r) + \lambda_{\ell j}U(r)\right]\right\}\eta_{\ell j} = 0 \qquad (16\text{-}2a)$$

with

$$k^2 = \frac{2\mu E}{\hbar^2} \qquad (16\text{-}2b)$$

$$\begin{aligned}\lambda_{\ell j} &= j(j+1) - \ell(\ell+1) - s(s+1)\\ &= \ell \qquad \text{when } j = \ell + \tfrac{1}{2}\\ &= -(\ell+1) \qquad \text{when } j = \ell - \tfrac{1}{2}\end{aligned} \qquad (16\text{-}2c)$$

The spin-orbit potential $\lambda_{\ell j} U(r)$ is seen to be the angular and spin expectation value of $2\boldsymbol{\ell}\cdot\mathbf{s}\,U(r)$ in the j^2, j_z, ℓ^2, and s^2 representation, and has different values for the two differnt modes of coupling ℓ with s. Consequently, the scattering solution will be characterized for each ℓ by two distinct phase shifts $\delta_{\ell j}$ which we call $\delta_{\ell+}$ and $\delta_{\ell-}$ for the two respective cases: $j = \ell + \frac{1}{2}$, $j = \ell - \frac{1}{2}$. The scattering solutions of (16-2a) have the asymptotic behavior

$$\eta_{\ell j}(r) \xrightarrow[r\to\infty]{} \sin\left(kr - \frac{\ell\pi}{2} + \delta_{\ell j}\right) \tag{16-3}$$

The incident wave is

$$\Psi_{in} = e^{i\mathbf{k}\cdot\mathbf{r}} \,|\, m_s\rangle \tag{16-4}$$

where the ket $|\, m_s\rangle$ represents the spin eigenstate of the incident particle: $s = \frac{1}{2}$, z component $m_s = \pm\frac{1}{2}$. We take the incident direction and the angular momentum quantization axis to coincide along the z axis. The incident wave has the usual angular decomposition

$$e^{i\mathbf{k}\cdot\mathbf{r}} \,|\, m_s\rangle = \sum_{\ell=0}^{\infty} \sqrt{4\pi(2\ell+1)}\, i^{\ell} j_{\ell}(kr)\, Y_{\ell 0}(\Omega) \,|\, m_s\rangle \tag{16-5}$$

The spin and angular part of (16-5) can be expanded in eigenstates of j^2, j_z, ℓ^2, s^2

$$Y_{\ell 0}(\Omega) \,|\, m_s\rangle = \sum_j |\, jm\rangle\langle jm \,|\, \ell 0 \tfrac{1}{2} m_s\rangle \tag{16-6}$$

Here $\langle jm \,|\, \ell m_\ell s m_s\rangle$ is the Clebsch-Gordon coefficient connecting the two representations.

Since we chose the incident direction and the angular momentum quantization axis to coincide along the z direction, we have $m_\ell = 0$ and $m_s = m$. Therefore in the following we shall frequently replace m_s by m.

The spherical Bessel function $j_\ell(kr)$ has the asymptotic form

$$j_\ell(kr) \xrightarrow[r\to\infty]{} \frac{\sin\left(kr - \frac{\ell\pi}{2}\right)}{kr} \tag{16-7}$$

Thus (16-5) has the asymptotic behavior

$$e^{i\mathbf{k}\cdot\mathbf{r}}|m\rangle \xrightarrow[r\to\infty]{} \frac{1}{2ikr}\sum_{\ell=0}^{\infty}\sqrt{4\pi(2\ell+1)}\, i^{\ell}\sum_{j}|jm\rangle\langle jm|\ell 0\tfrac{1}{2}m\rangle$$

$$\times\,[e^{i(kr-\ell\pi/2)} - e^{-i(kr-\ell\pi/2)}] \qquad (16\text{-}8)$$

The solution of (16-1) corresponding to the scattering problem can be written as

$$\Psi = \sum_{\ell=0}^{\infty}\frac{\sqrt{4\pi(2\ell+1)}}{kr}\, i^{\ell}\sum_{j} e^{i\delta_{\ell j}}\,\eta_{\ell j}(r)\,|jm\rangle\langle jm|\ell 0\tfrac{1}{2}m\rangle \qquad (16\text{-}9)$$

(16-3) gives the asymptotic behavior for Ψ

$$\Psi \xrightarrow[r\to\infty]{} \frac{1}{2ikr}\sum_{\ell=0}^{\infty}\sqrt{4\pi(2\ell+1)}\, i^{\ell}\sum_{j}|jm\rangle\langle jm|\ell 0\tfrac{1}{2}m\rangle\, e^{i\delta_{\ell j}}$$

$$\times\,[e^{i(kr-\ell\pi/2+\delta_{\ell j})} - e^{-i(kr-\ell\pi/2+\delta_{\ell j})}] \qquad (16\text{-}10)$$

$$= \frac{1}{2ikr}\sum_{\ell=0}^{\infty}\sqrt{4\pi(2\ell+1)}\, i^{\ell}\sum_{j}|jm\rangle\langle jm|\ell 0\tfrac{1}{2}m\rangle$$

$$\times\,\{[e^{i(kr-\ell\pi/2)} - e^{-(kr-\ell\pi/2)}]$$

$$+\, e^{i(kr-\ell\pi/2)}[e^{2i\delta_{\ell}} - 1]\} \qquad (16\text{-}11)$$

The term in the first square bracket is seen to be the asymptotic form of the incident wave (16-8). Recalling that the scattering amplitude f, is the coefficient of e^{ikr}/r in the asymptotic form of Ψ

$$\Psi \xrightarrow[r\to\infty]{} e^{i\mathbf{k}\cdot\mathbf{r}}|m\rangle + f\,\frac{e^{ikr}}{r} \qquad (16\text{-}12)$$

we obtain from (16-11)

$$f = \frac{1}{2ik}\sum_{\ell=0}^{\infty}\sqrt{4\pi(2\ell+1)}\sum_{j}(e^{2i\delta_{\ell j}} - 1)\,|jm\rangle\langle jm|\ell 0\tfrac{1}{2}m\rangle \qquad (16\text{-}13)$$

Obviously f has two components. We now expand $|jm\rangle$ in the ℓ^2, ℓ_z, s^2, s_z representation

$$|jm\rangle = \sum_{m_\ell} Y_{\ell m_\ell}(\Omega)\,|m - m_\ell\rangle\langle \ell m_\ell \tfrac{1}{2}\; m-m_\ell\,|\,jm\rangle \qquad (16\text{-}14)$$

Here as before $|m - m_\ell\rangle$ represents the eigenstate of s^2, s_z with s_z eigenvalue $m - m_\ell$. Since $s = \frac{1}{2}$ only two terms occur in the sum over m_ℓ. If $m = \frac{1}{2}$, then $m_\ell = 0, 1$, while if $m = -\frac{1}{2}$, $m_\ell = 0, -1$ or in general $m_\ell = 0, 2m$. Inserting (16-14) into (16-13)

$$f = \frac{1}{2ik} \sum_{\ell j m_\ell} \sqrt{4\pi(2\ell + 1)}\,(e^{2i\delta_{\ell j}} - 1)\,Y_{\ell m_\ell}(\Omega)\,|m - m_\ell\rangle$$

$$\times \langle \ell m_\ell \tfrac{1}{2} m-m_\ell\,|\,jm\rangle\langle jm\,|\,\ell 0 \tfrac{1}{2} m\rangle \qquad (16\text{-}15)$$

Let us for definiteness assume $m = \frac{1}{2}$, viz., the incident particle is in an "up" spin state. Then the summation over m_ℓ gives terms

$$Y_{\ell 0}(\Omega)\,|\tfrac{1}{2}\rangle\,|\langle \ell 0 \tfrac{1}{2} \tfrac{1}{2}\,|\,j \tfrac{1}{2}\rangle|^2$$

$$+ Y_{\ell 1}(\Omega)\,|-\tfrac{1}{2}\rangle\langle \ell\, 1 \tfrac{1}{2} -\tfrac{1}{2}\,|\,j \tfrac{1}{2}\rangle\langle j \tfrac{1}{2}\,|\,\ell 0\, \tfrac{1}{2} \tfrac{1}{2}\rangle \qquad (16\text{-}16)$$

The Clebsch-Gordan coefficients are given in equations (6-33) and (6-34), while the spherical harmonics are

$$Y_{\ell 0}(\Omega) = \sqrt{\frac{2\ell + 1}{4\pi}}\; P_\ell(\cos\theta)$$

$$Y_{\ell \pm 1}(\Omega) = \mp \frac{1}{\sqrt{\ell(\ell + 1)}}\; e^{\pm i\varphi}\, \sin\theta\, \frac{dY_{\ell 0}}{d\cos\theta} \qquad (16\text{-}17)$$

Thus f, which we designated in the case of $+\frac{1}{2}$ incident spin state by $f_\uparrow$, becomes

$$f_\uparrow = f_1(\theta)\,|\tfrac{1}{2}\rangle + f_2(\theta)\,e^{i\varphi}\,|-\tfrac{1}{2}\rangle \qquad (16\text{-}18a)$$

$$f_1(\theta) = \frac{1}{4ik} \sum_{\ell, j} (2j + 1)(e^{2i\delta_{\ell j}} - 1)\,P_\ell(\cos\theta) \qquad (16\text{-}18b)$$

$$f_2(\theta) = \frac{\sin\theta}{2ik} \sum_{\ell} (e^{2i\delta_{\ell +}} - e^{2i\delta_{\ell -}})\, \frac{d}{d\cos\theta}\, P_\ell(\cos\theta) \qquad (16\text{-}18c)$$

where $\delta_{\ell\pm} = \delta_{\ell, j=\ell\pm\frac{1}{2}}$.

An analogous derivation for the case of incident spin state $-\frac{1}{2}$ gives for $f_\downarrow$

$$f_\downarrow = f_1(\theta)|-\tfrac{1}{2}\rangle - f_2(\theta)e^{-i\varphi}|\tfrac{1}{2}\rangle \tag{16-19}$$

with the same f_1 and f_2 as before.

It is clear that f_1 represents the probability amplitude for scattering with conserved spin direction, whereas f_2 is the spin flip amplitude. The $e^{\pm i\varphi}$ occurring with f_2 represents total j_z conservation. The $\sin\theta$ in f_2 indicates that there is no spin flip in the forward or backward direction which also is a consequence of j_z conservation. The formulas (16-18) and (16-19) reduce to the usual nonspin dependent amplitudes in the case that U is 0, i.e., there is no spin-orbit interaction. In that instance $\delta_{\ell+} = \delta_{\ell-} \equiv \delta_\ell$ and $f_2 = 0$,, while

$$f_1(\theta) = \frac{1}{2ik}\sum_\ell (2\ell+1)(e^{2i\delta_\ell} - 1)P_\ell(\cos\theta) \tag{16-20}$$

In the Born approximation, the exponentials occurring in (16-18b,c) are replaced by the first two terms of their series expansion, thus f_1 and f_2 are real.

GENERALIZATION

If the spin of the incident particle points in an arbitrary direction, the incident particle will be in a superposition of ↑ and ↓ states, say

$$e^{i\mathbf{k}\cdot\mathbf{r}}|\xi\rangle = e^{i\mathbf{k}\cdot\mathbf{r}}[a|\tfrac{1}{2}\rangle + b|-\tfrac{1}{2}\rangle] \tag{16-21a}$$

$$|a|^2 + |b|^2 = 1 \tag{16-21b}$$

Then the superposition principle gives the total scattering amplitude f_ξ

$$\begin{aligned} f_\xi = {} & (af_1(\theta) - bf_2(\theta)e^{-i\varphi})|\tfrac{1}{2}\rangle \\ & + (bf_1(\theta) + af_2(\theta)e^{i\varphi})|-\tfrac{1}{2}\rangle \end{aligned} \tag{16-22}$$

Using a specific representation for the state $|\pm\frac{1}{2}\rangle$; i.e., $|\frac{1}{2}\rangle = \binom{1}{0}$ $|-\frac{1}{2}\rangle = \binom{0}{1}$, the general state $|\xi\rangle$ can be expressed as

$$|\xi\rangle = \begin{pmatrix} a \\ b \end{pmatrix} \tag{16-23}$$

and (16-22) can be written in matrix form

$$f_\xi = S\,|\,\xi\rangle \tag{16-24a}$$

$$S = \begin{bmatrix} f_1(\theta) & -f_2(\theta)\,e^{-i\varphi} \\ f_2(\theta)\,e^{i\varphi} & f_1(\theta) \end{bmatrix} \tag{16-24b}$$

Since S is a 2×2 matrix it can be expanded in terms of the complete set of four 2×2 matrices (I, σ) where I is the unit matrix and σ_i are the three Pauli matrices. It is found that

$$S = g(\theta)I + h(\theta)\,\sigma\cdot\hat{n} \tag{16-25}$$

with

$$\hat{n} = (-\sin\varphi,\ \cos\varphi,\ 0) \tag{16-26a}$$

$$g(\theta) = f_1(\theta) \tag{16-26b}$$

$$h(\theta) = -if_2(\theta) \tag{16-26c}$$

The unit vector $\hat{n}$ is perpendicular to the scattering plane and is given by[1]

$$\hat{n} = \frac{\mathbf{k}_i \times \mathbf{k}_f}{|\mathbf{k}_i \times \mathbf{k}_f|} \tag{16-27}$$

The differential cross section is obtained by the usual method of determining the flux of the scattered beam, and is

$$\frac{d\sigma}{d\Omega} = |f_\xi|^2 = \langle \xi\,|\,S^+S\,|\,\xi\rangle \tag{16-28}$$

---

[1]That $\hat{n}$ occurring in (16-25) is a (pseudo) vector follows from the fact that S is a scalar and σ a (pseudo) vector under rotations and reflections. Of course, we can write any 2×2 matrix as $gI + \Sigma_i\sigma_i h_i$. However, the conclusion that the three numbers h_i form a (pseudo) vector follows from the transformation properties of the S matrix. Similar considerations apply in the expansion of the density matrix, p. 288, below. Indeed many of the conclusions about the form of the various quantities we consider in this chapter: S matrix, density, matrix, cross section, follow from symmetry arguments. For a discussion of these points, as well as more complete treatment of spin in scattering problems see, M. L. Goldberger and K. M. Watson, *Collision Theory*, Krieger, Melbourne FL, 1975.

If the incident particle is described by $\begin{pmatrix} a \\ b \end{pmatrix}$, equation (16-23), then the quantities $|a|^2$ and $|b|^2$ are the probabilities that a measurement of the z component of the spin will result in $\frac{1}{2}$ and $-\frac{1}{2}$, respectively. However, there exists always some direction $\hat{\eta}$ such that

$$(\sigma \cdot \hat{\eta}) | \xi \rangle = | \xi \rangle \tag{16-29}$$

Thus if the spin component is measured in direction $\hat{\eta}$ the result will always be $+\frac{1}{2}$. We call a particle beam described by such a state, fully polarized. Most particle beams are not fully polarized; in fact the most usual case is that of an unpolarized beam. In this case, if the spin component is measured in *any* direction, we are equally likely to find $+\frac{1}{2}$ and $-\frac{1}{2}$.

To calculate the scattering of an unpolarized beam, we merely need to take the average of (16-28) for two states $\xi_{1,2}$ which represent opposite polarization, e.g., for one state ξ_1 such that $\sigma \cdot \hat{n} | \xi_1 \rangle = | \xi_1 \rangle$ and another such that $\sigma \cdot \hat{n} | \xi_2 \rangle = -| \xi_2 \rangle$. The result, according to (16-25), is

$$\left(\frac{d\sigma}{d\Omega} \right)_{\text{unpol.}} = | g(\theta) |^2 + | h(\theta) |^2 \tag{16-30}$$

It is also possible to calculate the scattering of partially polarized beams in this manner.

DENSITY MATRIX

It is, however, useful to introduce a more powerful method for the discussion of problems of partial polarization, the ***density matrix.*** We shall first introduce this concept for a ***pure state,*** viz., the state described by $| \xi \rangle = \begin{pmatrix} a \\ b \end{pmatrix}$. The density matrix ρ for the system described by ξ is defined as

$$\rho(\xi) = | \xi \rangle \langle \xi | = \begin{bmatrix} |a|^2 & ab^* \\ ba^* & |b|^2 \end{bmatrix} \tag{16-31}$$

Evidently ρ is a 2×2 Hermitian matrix, with $\text{Tr}\, \rho = 1$. The expectation value in the state $| \xi \rangle$ of any operator A can be obtained with the help of $\rho(\xi)$. Thus

$$\langle \xi | A | \xi \rangle = \sum_{ij} a_i^* A_{ij} a_j = \sum_{ij} a_j a_i^* A_{ij}$$

$$= \sum_{ij} \rho_{ji} A_{ij} = \mathrm{Tr}[\rho(\xi)A] \tag{16-32}$$

where we have represented $|\xi\rangle$ as $\begin{pmatrix} a_1 \\ a_2 \end{pmatrix}$ for convenience. Since ρ is a 2×2 Hermitian matrix, it can be expanded in the complete set (I, σ) with real coefficients. We write this expansion

$$\rho(\xi) = \tfrac{1}{2}(I + \mathbf{P} \cdot \sigma) \tag{16-33}$$

where

$$P_x = 2\,\mathrm{Re}\,(a^* b)$$

$$P_y = 2\,\mathrm{Im}\,(a^* b)$$

$$P_z = |a|^2 - |b|^2 \tag{16-34}$$

The real vector $\mathbf{P}$, called the *polarization vector*, has a simple physical significance. In our example it has unit magnitude. This follows from

$$\rho^2 = |\xi\rangle\langle\xi|\xi\rangle\langle\xi| = |\xi\rangle\langle\xi| = \rho$$

Therefore

$$\tfrac{1}{2}(I + \mathbf{P}\cdot\sigma) = \tfrac{1}{4}(I + \mathbf{P}\cdot\sigma)(I + \mathbf{P}\cdot\sigma) = \tfrac{1}{4}(I + 2\mathbf{P}\cdot\sigma + (\mathbf{P}\cdot\sigma)(\mathbf{P}\cdot\sigma))$$

$$= \frac{1}{2}\left(I\,\frac{[1 + P^2]}{2} + \mathbf{P}\cdot\sigma\right) \quad \text{or} \quad P^2 = 1 \tag{16-35}$$

where use has been made of the relation

$$(\mathbf{a}\cdot\sigma)(\mathbf{b}\cdot\sigma) = \mathbf{a}\cdot\mathbf{b} + i\mathbf{a} \times \mathbf{b}\cdot\sigma \tag{16-36}$$

Also using (16-32), $\langle\xi|\sigma|\xi\rangle = \mathrm{Tr}\,\rho\sigma = \mathbf{P}$,

$$\mathbf{P} = \langle\xi|\sigma|\xi\rangle \tag{16-37}$$

Thus $\mathbf{P}$ is the direction of the particle's spin. Furthermore, from (16-34) we see that P_z measures the ratio

$$\frac{|a|^2 - |b|^2}{|a|^2 + |b|^2} = \frac{\text{spin up} - \text{spin down}}{\text{spin up} + \text{spin down}}$$

where up and down refer to the z direction.

Applying (16-25), (16-28), and (16-32) to (16-37), we get

$$\frac{d\sigma}{d\Omega} = \mathrm{Tr}\, \rho(\xi)S^{+}S = \mathrm{Tr}\, \tfrac{1}{2}[I + \mathbf{P}\cdot\sigma][g^*(\theta) + h^*(\theta)\sigma\cdot\hat{n}]$$

$$\times\, [g(\theta) + h(\theta)\sigma\cdot\hat{n}]$$

$$= |g|^2 + |h|^2 + 2\,\mathrm{Re}\, g^* h\, \mathbf{P}\cdot\hat{n} \qquad (16\text{-}38)$$

We recall that $\mathbf{P}$ in (16-38) is the polarization vector for the *incident* particle.

Of particular interest is that incident spin state $|\chi\rangle$ which is not affected by the scattering, viz.,

$$S\,|\chi\rangle = \lambda\,|\chi\rangle \qquad (16\text{-}39)$$

The secular equation for (16-39) follows from (16-24b)

$$(f_1 - \lambda)^2 + f_2^2 = 0 \qquad (16\text{-}40a)$$

with solutions

$$\lambda = f_1 \mp i f_2 \qquad (16\text{-}40b)$$

and

$$|\chi\rangle = \frac{1}{\sqrt{2}}\begin{pmatrix} 1 \\ \pm i e^{i\varphi} \end{pmatrix} \qquad (16\text{-}40c)$$

Inserting this into (16-34) and using (16-26a), we obtain

$$\mathbf{P}_\chi = \pm\hat{n} \qquad (16\text{-}41)$$

where $\hat{n}$ is the unit vector perpendicular to the scattering plane given by (16-27). We conclude that when the direction of the spin is initially perpendicular to the scattering plane, it is not affected by the scattering. Using (16-38), the differential cross section for this case is

$$\frac{d\sigma}{d\Omega} = |g|^2 + |h|^2 \pm 2\,\mathrm{Re}\, g^* h \qquad (16\text{-}42a)$$

$$= \sigma_0(1 \pm P_f) \qquad (16\text{-}42b)$$

$$\sigma_0 = |g|^2 + |h|^2 \tag{16-42c}$$

$$P_f = \frac{2\,\mathrm{Re}\,g^*h}{|g|^2 + |h|^2} \tag{16-42d}$$

In the Born approximation g is real and h imaginary, equations (16-26) and (16-18), hence $P_f = 0$. The reason for calling (16-42d) by the name P_f will become apparent below.

We saw that the density matrix was a convenient formal expression in the analysis of scattering when the incident beam was described by a definite spin state $|\xi\rangle$. Its full utility is in the description of a beam which is made up of many (say, N) particles, each with its own spin state $|\xi\rangle_\alpha$ and density matrix ρ_α. The beam is no longer in a definite spin state. We can then define the average density matrix

$$\overline{\rho} = \frac{1}{N}\sum_{\alpha=1}^{N} \rho_\alpha \tag{16-43}$$

The average expectation value of an operator A is then

$$\langle\overline{A}\rangle = \frac{1}{N}\sum_{\alpha=1}^{N} {}_\alpha\langle\xi|A|\xi\rangle_\alpha = \frac{1}{N}\sum_{\alpha=1}^{N} \mathrm{Tr}\,\rho_\alpha A = \mathrm{Tr}\,\overline{\rho}A \tag{16-44}$$

Since $\overline{\rho}$ is still a 2×2 Hermitian matrix, it can be expanded in the complete set (I, σ) with real coefficients. Clearly,

$$\overline{\rho} = \frac{1}{N}\sum_{\alpha=1}^{N} \tfrac{1}{2}(I + \mathbf{P}_\alpha\cdot\sigma) = \tfrac{1}{2}(I + \overline{\mathbf{P}}\cdot\sigma) \tag{16-45}$$

where

$$\overline{\mathbf{P}} = \frac{1}{N}\sum_{\alpha=1}^{N} \mathbf{P}_\alpha \tag{16-46}$$

$\overline{\mathbf{P}}$ gives the average direction of the spin in the incident beam since

$$\langle\overline{\sigma}\rangle = \mathrm{Tr}\,\overline{\rho}\sigma = \overline{\mathbf{P}} \tag{16-47}$$

Its magnitude lies between 0 and 1. When $|\overline{\mathbf{P}}| = \overline{P} = 0$, we say the beam is unpolarized and $\overline{\rho} = \tfrac{1}{2}I$. When $\overline{P} = 1$, the beam is fully polarized and $\bar{\rho} = \begin{pmatrix} 1 & 0 \\ 0 & 0 \end{pmatrix}$ (in some basis). We see that when all the particles are in the same spin state, the beam is fully polarized.

The differential cross section must now be averaged over all the initial states present in the beam.

$$\frac{d\sigma}{d\Omega} = \frac{1}{N} \sum_{\alpha=1}^{N} {}_{\alpha}\langle \xi | S^{+}S | \xi \rangle_{\alpha} = \mathrm{Tr}\, \bar{\rho}\, S^{+}S$$

$$= |g|^2 + |h|^2 + 2\,\mathrm{Re}\, g^{*}h\, \overline{\mathbf{P}} \cdot \hat{\mathbf{n}} \tag{16-48}$$

When the beam is unpolarized, $\overline{\mathbf{P}} = 0$ and the cross section is given by

$$\frac{d\sigma}{d\Omega} = |g|^2 + |h|^2 \equiv \sigma_0 \tag{16-49}$$

After the scattering, the incident states $|\xi\rangle_{\alpha}$ become $S|\xi\rangle_{\alpha}$ and the density matrix of the scattered beam is

$$\bar{\rho}_{fu} = \frac{1}{N} \sum_{\alpha=1}^{N} S|\xi\rangle_{\alpha\,\alpha}\langle \xi | S^{+} = S\bar{\rho}_0 S^{+} \tag{16-50}$$

If the initial beam was unpolarized, $\bar{\rho}_0 = \frac{1}{2}I$, and

$$\bar{\rho}_{fu} = \tfrac{1}{2} SS^{+} = \tfrac{1}{2}(g + h\hat{\mathbf{n}}\cdot\sigma)(g^{*} + h^{*}\hat{\mathbf{n}}\cdot\sigma)$$

$$= \tfrac{1}{2}\sigma_0 (I + \overline{\mathbf{P}}_f \cdot \sigma) \tag{16-51a}$$

Here σ_0 is the cross section for an unpolarized incident beam (16-49). $\overline{\mathbf{P}}_f$ the polarization of the scattered beam is

$$\overline{\mathbf{P}}_f = P_f \hat{\mathbf{n}} \tag{16-51b}$$

where P_f is given[2] by (16-42d) and $\hat{n}$ by (16-27). The presence of $\sigma_0 = |g|^2 + |h|^2$ as an overall factor in (16-51a) means that the final states $S|\xi\rangle_{\alpha}$ are unnormalized; this is indicated by the subscript u; to normalize, divide by σ_0 and get

$$\bar{\rho}_f = \tfrac{1}{2}(I + P_f \hat{n}\cdot\sigma) \tag{16-51c}$$

If the beam is scattered a second time (see Figure 16-1), we have

---

[2]The reason for the notation P_f, (16-42d), final polarization, is now clear.

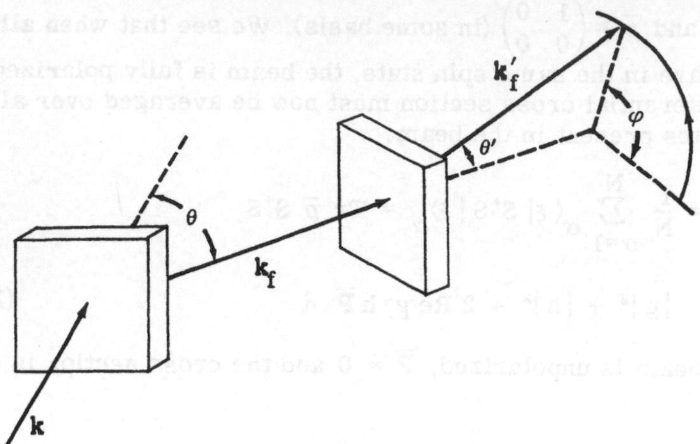

Figure 16-1

Geometry of the double-scattering experiment for the detection of polarization. The initially unpolarized beam is incident on the first target with wave vector **k**. It is scattered through the angle θ, taking the new direction $\mathbf{k}_f$. It is scattered a second time in the second target, through the angle θ' into the direction $\mathbf{k}'_f$, specified by θ' and by the azimuthal angle φ. Any polarization resulting from the first scattering is now analyzed by observing the dependence of the intensity of the twice-scattered beam upon the angle φ. (After Bethe and Morrison[3], p. 140.)

$$\bar{\rho}_{f'} = \frac{S' \bar{\rho}_f S'^{+}}{\sigma'_0} \tag{16-52a}$$

$$S' = (g' + h'\mathbf{n}' \cdot \boldsymbol{\sigma}) \tag{16-52b}$$

$$\hat{\mathbf{n}}' = \frac{\mathbf{k}_f \times \mathbf{k}'_f}{|\mathbf{k}_f \times \mathbf{k}'_f|} \tag{16-52c}$$

The differential cross section for the twice scattered beam is

$$\frac{d\sigma'}{d\Omega'} = \operatorname{Tr} \bar{\rho}_f S'^{+} S' \tag{16-53}$$

This has the same form as (16-18); thus

---

[3]H. A. Bethe and P. Morrison, *Elementary Nuclear Theory*, 2nd ed., Wiley, New York, 1956.

$$\frac{d\sigma'}{d\Omega'} = |g'|^2 + |h'|^2 + 2\,\mathrm{Re}(g'h'^*)\,\overline{\mathbf{P}}_f \cdot \hat{\mathbf{n}}' \qquad (16\text{-}54)$$

If the beam entering the second scatter were unpolarized, the second scatterer acting alone would produce a polarization which we call $\overline{\mathbf{P}}_f' = P_f'\hat{\mathbf{n}}'$

$$P_f' = \frac{2\,\mathrm{Re}\,(g'h'^*)}{|g'|^2 + |h'|^2} \qquad (16\text{-}55)$$

Using (16-55) and (16-51b) we may rewrite (16-54) as

$$\frac{d\sigma'}{d\Omega'} = (|g'|^2 + |h'|^2)(1 + P_f P_f' \hat{n}\cdot\hat{n}') \qquad (16\text{-}56)$$

If the experimental conditions are so arranged as to make P_f and P_f' equal (by choosing the angles θ and θ' appropriately—see Figure 16-1), then

$$\frac{d\sigma'}{d\Omega'} \sim 1 + P_f^2 \cos\varphi \qquad (16\text{-}57)$$

where φ is the azimuthal angle between $\mathbf{k}_f$ and $\mathbf{k}_f'$. Thus from the observed variation of $d\sigma'/d\Omega'$ with φ in a double scattering experiment, we can determine the polarization of the scattered beam.

PROBLEMS

1. Derive equations (16-28).
2. Show that the spin function

$$\chi = 2^{1/2}\begin{pmatrix} 1 \\ ie^{i\varphi} \end{pmatrix}$$

represents a spin pointing in the direction of the normal

$$\hat{\mathbf{n}} = \frac{\mathbf{k}_i \times \mathbf{k}_f}{|\mathbf{k}_i \times \mathbf{k}_f|}$$

The polar coordinate system is chosen such that $\mathbf{k}_i$ is the polar axis, and $\mathbf{k}_f$ has the angular coordinates θ, φ.

3. Prove that any 2×2 matrix can be expanded in terms of the complete set of four 2×2 matrices (I, σ).

4. Repeat the argument from (16-15) to (16-18) with $m = -\frac{1}{2}$, thus deriving (16-19).
5. Let ρ be an arbitrary 2×2 matrix. What are the restrictions on its elements such that $\mathrm{Tr}\,\rho = 1$, $\rho^2 = \rho$, and $\rho^\dagger = \rho$?

Chapter 17

INELASTIC SCATTERING AT HIGH ENERGIES

In the previous chapters we considered elastic scattering of the incident particle. This is characterized by the fact that the energy of the scattered particle does not change between the initial and final state. We assumed that to describe such scattering, the incident particle can be considered to be moving in an effective potential. The simplest was the high energy case, where we could use the Born approximation on a Hartree or Thomas-Fermi potential. As the energy of the particle decreases, we found that better potentials were required, such as the Hartree-Fock exchange potential and the polarization correction potential.

In this and subsequent chapters we shall study inelastic scattering, where the energy of the external particle, carrying charge ze, $e > 0$, changes between the initial and final state. The change in energy is absorbed by the Z electron atom, leading to excitation or ionization. As we shall need to describe these internal changes in the atom, we remain with the full many-body Hamiltonian describing the interactions of our system.

For the present we confine ourselves to high energy scattering where the Born approximation may be used, and derive the Born formula for inelastic scattering from first principles. As a special case, we shall arrive at the elastic scattering amplitude thus justifying the "physically reasonable" equivalent potential used in Chapter 13. As in Chapter 13 we assume the initial particle velocity to be between $(Z/137)c$ and $\frac{1}{2}c$.

The Hamiltonian for the entire problem is

$$H = H_{atom} + H_{particle} + H^I \tag{17-1}$$

H_{atom} is the Hamiltonian of the atom, $H_{particle}$ the free-particle Hamiltonian, and H^I the interaction Hamiltonian representing the interaction of the incident particle with the electrons and nucleus of the atom. From general collision theory[1] we recall that the Born approximation to the amplitude for transition from state α to state β is

$$T_{\alpha\beta} = \langle \psi_\beta | H^I | \psi_\alpha \rangle \tag{17-2}$$

where ψ is an eigenfunction of $H_{atom} + H_{particle}$, thus

$$\psi = \varphi(A)\, e^{i\mathbf{k}\cdot\mathbf{r}} \tag{17-3}$$

$\varphi(A)$ is an atomic eigenstate, with A representing the Z spatial coordinates of the atomic electrons (we are ignoring spin). The exponential is the incident particle eigenstate with momentum $\hbar\mathbf{k}$. In the case of our interest

$$\psi_\alpha = \varphi_0\, e^{i\mathbf{k}_0\cdot\mathbf{r}} \tag{17-4a}$$

$$\psi_\beta = \varphi_n\, e^{i\mathbf{k}_n\cdot\mathbf{r}} \tag{17-4b}$$

since initially the atom is in its ground state φ_0, and after the collision the atom is left in the nth excited state φ_n. The initial and final energies of the atom are E_0 and E_n, respectively; the initial and final momenta of the scattered particle are $\mathbf{p}_0 = \hbar\mathbf{k}_0$ and $\mathbf{p}_n = \hbar\mathbf{k}_n$, respectively. By energy conservation, we must have

$$\hbar^2(k_0^2 - k_n^2) = 2m(E_n - E_0) \tag{17-5a}$$

$$= -2m(W_n - W_0) \tag{17-5b}$$

where W_n is the total energy (including the rest mass) of the *incident* particle. The momentum change of the scattered particle is

$$\hbar\mathbf{q} = \hbar(\mathbf{k}_0 - \mathbf{k}_n) \tag{17-6a}$$

$$q^2 = (k_0 - k_n)^2 + 4k_0 k_n \sin^2 \tfrac{1}{2}\theta \tag{17-6b}$$

$$\cos\theta = \frac{\mathbf{k}_0\cdot\mathbf{k}_n}{k_0 k_n} \tag{17-6c}$$

Thus the Born approximation to the transition amplitude is

---

[1]For example, L. I. Schiff, *Quantum Mechanics,* 3rd ed., McGraw-Hill, New York, 1968, p. 324.

$$T_{on} = \int e^{i\mathbf{q}\cdot\mathbf{r}} \varphi_n^*(A) H^I(\mathbf{r}, A) \varphi_0(A)\, d\tau\, dA \tag{17-7a}$$

$$= -\frac{1}{q^2}\int e^{i\mathbf{q}\cdot\mathbf{r}} \nabla^2 V_n(\mathbf{r})\, d\tau \tag{17-7b}$$

where V_n is the potential.

$$V_n(\mathbf{r}) = \int \varphi_n^*(A) H^I(\mathbf{r}, A) \varphi_0(A)\, dA \tag{17-8}$$

In going from (17-7a) to (17-7b), we have used the transformation discussed on p. 244.

The interaction Hamiltonian is, of course,

$$H^I(\mathbf{r}, A) = ze\left[\frac{Ze}{r} - \sum_{i=1}^{Z} \frac{e}{|\mathbf{r} - \mathbf{r}_i|}\right] \tag{17-9}$$

Thus

$$\nabla^2 V_n(\mathbf{r}) = -4\pi e^2 z\left[Z\,\delta_{no}\,\delta(\mathbf{r}) - \sum_{i=1}^{Z}\int \varphi_n^*(A)\varphi_0(A)\,\delta(\mathbf{r} - \mathbf{r}_i)\, dA\right] \tag{17-10}$$

The δ_{no} arises from the orthonormality of φ_n and φ_0. We define

$$\rho_n(\mathbf{r}) = \sum_{i=1}^{Z}\int \varphi_n^*(\mathbf{r}_1, \ldots, \mathbf{r}_{i-1}, \mathbf{r}, \mathbf{r}_{i+1}, \ldots, \mathbf{r}_Z)$$

$$\times\, \varphi_0(\mathbf{r}_1, \ldots, \mathbf{r}_{i-1}, \mathbf{r}, \mathbf{r}_{i+1}, \ldots, \mathbf{r}_Z) \prod_{j\neq i} d\tau_j$$

$$= \sum_{i=1}^{Z}\int \varphi_n^*(\mathbf{r}_1, \ldots, \mathbf{r}_Z)\,\varphi_0(\mathbf{r}_1, \ldots, \mathbf{r}_Z)\,\delta(\mathbf{r} - \mathbf{r}_i)\prod_j d\tau_j \tag{17-11}$$

Then (17-10) can be written

$$\nabla^2 V_n(\mathbf{r}) = -4\pi e^2 z\,[Z\,\delta_{no}\,\delta(\mathbf{r}) - \rho_n(\mathbf{r})] \tag{17-12}$$

Thus the equivalent potential is given by

$$V_n(\mathbf{r}) = e^2 z\left[\frac{Z\delta_{no}}{r} - \int \frac{\rho_n(\mathbf{r}')}{|\mathbf{r} - \mathbf{r}'|}\, d\tau\right] \tag{17-13}$$

In particular, for elastic scattering we have n = 0, and

$$V_0(\mathbf{r}) = e^2 z \left[\frac{Z}{r} - \int \frac{\rho(\mathbf{r}')}{|\mathbf{r} - \mathbf{r}'|} \, d\tau' \right] \tag{17-14a}$$

$$\rho(\mathbf{r}) = \sum_{i=1}^{Z} \int |\phi_0(\mathbf{r}_1, \ldots, \mathbf{r}_{i-1}, \mathbf{r}, \mathbf{r}_{i+1}, \ldots, \mathbf{r}_Z)|^2 \prod_{j \neq i} d\tau_j \tag{17-14b}$$

which verifies (13-2) and (13-5).

For inelastic scattering ($n \neq 0$), combining (17-7) and (17-12)

$$T_{on} = -\frac{4\pi e^2 z}{q^2} \int e^{i\mathbf{q}\cdot\mathbf{r}} \rho_n(\mathbf{r}) \, d\tau \tag{17-15a}$$

$$= -\frac{4\pi e^2 z}{q^2} F_n(\mathbf{q}) \tag{17-15b}$$

where we have defined the transition form factor

$$F_n(\mathbf{q}) = \int e^{i\mathbf{q}\cdot\mathbf{r}} \rho_n(\mathbf{r}) \, d\tau$$

$$= \int \varphi_n^*(A) \sum_i e^{i\mathbf{q}\cdot\mathbf{r}_i} \varphi_0(A) \, dA \tag{17-16}$$

The differential cross section for inelastic scattering into the final state n is obtained from the transition amplitude by Fermi's golden rule, viz.,

$$\frac{d\sigma_n}{d\Omega} = \frac{2\pi}{\hbar v_0} |T_{on}|^2 \rho(W_n) \tag{17-17a}$$

$\rho(W_n)$ is the density of final states per unit energy per unit solid angle, which equals $p_n^2 \, dp_n/(2\pi\hbar)^3 \, dW_n$. (The usual volume term has been absorbed in the normalization of the final state wave function.) The initial and final velocities are v_0 and v_n, respectively. With $dW_n/dp_n = v_n$ and

$$p_n = \left(\frac{W_n}{c^2} \right) v_n \tag{17-17b}$$

we obtain for the differential cross section.

$$\frac{d\sigma_n}{d\Omega} = \left[\frac{2e^2 z W_n}{\hbar^2 c^2 q^2} \right]^2 \frac{v_n}{v_0} |F_n(\mathbf{q})|^2 \tag{17-18a}$$

or explicitly writing out the form factor

$$\frac{d\sigma_n}{d\Omega} = \left[\frac{2e^2 z W_n}{\hbar^2 c^2 q^2}\right]^2 \frac{v_n}{v_0} \left| \int \sum_{j=1}^{Z} \varphi_n^* \varphi_0 e^{i\mathbf{q}\cdot\mathbf{r}_j} d\tau_1 \ldots d\tau_Z \right|^2 \tag{17-18b}$$

The total cross section, in the Born approximation, for inelastic scattering into the final state n is

$$\sigma_n = \int \frac{d\sigma_n}{d\Omega} d\Omega$$

$$= \left[\frac{2e^2 z W_n}{\hbar^2 c^2}\right]^2 \frac{v_n}{v_0} \int d\Omega \frac{|F_n(q)|^2}{q^4} \tag{17-19}$$

To evaluate this, we consider

$$\frac{v_n}{v_0} d\Omega = \frac{v_n}{v_0} \sin\theta \, d\theta \, d\varphi \tag{17-20a}$$

From (17-6b)

$$q\,dq = k_n k_0 \sin\theta \, d\theta \tag{17-20b}$$

According to the relativistic formula (17-17b)

$$k_n = \frac{W_n}{\hbar c^2} v_n \tag{17-20c}$$

and

$$\frac{v_n}{v_0} d\Omega = \frac{1}{k_0^2} \frac{v_n}{k_n} \frac{k_0}{v_0} q\,dq\,d\varphi = \frac{W_0}{W_n} \frac{q\,dq\,d\varphi}{k_0^2} \tag{17-20d}$$

Then (17-19) gives for the total cross section

$$\sigma_n = 2\pi \left[\frac{2e^2 z W_0}{\hbar c^2 p_0}\right]^2 \frac{W_n}{W_0} \int_{q_{min}}^{q_{max}} \frac{F_n^2(q)}{q^3} dq \tag{17-21a}$$

with q_{max} and q_{min} given by $k_0 + k_n$ and $|k_0 - k_n|$, respectively. We have assumed spherical symmetry for F. In nearly all practical cases, $W_n = W_0$ very nearly; therefore using (17-20c) again,

$$\sigma_n = 2\pi\left(\frac{2e^2 z}{\hbar v_0}\right)^2 \int_{q_{min}}^{q_{max}} \frac{F_n^2(q)}{q^3}\, dq \tag{17-21b}$$

We shall show in Chapter 19 that this formula remains valid when some of the conditions for the Born approximation are no longer satisfied.

We recall that φ_n, φ_0 describe the electron configuration of the atom. In particular, if they are approximated by Slater determinants, the observation that $\Sigma_{j=1}^{Z}\, e^{i\mathbf{q}\cdot\mathbf{r}_j}$ is a one-electron operator leads to the conclusion that no more than one orbital can change between the initial and final states.

We estimate the limits q_{min} and q_{max}

$$q^2 = k_0^2 + k_n^2 - 2k_0 k_n \cos\theta$$

$$q_{min} = k_0 - k_n = \frac{p_0 - p_n}{\hbar}$$

$$= \frac{\Delta p}{\hbar} \approx \frac{\Delta W}{\hbar v_0} \qquad \text{since } \frac{dW}{dp} = v$$

$$q_{min} = \frac{E_n - E_0}{\hbar v_0} \neq 0 \tag{17-22a}$$

Assume $E_n - E_0 \approx Ry \approx e^2/a$ where a is the radius of the atom; $e^2/\hbar = u_0$ is the velocity of the electron in the first Bohr orbit. Therefore,

$$q_{min} \approx \frac{u_0}{v_0 a} \qquad q_{min}\, a \approx \frac{u_0}{v_0} \ll 1 \tag{17-22b}$$

because the Born approximation is supposed to be valid. On the other hand

$$q_{max} = k_n + k_0 \approx 2k_0 \qquad q_{max}\, a \gg 1 \tag{17-22c}$$

Hence as q ranges from $k_n - k_0$ to $k_n + k_0$, qa goes from much less than one to much greater than one.

We now examine the behavior of the integral occurring in (17-18b), viz., of the transition form factor in the two limits $qa \ll 1$ and $qa \gg 1$.

For $qa \ll 1$ we expand the exponential in (17-18b). The zero-order term vanishes because of the orthogonality of the wave functions. The first nonvanishing term leads to the integral

$$\mathbf{iq} \cdot \int \varphi_n^* \varphi_0 \sum_j \mathbf{r}_j \, d\tau \tag{17-23}$$

which is the dipole moment for the transition $0 \to n$. Since we know that the cross section contains a factor q^{-4}, small q are most frequent, and for these we see that the collisions mostly cause transitions which are also optically allowed. The result (17-23) does not depend on any Hartree-Fock approximation. For $qa \approx 1$, transitions to any state n are possible, not only dipole transitions.

For $qa \gg 1$, the integral occurring in (17-18b) will become quite small due to the rapid oscillation of the exponential unless φ_n varies the same way. If we consider the determinantal approximation, then the integral reduces to

$$\int e^{i\mathbf{q} \cdot \mathbf{r}} u_n^*(\mathbf{r}) u_0(\mathbf{r}) \, d\tau \tag{17-24}$$

u_n being the nth state orbital. This can be large only if u_n varies as $e^{i\mathbf{q}\cdot\mathbf{r}}$, i.e., if the final state of the excited atomic electron has momentum $\approx \mathbf{q}$. This leads to the conclusion that there is approximate conservation of momentum between incident particle and excited electron; i.e., the nucleus gets little momentum in the case $qa \gg 1$.

This result has an important influence on q_{max} because $\hbar q_{max}$ is now the maximum momentum which can be transferred by the incident particle (mass m) to an electron (mass m_e) in a collision which conserves momentum and energy between these two particles. By elementary classical mechanics this is in the nonrelativistic case

$$\hbar q_{max} = \frac{2m_e m v_0}{m_e + m} \tag{17-25a}$$

$$\hbar q_{max} \approx 2m_e v_0 \qquad \text{if } m \gg m_e \tag{17-25b}$$

$$\hbar q_{max} \approx m_e v_0 \qquad \text{if } m \approx m_e \tag{17-25c}$$

In the latter case (incident electron) it is customary to consider the slower of the two resulting electrons as that ejected from the atom which modifies (17-25c) to

$$\hbar q_{max} \approx \frac{m_e v_0}{\sqrt{2}} \tag{17-25d}$$

Moreover, there is in this case an exchange term between incident and atomic electron which reduces the cross section when q is of the order of q_{max}.

RESULTS OF CALCULATION AND COMPARISON WITH EXPERIMENT

In Figure 17-1 we exhibit the results of the Born formula calculation for excitation of helium at 200 eV. The elastic cross section is also indicated. The general predictions made above are verified, viz., at large angles, the scattering is mainly elastic; at small angles optically allowed transitions $2\,^1P$, $3\,^1P$ dominate the optically forbidden transition $2\,^1S$. At intermediate angles, the forbidden transition dominates the allowed transitions.

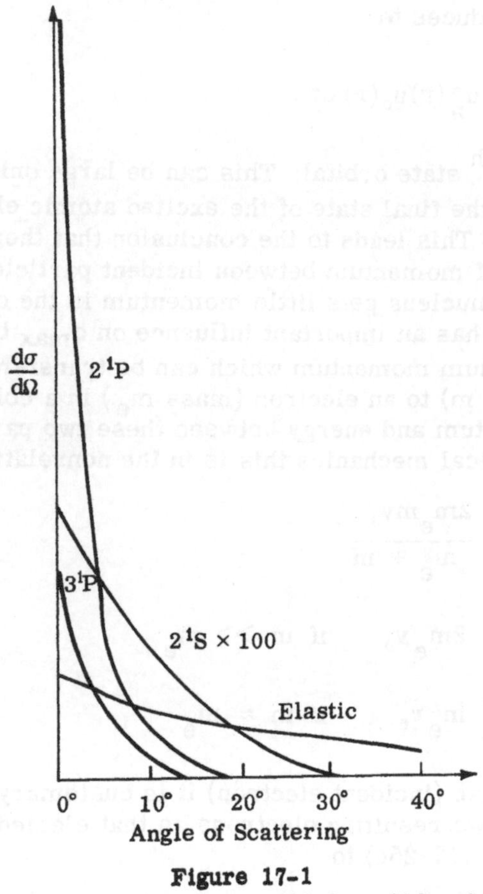

Figure 17-1

Angular distribution of 200 eV electrons off helium (After Mott and Massey[2], p. 483).

---

[2]N. F. Mott and H. S. W. Massey, *The Theory of Atomic Collisions*, 3rd ed., Oxford University Press, Oxford, 1965.

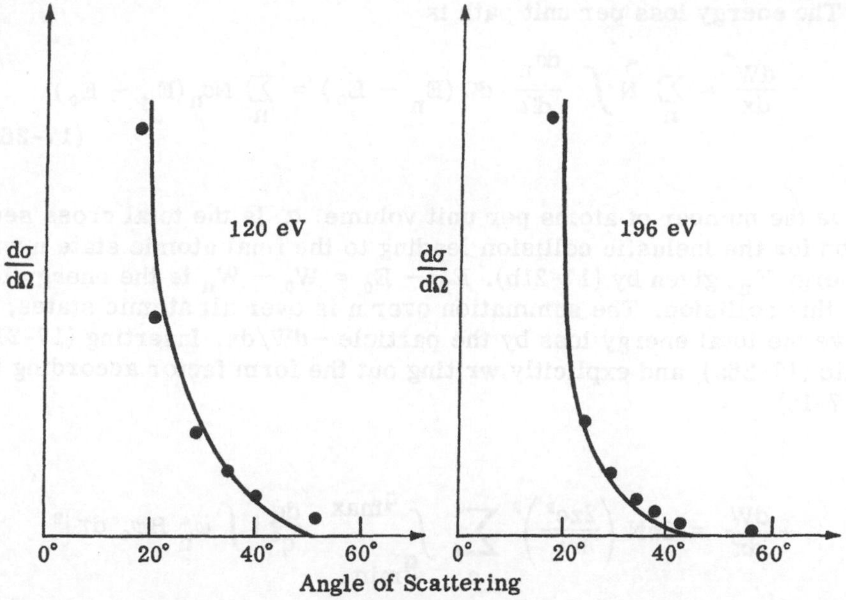

Figure 17-2

Angular distribution of inelastic electrons off helium (After Mott and Massey[3] p. 231). Observed points on a curve calculated from Born's formula.

In Figure 17-2 we exhibit comparison between experiment and Born formula calculation for the angular distribution of *all* the inelastically scattered electrons off helium. It is seen that the agreement is good. The large angle discrepancy is similar to that for elastic scattering and can be understood in the same fashion (see end of Chapter 13).

ENERGY LOSS OF THE INCIDENT PARTICLE

When a charged particle passes through matter, it makes many collisions, the cross section for each being given by (17-18). Inelastic collisions occur even at large atomic distances between particle and atom, the largest being given by $1/q_{min}$, which is about 100 atomic radii when $v \to c$. (In the relativistic region, this distance increases further, as p.) The particle loses its kinetic energy by such collisions and eventually is brought to a stop.

[3]Ibid, 2nd ed., 1949.

The energy loss per unit path is

$$-\frac{dW}{dx} = \sum_n N \int \frac{d\sigma_n}{d\Omega}\, d\Omega\, (E_n - E_0) = \sum_n N\sigma_n (E_n - E_0) \tag{17-26a}$$

N is the number of atoms per unit volume. σ_n is the total cross section for the inelastic collision leading to the final atomic state n with energy E_n, given by (17-21b). $E_n - E_0 = W_0 - W_n$ is the energy loss in this collision. The summation over n is over all atomic states, to give the total energy loss by the particle $-dW/dx$. Inserting (17-21b) into (17-26a), and explicitly writing out the form factor according to (17-16)

$$-\frac{dW}{dx} = 2\pi N \left(\frac{2ze^2}{\hbar v_0}\right)^2 \sum_n \int_{q_{min}}^{q_{max}} \frac{dq}{q^3} \left| \int \varphi_n^* B \varphi_0 \, d\tau \right|^2 \times (E_n - E_0) \tag{17-26b}$$

where

$$B = \sum_i e^{i\mathbf{q}\cdot\mathbf{r}_i}$$

We now make the approximation that we can interchange the summation over n with the integration over q. This is an approximation, since q_{max} and q_{min} both depend on E_n. Thus we are led to rewrite (17-26b) as

$$-\frac{dW}{dx} = 2\pi N \left(\frac{2ze^2}{\hbar v_0}\right)^2 \int_{\bar{q}_{min}}^{\bar{q}_{max}} \frac{dq}{q^3} \sum_n \left| \int \varphi_n^* B \varphi_0 \, d\tau \right|^2 \times (E_n - E_0) \tag{17-27}$$

where the bar over the limits of the q integration indicates that they have been replaced by suitable average values which are independent of n. The effect of this averaging will be examined below.

The sum over n can be evaluated exactly.[4] It is first written as

[4]It is recognized that this sum rule is a special case of the general sum rule obtained in Chapter 11.

$$\sum_n |B_{no}|^2 (E_n - E_0) = \sum_n (B^*_{no}) B_{no} (E_n - E_0)$$

$$= \sum_n (B^\dagger)_{on} [H, B]_{no}$$

$$= (B^\dagger [H, B])_{00} \tag{17-28}$$

We evaluate this in the nonrelativistic case.[5]

$$H \equiv H_{atom} = V + \sum_j \frac{p_j^2}{2m_e} \tag{17-29}$$

V clearly commutes with B. Hence

$$[H, B] = \sum_{ij} -\frac{\hbar^2}{2m_e} [\nabla_j^2, e^{i\mathbf{q}\cdot\mathbf{r}_i}]$$

$$= \sum_{ij} -\frac{\hbar^2}{2m_e} (-q^2 e^{i\mathbf{q}\cdot\mathbf{r}_i} + 2i\, e^{i\mathbf{q}\cdot\mathbf{r}_i}\, \mathbf{q}\cdot\nabla_j)\, \delta_{ij}$$

$$= \frac{\hbar^2}{2m_e} \sum_i e^{i\mathbf{q}\cdot\mathbf{r}_j} (q^2 - 2i\mathbf{q}\cdot\nabla_j) \tag{17-30}$$

$$(B^\dagger [H, B])_{00} = \frac{\hbar^2}{2m_e} \sum_{ij} \int \varphi_0^* \, e^{-i\mathbf{q}\cdot\mathbf{r}_i} e^{i\mathbf{q}\cdot\mathbf{r}_j}$$

$$\times (q^2 - 2i\mathbf{q}\cdot\nabla_j)\, \varphi_0 \, d\tau \tag{17-31}$$

We can always take[6] φ_0, the ground state, to be real, so that we can write (17-31) as

$$\frac{\hbar^2}{2m_e} \sum_{i\neq j} \int e^{-i\mathbf{q}\cdot(\mathbf{r}_i - \mathbf{r}_j)} [q^2 \varphi_0^2 - i\mathbf{q}\cdot\nabla_j \varphi_0^2]\, d\tau$$

$$= \frac{\hbar^2}{2m_e} \sum_i \int [q^2 \varphi_0^2 - i\mathbf{q}\cdot\nabla_i \varphi_0^2]\, d\tau \tag{17-32}$$

---

[5]"Nonrelativistic" refers here to the atomic electrons; the incident particle is permitted to be relativistic.

[6]Except for a factor like $e^{im\varphi}$ representing the orbital angular momentum but this can easily be seen to make no difference.

The second term in the first integral can be integrated by parts, the surface term is dropped, and the integrated term just cancels the first term in the first integral. The second term in the second integral is zero, as can be seen by performing partial integration. Therefore, (17-32) becomes

$$\frac{\hbar^2}{2m_e} \sum_i \int q^2 \varphi_0^2 \, d\tau = \frac{\hbar^2}{2m_e} Zq^2 \tag{17-33}$$

and

$$\sum_n \left| \int \varphi_n^* \sum_i e^{i\mathbf{q}\cdot\mathbf{r}_i} \varphi_0 \, d\tau \right|^2 (E_n - E_0) = \frac{\hbar^2}{2m_e} Zq^2 \tag{17-34}$$

This is a remarkable generalization of the many-electron f (oscillator strength) sum rule (11-23).

What is even more remarkable is that the dipole approximation to $e^{i\mathbf{q}\cdot\mathbf{r}_i}$ already gives the correct answer for (17-28). To see this we note that the right-hand side (17-34) is proportional to q^2. Hence, if we expand the left-hand side in powers of q, we need to keep only terms up to order q^2 to obtain the correct result. To this order we have

$$\begin{aligned} B_{no} &= Z\delta_{no} + i \sum_i \int d\tau \, \varphi_n^* \, \mathbf{q}\cdot\mathbf{r}_i \, \varphi_0 \\ &\quad - \tfrac{1}{2} \sum_i \int d\tau \, \varphi_n^* (\mathbf{q}\cdot\mathbf{r}_i)^2 \varphi_0 + O(q^3) \end{aligned} \tag{17-35}$$

$$\begin{aligned} |B_{no}|^2 &= Z\delta_{no} \Big(Z + 2\,\mathrm{Re}\Big[i \sum_i \int d\tau \, \varphi_n^* \, \mathbf{q}\cdot\mathbf{r}_i \, \varphi_0 \\ &\quad - \tfrac{1}{2} \sum_i \int d\tau \, \varphi_n^* (\mathbf{q}\cdot\mathbf{r}_1)^2 \varphi_0 \Big]\Big) \\ &\quad + \Big| \sum_i \int d\tau \, \varphi_n^* \, \mathbf{q}\cdot\mathbf{r}_i \, \varphi_0 \Big|^2 + O(q^3) \end{aligned} \tag{17-36}$$

The term proportional to δ_{no} does not contribute to the sum because of the $E_n - E_0$ factor. What we are left with for the sum is only

$$\sum_n |B_{no}|^2 (E_n - E_0) = \sum_n (\mathbf{q}\cdot\mathbf{D}_{on})(\mathbf{q}\cdot\mathbf{D}_{no})(E_n - E_0)$$

$$\mathbf{D}_{no} = \sum_i \int d\tau \, \varphi_n^* \, \mathbf{r}_i \, \varphi_0 \tag{17-37}$$

which is just the result obtained from the dipole approximation to $e^{iq \cdot r_i}$. That the right-hand side of (17-37) does indeed sum to $(\hbar^2/2m_e) Zq^2$ can be easily verified by using the oscillator strength sum rule (11-23).

Returning to (17-27)

$$-\frac{dW}{dx} = \frac{4\pi z^2 e^4}{m_e v_0^2} NZ \log \frac{\bar{q}_{max}}{\bar{q}_{min}} \tag{17-38}$$

The quantity NZ is seen to be the total number of electrons per unit volume. For q_{max} we can use (17-25b) which is independent of energy. (We assume the mass of the incident particle to be much greater than the electron mass.) For q_{min} we use an average of (17-22a)

$$\frac{\bar{q}_{max}}{\bar{q}_{min}} = \frac{2m_e v_0^2}{I_{AV}} \qquad I_{AV} = (E_n - E_0)_{AV} \tag{17-39}$$

(The precise definition of the averaging process is given below.) This is the form commonly used for the stopping-power formula.

$$-\frac{dW}{dx} = \frac{4\pi z^2 e^4}{m_e v_0^2} NZ \log \frac{2m_e v_0^2}{I_{AV}} \tag{17-40}$$

Thus we have expressed the stopping power in terms of one free parameter I_{AV}, which depends on the atom in question.

We may obtain a precise definition of I_{AV} by recalling that it is defined in order to allow the interchange of integration and summation in (17-26), viz.,

$$\sum_n \int_{(E_n - E_0)/\hbar v_0}^{2m_e v_0/\hbar} \frac{dq}{q^3} |B_{no}|^2 (E_n - E_0)$$

$$= \int_{I_{AV}/\hbar v_0}^{2m_e v_0/\hbar} \frac{dq}{q^3} \sum_n |B_{no}|^2 (E_n - E_0) \tag{17-41}$$

This may be rewritten

$$\sum \int_{I_{AV}/\hbar v_0}^{(E_n - E_0)/\hbar v_0} \frac{dq}{q^3} |B_{no}|^2 (E_n - E_0) = 0 \tag{17-42}$$

Both limits of the integral are small in the sense that $qa \ll 1$, cf. (17-27b). Therefore (and because elastic collisions do not contribute), we may use the dipole approximation (17-35); thus

$$|B_{no}|^2 = |\mathbf{q} \cdot \mathbf{D}_{no}|^2 \tag{17-43}$$

If we average over all magnetic substates of the excited state n, we may further write

$$|\mathbf{q} \cdot \mathbf{D}_{no}|^2 = q^2 |D^z_{no}|^2 \tag{17-44}$$

Then we may use the definition of the oscillator strength

$$f^z_{no} = \frac{2m}{\hbar^2}(E_n - E_0)|D^z_{no}|^2 \tag{17-45a}$$

Inserting all this into (17-42) yields

$$\sum_n f^z_{no} \ln \frac{E_n - E_0}{I_{AV}} = 0 \tag{17-45b}$$

Remembering the sum rule of oscillator strengths, (11-21), we get

$$\ln I_{AV} = Z^{-1} \sum_n f^z_{no} \ln (E_n - E_0) \tag{17-46}$$

which is the desired definition of I_{AV}.

One can go further and obtain the dependence of I_{AV} on Z from the Thomas-Fermi model. Bloch has done this and found that

$$I_{AV} = CZ \tag{17-47}$$

where C is some constant. Experiment verifies this law and an empirical value of about 10 eV is found for C. The final stopping-power formula, for heavy incident, nonrelativistic particles, is then

$$-\frac{dW}{dx} = \frac{4\pi z^2 e^4}{m_e v_0^2} NZ \ln \frac{2m_e v_0^2}{CZ} \tag{17-48}$$

This depends only on the velocity of the particle, not on its mass. For further evaluation see, e.g., Bethe and Ashkin,[7] in Segrè's book *Experimental Nuclear Physics*.

---

[7] H. A. Bethe and J. Ashkin, "Passage of Radiation through Matter" in *Experimental Nuclear Physics,* Vol. 1, E. Segrè (ed.), Wiley, New York, 1953.

PROBLEMS

1. The atoms of a gas have an ionization potential I. A particle of mass M and velocity v traverses the gas, and has an interaction with the atomic electrons of the form $V = \alpha/r^2$. Calculate the energy loss per centimeter of the particle.
2. Justify the linear dependence of I_{AV} on Z, equation (17-47), on the basis of the Thomas-Fermi model. Hint: set $I_{AV} = \hbar\omega_{AV}$, where ω_{AV} is the average excitation frequency; the average frequency is in turn proportional to the average electronic velocity divided by the radius of the atom. The Z dependence of these two quantities in the Thomas-Fermi atom is readily determined. Specifically, the average velocity is taken to be proportional to the square root of the energy per electron, while the energy may be expressed as an integral over the charge density.
3. Verify the steps from (17-18) to (17-21).
4. Calculate $F_n(q)$ for hydrogen, where the final state n is the isotropic state with principal quantum number 2.
5. Using the oscillator strength sum rule (11-23), verify that the right-hand side of (17-37) is equal to $(\hbar^2/2m_e)Zq^2$.
6. A particle "a," of mass m_1, is bound to an infinitely heavy center by a potential $U = \frac{1}{2}m_1\omega^2 r_a{}^2$, and is in the ground state in this potential. Another particle "b," of mass m_2 and velocity v, interacts with particle "a" with the potential $V = B\,e^{-\mu r}$, $r = |r_a - r_b|$. What is the differential and total cross section in Born approximation for scattering of particle "b", with excitation of particle "a" to the first excited state, which has angular momentum quantum numbers $\ell = 1$, $m = 0$? (You may leave the formula for the total cross section as a parametric integral.)
7. A particle "a," governed by an attractive separable nonlocal potential (as in Chapter 4, Problem 4 and Chapter 15, Problem 1, with $\lambda < 0$) interacts with another particle "b" through the potential $V(r_a - r_b)$. Recall that there is one bound state of particle "a" in the separable potential. Compute the differential scattering cross section for the elastic scattering of particle "b" on particle "a" in the latter's bound state, and for the release of "a" from the bound state. Treat the a$\leftrightarrow$b interaction in the Born approximation.

Chapter 18

INELASTIC SCATTERING AT LOW ENERGIES

At low incident energies, the Born approximation is not valid, and we need more accurate methods of calculating. The problem is more difficult than in the elastic scattering case since an inelastically scattered particle cannot be considered to be moving in an equivalent static potential. Thus we must remain with the many-body Hamiltonian (17-1) and develop approximate methods of solution. Such approximate methods exist, and are discussed in Mott and Massey,[1] Chapters 12–15, 17. We shall confine ourselves to the simplest of these, the *distorted wave approximation*.

It is clear that the Born approximation (17-2) can be improved by replacing the initial and final free wave functions, with wave functions which are closer to the exact solution. We shall now derive an approximation to the transition amplitude which makes use of this idea.

We recall from general scattering theory that the transition amplitudes between states a and b, in two complex systems described by Hamiltonians H and $\hat{H}$, are related by the formula[2]

$$\langle b | T | a \rangle = \langle b | \hat{T} | a \rangle + \langle \hat{\Psi}_b^{(-)} | (H - \hat{H}) | \Psi_a^{(+)} \rangle \tag{18-1}$$

It is assumed that the kinetic energy parts of H and $\hat{H}$ are the same. The $\Psi_a^{(+)}$ is the complete scattering eigenstate of H with outgoing waves and initial conditions a. Similarly $\hat{\Psi}_b^{(-)}$ is the complete scattering eigenstate of $\hat{H}$ with incoming waves and initial conditions b.

---

[1] N. F. Mott and H. S. W. Massey, *The Theory of Atomic Collisions*, 3rd ed., Oxford University Press, Oxford, 1965.

[2] L. I. Schiff, *Quantum Mechanics*, 3rd ed., McGraw-Hill, New York, 1968, p. 327.

We first make the following choices for H and $\hat{H}$ in (18-1). For $\hat{H}$ we take the exact Hamiltonian of our problem (17-1):

$$\hat{H} = H^o + H^I \tag{18-2a}$$

$$H^o = H_{atom} + H_{particle} \tag{18-2b}$$

$$H^I = ze\left[\frac{Ze}{r} - \sum_{i=1}^{Z} \frac{e}{|\mathbf{r} - \mathbf{r}_i|}\right] \tag{18-2c}$$

Thus $\hat{\Psi}_b^{(-)}$ is that eigenfunction of $H^o + H^I$ which has the asymptotic behavior in r

$$\hat{\Psi}_b^{(-)} \xrightarrow[r \to \infty]{} \varphi_\beta \left[e^{i\mathbf{k}_b \cdot \mathbf{r}} + f_b^{(-)}(\Omega)\frac{e^{-ik_b r}}{r}\right] + \sum_{\gamma \neq \beta} \varphi_\gamma f_\gamma^{(-)}(\Omega)\frac{e^{-ik_\gamma r}}{r} \tag{18-3}$$

Here the φ_γ are the atomic states. The label b specifies the conditions: atomic state φ_β and incident particle with momentum $\mathbf{k}_b$. Thus we write $b = (\beta, \mathbf{k}_b)$.

For H we take

$$H = H^o + U_a \tag{18-4a}$$

$$U_a(\mathbf{r}) = ze\left[\frac{Ze}{r} - \int \frac{\rho_\alpha(\mathbf{r}')e}{|\mathbf{r} - \mathbf{r}'|} d\tau'\right] \tag{18-4b}$$

$$\rho_\alpha(\mathbf{r}) = \sum_{i=1}^{Z} \int |\varphi_\alpha(\mathbf{r}_1, \ldots, \mathbf{r}_{i-1}, \mathbf{r}, \mathbf{r}_{i+1}, \ldots, \mathbf{r}_Z)|^2 \prod_{j \neq i} d\tau_j \tag{18-4c}$$

U_a is the Hartree potential with the charge density corresponding to the initial state, α of the atom. This potential depends only on $\mathbf{r}$, the incident particle's coordinate, and the associated scattering problem can be solved.

Let us denote the transition amplitude in this potential by $\langle b | T^X | a \rangle$ and the scattering solutions by $X_a^{(+)}$. Then from (18-1)

$$\langle b | \hat{T} | a \rangle = \langle b | T^X | a \rangle + \langle \hat{\Psi}_b^{(-)} | H^I - U_a | X_a^{(+)} \rangle \tag{18-5}$$

We first note that $X_a^{(+)}$ factors into a function of the internal atomic coordinates only, and a function of $\mathbf{r}$ the position of the particle. This is because U_a, equation (18-4b), depends only on $\mathbf{r}$, not on the internal coordinates A. Because of the initial condition that the atom is in state α, we have

$$X_a^{(+)} = \varphi_\alpha(A)\,\chi_a^{(+)}(\mathbf{r}) \tag{18-6}$$

Then the transition amplitude T^X in the potential U_a is zero for all inelastic transitions, for we have

$$\langle b\,|\,T^X\,|\,a\rangle = \langle \varphi_\beta(A)\,e^{i\mathbf{k}_b\cdot\mathbf{r}}\,|\,U_a(\mathbf{r})\,|\,\varphi_\alpha(A)\,\chi_a^{(+)}(\mathbf{r})\rangle \tag{18-7a}$$

Since U_a is independent of the internal coordinates A, this vanishes for $\beta \neq \alpha$ and we have

$$\langle b\,|\,T^X\,|\,a\rangle = \langle a'\,|\,T^X\,|\,a\rangle\,\delta_{\alpha\beta} \tag{18-7b}$$

where the state $|\,a'\rangle$ contains the same atomic state as $|\,a\rangle$, but may differ from $|\,a\rangle$ in the momentum of the external particle.

We suppress the carets in (18-5) and define

$$W_a = H^I - U_a = -ze^2\left[\sum_{i=1}^{Z}\frac{1}{|\mathbf{r}-\mathbf{r}_i|} - \int\frac{\rho_\alpha(\mathbf{r}')}{|\mathbf{r}-\mathbf{r}'|}d\tau'\right] \tag{18-8a}$$

Then (18-5) simplifies to

$$\langle b\,|\,T\,|\,a\rangle = \langle a'\,|\,T^X\,|\,a\rangle\,\delta_{\alpha\beta} + \langle \Psi_b^{(-)}\,|\,W_a\,|\,X_a^{(+)}\rangle \tag{18-8b}$$

This is still exact. Now we replace $\Psi_b^{(-)}$ by the approximation corresponding to (18-4). Thus we define a potential U_b which is the same as U_a except that φ_α is replaced by φ_β, the atomic wave function in the final state β. We designate the scattering wave function in this potential by $Y_b^{(-)}$. According to (18-6) $Y_b^{(-)}$ has the form $\varphi_\beta(A)\,\chi_b^{(-)}(\mathbf{r})$. $Y_b^{(-)}$ is a good approximation to $\Psi_b^{(-)}$ for our purposes because U_b, according to (18-4), is simply the average of the exact H^I, equation (18-2), over the wave function φ_β of the atom. We then obtain

$$\langle b\,|\,T\,|\,a\rangle \approx \langle a'\,|\,T^X\,|\,a\rangle\,\delta_{\alpha\beta} + \langle \varphi_\beta\chi_b^{(-)}\,|\,W_a\,|\,\varphi_\alpha\chi_a^{(+)}\rangle \tag{18-9}$$

It follows from a similar argument that W_b could be substituted for W_a in (18-9). As will be shown below, it makes no difference whether W_a or W_b is used.

The difference between U_a and U_b is also small. In general only one of the atomic electrons changes its state (see Chapter 17), so that the charge density (18-4c) changes by about 1 part in Z, or less. It is therefore a good approximation to use U_a also for the calculation of $\chi_b^{(\pm)}$, changing only the value of the particle energy, i.e., replacing k_a by k_b.

Equation (18-9) is the ***distorted wave approximation.*** It is derived in Mott and Massey,[1] p. 349, by a somewhat different method.[3] As in the Born approximation, the wave functions of initial and final states are simply products of an atomic function and a particle function. However, the latter is not a plane wave but a complete solution of the scattering of the particle in the potential U_a (or U_b if we wish to distinguish the two), including outgoing or incoming spherical waves, respectively.

Again because of the orthogonality of φ_α and φ_β, for inelastic scattering, the term U_a in W_a, equation (18-8a), does not give any contribution to transition amplitude (18-9), only the interaction H^I does. (Recall that U_a depends only on the particle coordinates, and is independent of the atomic coordinates.) Specializing to the case where α is the ground state of the atom, $\alpha = 0$ and $\beta = n \neq 0$, an excided state, the transition amplitude may then be written

$$\langle n | T | 0 \rangle \approx \langle \varphi_n \chi_b^{(-)} | H^I | \varphi_0 \chi_a^{(+)} \rangle \tag{18-10a}$$

$$= \int V_n(\mathbf{r}) \chi_b^{(-)*}(\mathbf{r}) \chi_a^{(+)}(\mathbf{r}) \, d\tau \tag{18-10b}$$

where

$$V_n(\mathbf{r}) = -ze^2 \int \varphi_n^* \varphi_0 \sum_{i=1}^{N} \frac{1}{|\mathbf{r} - \mathbf{r}_i|} \, d\tau_1, \ldots, d\tau_Z$$

$$= -ze^2 \int \frac{\rho_n(\mathbf{r}')}{|\mathbf{r} - \mathbf{r}'|} \, d\tau' \tag{18-10c}$$

Obviously (18-10c) is exactly the same as in the Born approximation, only the particle wave functions in (18-10b) are now more complicated. Clearly, the simple expression $e^{i\mathbf{q}\cdot\mathbf{r}}$ for the product of initial and final wave functions is no longer valid, and therefore we can

[3] It is also derived and discussed by M. L. Goldberger and K. M. Watson *Collision Theory*, Krieger, Melbourne FL. 1975, p. 818 ff., under the name "Single Inelastic Scattering Approximation," and by Schiff,[2] p. 327 ff.

no longer carry out the integration over $\mathbf{r}$ before that over the atomic coordinates, thus T can no longer be reduced to a form factor of the transition charge distribution.

We note in passing that for *elastic* scattering, the matrix element of W_a in (18-9) vanishes for the following reason. From the definition (18-4b)

$$U_a(\mathbf{r}) = \langle \varphi_\alpha | H^I | \varphi_\alpha \rangle \tag{18-11a}$$

and hence

$$\langle \varphi_\alpha | W_a | \varphi_\alpha \rangle = \langle \varphi_\alpha | H^I - U_a | \varphi_\alpha \rangle = 0 \tag{18-11b}$$

Therefore

$$\langle 0' | T | 0 \rangle \approx \langle 0' | T^X | 0 \rangle \tag{18-12}$$

The elastic scattering can be calculated simply as the scattering from the potential U_a, as we have known before.

VALIDITY OF DISTORTED WAVE APPROXIMATION

To evaluate the validity of this distorted wave approximation, let us examine the exact scattering wave function $\Psi_b^{(-)}$ which we have approximated by $Y_b^{(-)}$. We recall that $\Psi_b^{(-)}$ is a scattering solution of $H = H^0 + H^I$, and it is a function of the atomic coordinates A, and the particle coordinate $\mathbf{r}$. We may expand $\Psi_b^{(-)}$ in the atomic wave functions with coefficients which are functions of $\mathbf{r}$.

$$\Psi_b^{(-)}(A, \mathbf{r}) = \sum_\gamma \varphi_\gamma(A)\, \psi_\gamma^{(-)}(\mathbf{r}) \tag{18-13}$$

According to (18-3), the ψ_γ satisfy the boundary conditions at large r

$$\psi_\beta^{(-)} \xrightarrow[r \to \infty]{} e^{i\mathbf{k}_b \cdot \mathbf{r}} + f_b^{(-)}(\Omega)\, \frac{e^{-ik_b r}}{r}$$

$$\psi_{\gamma \neq \beta}^{(-)} \xrightarrow[r \to \infty]{} f_\gamma^{(-)}(\Omega)\, \frac{e^{-ik_\gamma r}}{r}$$

The individual ψ_γ satisfy equations which may be obtained as follows. $\Psi_b^{(-)}$ satisfies $(H - E)\, \Psi_b^{(-)} = 0$ or

$$\sum_\gamma (H^0 + H^I - E)\, \varphi_\gamma(A)\, \psi_\gamma^{(-)}(\mathbf{r}) = 0 \tag{18-14a}$$

Multiplying by $\varphi_{\gamma'}(A)$ and integrating over A gives

$$\sum_{\gamma} (\langle \varphi_{\gamma'} | H^0 | \varphi_{\gamma} \rangle + \langle \varphi_{\gamma'} | H^I | \varphi_{\gamma} \rangle) \psi_{\gamma}^{(-)}(r) = E\psi_{\gamma'}^{(-)}(r) \qquad (18\text{-}14b)$$

Now $H^0 = H_{atom} + H_{particle}$; where the former operating on φ_γ gives E_γ, and the latter does not affect φ_γ since $H_{particle}$ does not depend on A. Thus $\langle \varphi_{\gamma'} | H^0 | \varphi_{\gamma} \rangle = (E_{\gamma'} + H_{particle})\delta_{\gamma\gamma'}$. Therefore from (18-14b)

$$H_{particle}\psi_{\gamma}^{(-)} + \sum_{\gamma'} \langle \varphi_{\gamma} | H^I | \varphi_{\gamma'} \rangle \psi_{\gamma'}^{(-)} = \epsilon_{\gamma}\psi_{\gamma}^{(-)}$$

$$\epsilon_{\gamma} + E_{\gamma} = E \qquad (18\text{-}14c)$$

These equations together with the boundary conditions given above determine $\psi_{\gamma}^{(-)}$. Note that the momentum k_γ appearing in (18-3) is given by $\hbar^2 k^2_\gamma/2m = \epsilon_\gamma$.

The scattering wave function $Y_b^{(-)}$ on the other hand satisfies

$$(H^0 + \langle \varphi_\beta | H^I | \varphi_\beta \rangle - E) Y_b^{(-)} = 0 \qquad (18\text{-}15a)$$

As stated above $Y_b^{(-)}$ factors into $\varphi_\beta \chi_b^{(-)}$ and we obtain

$$(H_{particle} + \langle \varphi_\beta | H^I | \varphi_\beta \rangle)\chi_b^{(-)} = \epsilon_b \chi_b^{(-)} \qquad (18\text{-}15b)$$

$$\chi_b^{(-)} \xrightarrow[r \to \infty]{} e^{ik_b \cdot r} + f_b^{(-)}(\Omega)\frac{e^{-ikr}}{r}$$

where

$$H_{atom}\varphi_\beta = E_b \varphi_\beta$$

$$E_b + \epsilon_b = E \qquad (18\text{-}15c)$$

Comparing (18-14c) to (18-15b) we see that the approximation of replacing $\Psi_b^{(-)}$ by $Y_b^{(-)}$ consists of the following two approximations. First, we replace $\Psi_b^{(-)}$ by just the elastic scattering member in its expansion (18-13), i.e., by $\varphi_\beta \psi_\beta^{(-)}$. Then we simplify the equation determining $\psi_\beta^{(-)}$ by neglecting all off-diagonal matrix elements of H^I; i.e., we replace $\psi_\beta^{(-)}$ by $\chi_b^{(-)}$.

We note however that one member, $\gamma = \alpha$, in the expansion of $\Psi_b^{(-)}$ is rigorously absent in (18-8b). This term contributes a quantity which contains the factor

$$\langle \varphi_\alpha | H^I - U_a | \varphi_\alpha \rangle$$

which is zero by (18-11b). (This is really the reason for the choice of U_a in (18-4).) However the term $\varphi_\alpha \psi_\alpha^{(-)}$ does contribute in an exact calculation to the equations determining the individual $\psi_\gamma^{(-)}$ in the form of the matrix element $\langle \varphi_\alpha | H^I | \varphi_\gamma \rangle$.

It is seen therefore that the distorted wave approximation consists in assuming that the various channels $\gamma \neq \beta$, viz., the inelastic channels other than that corresponding to the final state; are closed in the reaction. This amounts in general to neglecting the *square* of the inelastic scattering. However, there can be coherent effects between the various channels γ; one such effect is the absorptive effect which we discussed in Chapter 15. We saw at that time that its effect on the elastic scattering could be represented by parametrizing the partial wave amplitude η_ℓ as $a_\ell e^{2i\delta\ell}$ with $a_\ell < 1$—or if we wish to find the effect on the whole wave function $\chi_a^{(+)}$, we might use an imaginary potential, such as that used by Mott and Massey,[4] p. 222. It is likely that the most important effect of the "intermediate channels" can be expressed in terms of such a "quasi-potential" which will modify the wave functions $\chi_a^{(+)}$ and $\chi_b^{(-)}$. Theoretical considerations and literature on the quasi-potential are given in Goldberger and Watson,[3] Section 11.7.

The effect of the quasi-potential on the wave function in turn may be measured by $1 - a_\ell$, in the notation of Chapter 15. Using (15-18),

$$1 - a_\ell \approx \frac{2}{3}\left(\frac{e^2}{\hbar v b}\right)^2 \sum_\alpha z_\alpha r_\alpha^2 \tag{18-16a}$$

For a given impact parameter b, we should include in the sum only the electron shells $r_\alpha < b$, because of the expansion (15-12) which makes this assumption; the outer shells $r_a > b$ contribute little. Of the inner shells, however, the main contribution comes from that with r_α closest to b. Thus we find approximately

$$1 - a_\ell \leq \left(\frac{e^2}{\hbar v}\right)^2 z_b \tag{18-16b}$$

---

[4]N. F. Mott and H. S. W. Massey, *Theory of Atomic Collisions*, 2nd ed., Oxford University Press, London, 1949.

where z_b is the number of electrons in that shell whose radius is closest to b. For the distorted wave approximation to be good, we must require that (18-16b) be small compared with one, or

$$\frac{\hbar v}{e^2} > \sqrt{z_b} \leq 4 \tag{18-17a}$$

The numerical estimate is based on the fact that electron shells in general do not contain more than 14 to 18 electrons (the 4f shell is sufficiently separated from 4d as well as 5s + p). Obviously, (18-17a) is a very much weaker condition than the usual one for the Born approximation

$$\frac{\hbar v}{e^2} > Z$$

The distorted wave approximation should thus be applicable to any atom for

$$E \geq 250 \text{ eV} \tag{18-17b}$$

For light atoms, up to Z = 20, the validity should start at even lower energy. We have used the corresponding theory for elastic scattering, viz., scattering by a potential, in Chapter 15. Obviously, the distorted wave approximation is enormously better than the simple Born approximation.

A further improvement in the approximation should be obtainable by using the quasi-potential consistently for the determination of the wave functions $\chi_a^{(+)}$ and $\chi_b^{(-)}$. However, while that potential can be defined fairly well for large r, its definition for small r is rather arbitrary (see Goldberger and Watson,[3] p. 853). No precise theory can be expected from such a procedure. However, in nuclear reaction theory where the absorption of incident and final state waves is enormous, it must be included in the distorted wave calculation (see Goldberger and Watson,[3] p. 823).

RESULTS OF CALCULATION AND COMPARISON WITH EXPERIMENT

The distorted wave approximation to the transition amplitude can be evaluated once the charge density for the atom and its excited state wave function φ_β is known from a Hartree or Hartree-Fock calculation. The distorted waves are obtained from a (numerical) solution of the scattering problem and then the transition amplitude (18-10) is

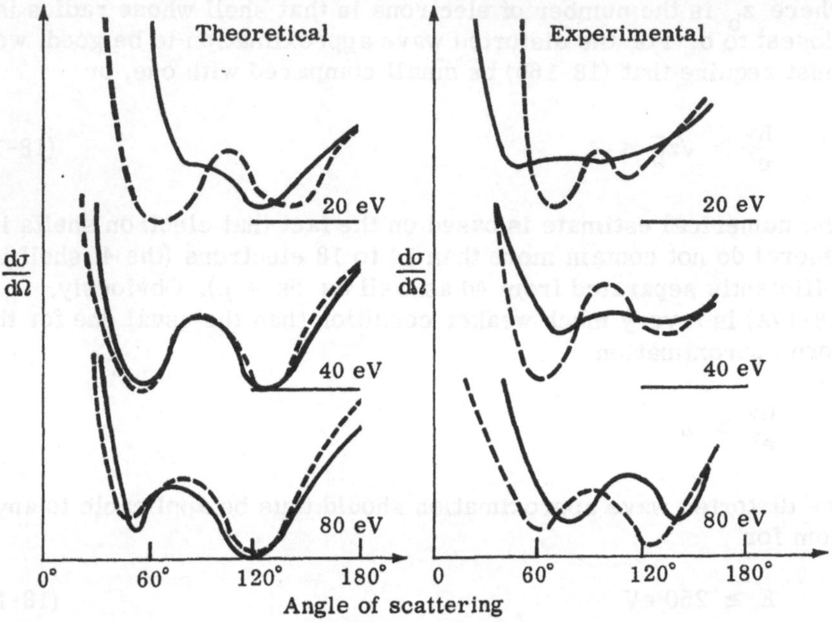

Figure 18-1
Theoretical and experimental differential cross sections for electrons off argon (After Mott and Massey[1], p. 593). Dashed line, elastically scattered electrons; solid line, electrons which have excited the argon resonance level.

calculated. Such calculations have been performed and Figure 18-1 exhibits the theoretical and experimental results. From the plots we see that the experimental data are in qualitative agreement with the calculation. It is striking that the elastic and inelastic angular distributions are roughly the same, both experimentally and theoretically. This can be understood by observing that the angular dependence of the cross section is contained in the distorted waves $\chi_0^{(+)}$, $\chi_n^{(-)*}$. The distorted waves measure the distortion of the wave function of the scattered particle by the normal and excited atom respectively. However at high energies of the incident particle (sufficiently greater than ionization energy) the distortion due to the excited atom is not markedly different from that due to a normal atom. The observed similarity of angular distribution, Figure 18-1, then means that the factor $V_n(\mathbf{r})$ in the matrix element (18-10b) does not have a great influence on the angular distribution but that the latter is mainly given by the wave functions $\chi_0^{(+)}$ and $\chi_n^{(-)*}$.

FURTHER CORRECTIONS

We have already pointed out that a better approximation could be obtained if the Hartree potential acting on the scattered electron were replaced by a quasi-potential which takes into account the absorption of the waves $\chi_a^{(+)}$ and $\chi_b^{(-)}$ by inelastic scattering. The "absorption potential" added for this purpose is purely imaginary. We may also add a real potential to represent the "polarization" of the atom by the scattered particle, similar to our discussion of the Ramsauer effect. (Polarization and absorption are presumably related by a dispersion relation.) All these corrections become more important at low energies. For electrons, also exchange must be taken into account. In particular, the scattered electron may go into the excited atomic state while the atomic electron leaves in the state $\mathbf{k}_a$ The reader is referred to Mott and Massey,[1] and to Goldberger and Watson,[3] for a partial account of all these corrections.

CLOSE COUPLING APPROXIMATION

Our discussion of the validity of the distorted wave approximation, p. 313 makes it evident that whenever some of the off-diagonal matrix elements, $\langle \varphi_\gamma | H^I | \varphi_{\gamma'} \rangle$, are large, the distorted wave approximation fails. One may then proceed by including the contribution of these off-diagonal matrix elements.

Returning to the exact expression for the inelastic scattering amplitude, we have from (18-8b)

$$\langle b | T | a \rangle \underset{a \neq b}{=} \langle \Psi_b^{(-)} | W_a | X_a^{(+)} \rangle$$

The scattering wave function $\Psi_b^{(-)}$ is given by equations (18-13) and (18-14c). It may happen that of the off-diagonal matrix elements only $\langle \varphi_\beta | H^I | \varphi_\alpha \rangle$ is large. Thus we replace the exact series for $\Psi_b^{(-)}$, (18-13), by

$$\Psi_b^{(-)}(A, \mathbf{r}) = \varphi_\alpha(A)\, \psi_a^{(-)}(\mathbf{r}) + \varphi_\beta(A)\, \psi_\beta^{(-)}(\mathbf{r}) \tag{18-18}$$

$\psi_\alpha^{(-)}$ and $\psi_\beta^{(-)}$ then satisfy the coupled equations

$$[H_{\text{particle}} + \langle \varphi_\beta | H^I | \varphi_\beta \rangle - \epsilon_\beta]\, \psi_\beta^{(-)} = -\langle \varphi_\beta | H^I | \varphi_\alpha \rangle\, \psi_\alpha^{(-)} \tag{18-19a}$$

$$[H_{\text{particle}} + \langle \varphi_\alpha | H^I | \varphi_\alpha \rangle - \epsilon_\alpha]\, \psi_\alpha^{(-)} = -\langle \varphi_\alpha | H^I | \varphi_\beta \rangle\, \psi_\beta^{(-)} \tag{18-19b}$$

$$\psi_\beta^{(-)} \xrightarrow[r\to\infty]{} e^{i\mathbf{k}_b\cdot\mathbf{r}} + f_b^{(-)}(\Omega)\,\frac{e^{-ik_b r}}{r}$$

$$\psi_\alpha^{(-)} \longrightarrow f_\alpha^{(-)}(\Omega)\,\frac{e^{-ik_a r}}{r} \tag{18-19c}$$

The term $\varphi_\alpha(A)\,\psi_\alpha^{(-)}(\mathbf{r})$ again does not contribute to $\langle \Psi_b^{(-)} | W_a | X_a^{(+)} \rangle$ and the transition amplitude, in this, the *close coupling approximation*, is given by

$$\langle b | T | a \rangle \approx \langle \varphi_\beta \psi_\beta^{(-)} | H^I | \varphi_\alpha \chi_a^{(+)} \rangle \tag{18-20}$$

This last expression is of the same form as in the distorted wave approximation (18-10), except that $\psi_\beta^{(-)}$ is now given by the two equations (18-19a) and (18-19b) instead of just (18-15b). Indeed, if in (18-19a) we were to ignore the right-hand term, we would regain the distorted wave approximation.

It should be pointed out that although the scattering amplitude which occurs in the first equation (18-19c) is the same as (18-20), for practical calculations the latter formula is much more useful. In order to obtain the scattering amplitude from (18-19c) one would have to know the solution of (18-19a) and (18-19b) at large r, while (18-20) requires the knowledge of $\psi_\beta^{(-)}$ only in the region of interaction.

The close coupling approximation is discussed by Mott and Massey,[1] p. 357, and recent applications have been summarized by Burke and Smith.[5]

COLLISION COMPLEX

The method of distorted waves is only applicable if inelastic scattering is relatively rare, i.e.,

$$1 - a_\ell^2 \ll 1 \qquad \text{for all } \ell \tag{18-21}$$

This is usually the case for atomic collisions, as we have shown above, but not for nuclear collisions. In the opposite case, which is usually fulfilled for nuclear collisions at moderate energies ($\lesssim 30$ MeV), many inelastic scatterings take place before the particle leaves the scatterer. In this case, the concept of the "collision complex" is applicable. For a discussion, see Mott and Massey,[1] p. 437.

[5] P. J. Burke and K. Smith, *Rev. Mod. Phys.* **34**, 458 (1962).

PROBLEMS

1. Using the general formula (18-1) and arguing as in the text up to equation (18-9), show that W_b could be substituted for W_a in (18-9).
2. Verify that (18-14c) follows from (18-14a).
3. Consider (18-1) applied to the scattering of a single particle in a potential. Obtain the well-known formula relating the transition amplitude $\langle b | T | a \rangle$ to a matrix element of the potential.
4. Outline a distorted wave approximation calculation for scattering off hydrogen where the final hydrogenic state is the isotropic state with principal quantum number 2. Perform as much of the calculation as feasible.
5. Outline a close coupling approximation calculation for the same process as described in Problem 4, above.

Chapter 19
SEMICLASSICAL TREATMENT OF INELASTIC SCATTERING

In our previous discussion of shadow scattering (Chapter 15) we used the results of a semiclassical method to evaluate the contribution of inelastic processes to the elastic scattering amplitude. We shall now examine this semiclassical method in detail. We shall also obtain the useful result that the Born approximation and the semiclassical method give the same formula for the inelastic cross section integrated over angles.

The basis for the semiclassical treatment is the assumption that the incident particle can be treated as a classical particle moving in a straight line. This is justified if (1) the particle has high momentum, such that

$$ka \gg 1 \tag{19-1}$$

(a = atomic radius), and if (2) the particle never encounters a potential energy greater than its total energy. The first condition is necessary in order that a classical trajectory can be defined, notwithstanding the uncertainty principle. (If the particle passes at an impact parameter $b > a$ from the atom, $kb \gg 1$ is sufficient.) The second condition ensures that the particle is not deflected by a large angle: If the particle moves initially in the z direction with velocity v, and if its orbit is in the xz plane, then the atomic potential V will cause a momentum change in the x direction which is approximately (in classical mechanics)

$$p_x \approx \frac{1}{v} \int \frac{\partial V}{\partial x}\, dz \approx \frac{V_{max}}{v} \tag{19-2a}$$

The last member of the equation follows from the assumption that $\partial V/\partial x$ and $\partial V/\partial z$ are of the same order of magnitude. For small deflection we must require

$$p_x \ll p \qquad V_{max} \ll pv \approx E \tag{19-2b}$$

This condition is always fulfilled for sufficiently large impact parameters b; for b of order of the atomic radius a, it is fulfilled only if E is large compared with the ionization energy, i.e., about

$$E \gg \text{Ry} \tag{19-2c}$$

Condition (19-1) is automatically fulfilled if (19-2c) is, and if the particle is heavy; for electrons, much higher energies are required to fulfill (19-1).

The external particle interacts with the atom with a potential energy

$$V_N(\mathbf{R}) + \sum_{i=1}^{Z} V(\mathbf{R} - \mathbf{r}_i) \tag{19-3}$$

Here $\mathbf{R}(t)$ is the (classical) position of the external particle which according to the assumption of rectilinear motion is

$$\mathbf{R}(t) = \mathbf{b} + \mathbf{v}t \qquad \mathbf{b}\cdot\mathbf{v} = 0 \tag{19-4}$$

where b is the impact parameter. In (19-3) the first term is the nuclear interaction and the second term is the interaction of the external particle with the Z atomic electrons. We shall see that V_N plays no role in the theory. We have taken the interaction V to be arbitrary but for an atom it is, of course, the Coulomb potential

$$V(\mathbf{r}) = -\frac{ze^2}{r} \tag{19-5}$$

(The external particle carries charge ze.)

The transition amplitude for the atomic system perturbed by this time-dependent potential is, apart from factors of unit magnitude

$$T_{nm} = \hbar^{-1} \int_{-\infty}^{\infty} e^{i\omega_{nm}t}\, V_{nm}(t)\, dt \tag{19-6}$$

$$\begin{aligned} V_{nm}(t) &= \int \psi_n^*(A)\psi_m(A)\, dA\, [V_N(\mathbf{R}) + \sum_i V(\mathbf{R} - \mathbf{r}_i)] \\ &= \int \psi_n^*(A)\psi_m(A)\, dA \sum_i V(\mathbf{R} - \mathbf{r}_i) \end{aligned} \tag{19-7}$$

The atomic states are designated by $\psi(A)$, A standing for the space variables of the atomic electrons. The second equation in (19-7) follows from the first by noting that ψ_n is orthogonal to ψ_m for inelastic scattering and $V_N(\mathbf{R})$ does not depend on A.

For large impact parameters, $b \gg a$, we can expand V in powers of r_i/R

$$V(\mathbf{R} - \mathbf{r}_i) = V(\mathbf{R}) - \nabla V(\mathbf{R}) \cdot \mathbf{D} \tag{19-8a}$$

$$\mathbf{D} = \sum_i \mathbf{r}_i \tag{19-8b}$$

The first term does not contribute to V_{nm}, due to the orthogonality of the final and initial states. (For an atom we also have $V_N(\mathbf{R}) + V(\mathbf{R}) =$ Thus

$$V_{nm}(t) = -\nabla V(\mathbf{R}) \cdot \mathbf{D}_{nm} \tag{19-9}$$

Specializing now to our case of interest; m the ground state, n the nth excited state and V given by (19-5), we get apart from factors of unit magnitude:

$$V_{no}(t) = \frac{ze^2 \mathbf{R} \cdot \mathbf{D}_{no}}{R^3} \tag{19-10}$$

We shall see that only the **b** component in the numerator **R** contributes appreciably. Choosing **b** in the x direction and **v** in the z direction, gives

$$V_{no}(t) = \frac{ze^2}{(b^2 + v^2 t^2)^{3/2}} (bD^x_{no} + vtD^z_{no}) \tag{19-11}$$

The integration over t gives

$$T_{no} = \hbar^{-1} ze^2 \int_{-\infty}^{\infty} dt \, \frac{e^{i\omega_{no} t}}{(b^2 + v^2 t^2)^{3/2}} (bD^x_{no} + vtD^z_{no})$$

$$= \frac{2ze^2}{\hbar bv} [\beta K_1(\beta) D^x_{no} + i\beta K_0(\beta) D^z_{no}] \tag{19-12}$$

where $\beta = \omega_{no} b/v$ and K is the Bessel function of imaginary argument.[1] Using the behavior of K for small and large argument, or simply evaluating the integral in (19-12) approximately, we can write this as

$$T_{no} = \frac{2e^2 z}{\hbar v b} D^x_{no} \qquad b < \frac{v}{\omega_{no}}$$

$$= \quad 0 \qquad\qquad b > \frac{v}{\omega_{no}} \tag{19-13}$$

We should really have $\ll$ and $\gg$, but (19-13) is sufficiently accurate with $<$ and $>$. It is seen from (19-13) that the **v**t component of **R** does not contribute in this approximation.

The probability of all inelastic transitions is then

$$\sum_n |\langle n | T | 0 \rangle|^2 = \left(\frac{2e^2 z}{\hbar v b}\right)^2 \sum_n |\langle n | D^x | 0 \rangle|^2 \tag{19-14}$$

For a given b, the condition (19-13) $\omega_{no} < v/b$ permits only excited states up to a certain energy. However, the matrix element $\langle n | D^x | 0$ is very small for high excitation, therefore the sum in (19-14) may be extended over all states n and evaluated by closure,

$$\sum_n |\langle n | D^x | 0 \rangle|^2 = \langle 0 | (D^x)^2 | 0 \rangle = \tfrac{1}{3} \langle 0 | D^2 | 0 \rangle$$

$$= \tfrac{1}{3} \langle 0 | \sum_i r_i^2 + 2 \sum_{j<i} \mathbf{r}_i \cdot \mathbf{r}_j | 0 \rangle \tag{19-15a}$$

In the above we are taking the ground state to be isotropic. Neglecting the correlation $\mathbf{r}_i \cdot \mathbf{r}_j$ and assuming that the state of the atom can be described by electron orbitals, we have

$$\sum_n |\langle n | D^x | 0 \rangle|^2 = \tfrac{1}{3} \sum_\alpha z_\alpha r_\alpha^2 \tag{19-15b}$$

Here z_α is the number of electrons in shell α, and r_α^2 is the mean square radius of the shell. Thus we see that the transition probability for transitions from the ground state, is

$$\sum_n |\langle n | T | 0 \rangle|^2 = \left(\frac{2e^2 z}{\hbar v b}\right)^2 \sum_n |\langle n | D^x | 0 \rangle|^2$$

$$a < b < \frac{v}{\overline{\omega}} \tag{19-16}$$

---

[1] G. N. Watson, *A Treatise on the Theory of Bessel Functions*, 2nd ed., Cambridge University Press, Cambridge, 1952, p. 78.

where $\hbar\bar{\omega}$ is an average excitation energy for those excited states n for which $|\langle n|D^x|0\rangle|^2$ is sizeable. This verifies (15-14), (15-15), (15-16), and (15-20). Thus we see that the transition probability from the ground state varies as $1/b^2$ when b lies between the limits a and $v/\bar{\omega} \approx (v/u)a$ where u is the orbital velocity of the atomic electrons. If the particle is an electron, this is (ka)a [see (15-20), (15-21), (15-22)]. Thus the atom can be excited by a fast charged particle passing by the atom at distances much larger than the extent of the atom.

This simple result has to be corrected for the following two effects. The first is the *density effect*. It arises from the fact that the interaction between an atom and an itinerant charged particle is modified (usually weakened) by the pressence of other atoms between the charged particle and the atom. To account for this, the dielectric properties of the material, as a function of density and frequency, have to be included. This reduces b_{max}.

The second modification arises from a relativistic effect. This effect can be included in our nonrelativistic theory phenomenologically by considering the electric field of the charged particle as being contracted in the direction of motion by the Lorentz contraction factor γ. Then the condition (19-13) is modified to read

$$\frac{b}{\gamma} < \frac{v}{\bar{\omega}} \tag{19-17}$$

which has the effect of increasing b_{max} by the factor γ. Thus a highly relativistic, charged particle ($\gamma \gg 1$) can produce ions at distance of the order of microns away from its path.

The two effects are seen to work against each other. It turns out that the density effect is the stronger of the two, and cuts the ionization distance off at a definite limit for very high particle energy. For details of both effects, the reader is referred to Bethe and Ashkin.[2]

The theory outlined here is valid only if

$$b > a \tag{19-18}$$

For smaller impact parameters, the expansion (19-8) is no longer justified, hence the following equations, particularly (19-13), no longer hold. However, (19-7) remains valid as long as

$$ka \gg 1 \qquad E \gg V(b) \tag{19-19}$$

---

[2] H. A. Bethe and J. Ashkin, "Passage of Radiation through Matter" in *Experimental Nuclear Physics*, Vol. 1, E. Segrè (ed.), Wiley, New York, 1953, pp. 166-190.

EQUIVALENCE WITH BORN APPROXIMATION

We now demonstrate the equivalence of the semiclassical and Born approximation in the calculation of the inelastic cross section *integrated over angles*. We now take the impact parameter to be arbitrary (i.e., we no longer assume b >> a), so as to derive the total cross section for the transition m → n. We can therefore no longer use the expansion (19-8) but must use (19-7) directly. To evaluate this, we introduce the Fourier transform of V,

$$\mathcal{V}(\mathbf{q}) = \int V(\mathbf{r})\, e^{-i\mathbf{q}\cdot\mathbf{r}}\, d\tau \qquad (19\text{-}20)$$

Then we have

$$V(\mathbf{R} - \mathbf{r}_i) = (2\pi)^{-3} \int d^3q\, \mathcal{V}(\mathbf{q})\, e^{i\mathbf{q}\cdot(\mathbf{R}-\mathbf{r}_i)} \qquad (19\text{-}21)$$

This permits a separation of the coordinates **R** and $\mathbf{r}_i$ and we may write

$$V_{nm}(t) = (2\pi)^{-3} \int d^3q\, \mathcal{V}(\mathbf{q})\, e^{i\mathbf{q}\cdot\mathbf{R}}\, F_{nm}(\mathbf{q}) \qquad (19\text{-}22)$$

where

$$F_{nm}(\mathbf{q}) = \int \sum_i e^{-i\mathbf{q}\cdot\mathbf{r}_i}\, \psi_n^*(A)\, \psi_m(A)\, dA \qquad (19\text{-}23)$$

is the form factor for the transition m → n which is familiar from the Born approximation. The relation to that approximation is beginning to emerge.

We put the z axis in the direction of the velocity **v**, and find for (19-6), using (19-21),

$$T_{nm} = (2\pi)^{-3}\hbar^{-1} \int d^3q\; \mathcal{V}(\mathbf{q})\, F_{nm}(\mathbf{q})\, e^{i(q_xX+q_yY)} \times \int_{-\infty}^{\infty} dt\, e^{iq_zvt+i\omega_{nm}t} \qquad (19\text{-}24)$$

where X, Y are the components of the vector **b**. The integral over t is well known and gives

$$\frac{2\pi}{v}\,\delta\!\left(q_z + \frac{\omega_{nm}}{v}\right) \qquad (19\text{-}25)$$

This means that q_z is fixed and has the value

$$q_z = -\frac{E_n - E_m}{\hbar v} \qquad (19\text{-}26)$$

This result also follows from energy and momentum conservation, if we make our assumptions (1) straight line motion of the particle and

(2) energy loss of the particle a small fraction of its initial energy. The second assumption gives

$$k_n = k_m - \frac{1}{\hbar}\frac{dp}{dE}(E_n - E_m) = k_m - \frac{1}{\hbar v}(E_n - E_m) \qquad (19\text{-}27)$$

By definition of **q**, for *any* deflection of the particle,

$$k_n^2 = (\mathbf{k}_m + \mathbf{q})^2 = k_m^2 + 2k_m q_z + q_x^2 + q_y^2$$

$$q_z \approx k_n - k_m - \frac{q_x^2 + q_y^2}{2k_m} = -\frac{E_n - E_m}{\hbar v} - \frac{q_x^2 + q_y^2}{2k_m} \qquad (19\text{-}28)$$

(We have dropped q_z^2 which is clearly small compared to $2\mathbf{k}_m \cdot \mathbf{q}$.) The first term is identical with (19-26). The second term is small if the particle is not appreciably deflected. However, even if the second term is taken into account, the result is not substantially changed: (19-25) will be replaced by

$$\frac{2\pi}{v}\,\delta(q_z + c)$$

where c is some quantity, small compared with q or 1/a whichever is larger. Integration of the δ function in (19-25) over q_z therefore gives unity, and in $F_{nm}(q)$ in (19-24) it is permissible to set

$$q^2 = q_x^2 + q_y^2$$

The cross section in the classical case is now

$$\sigma_{m\to n} = \iint_{-\infty}^{\infty} dX\, dY\, |T_{nm}|^2 \qquad (19\text{-}29)$$

Inserting (19-24) and (19-25),

$$\sigma_{m\to n} = (2\pi)^{-4}(\hbar v)^{-2}\int dX\, dY \int dq_x\, dq_y\, dq'_x\, dq'_y$$

$$\times\, \mathcal{V}(q)\,\mathcal{V}^*(q')\,F_{nm}(q)\,F^*_{nm}(q')\, e^{i(q_x - q'_x)X + i(q_y - q'_y)Y} \qquad (19\text{-}30)$$

All integrals are from $-\infty$ to ∞. We integrate first over X and Y:

$$\int dX\, e^{i(q_x - q'_x)X} = 2\pi\delta(q_x - q'_x) \qquad (19\text{-}31)$$

Then, since q_z is fixed we write $dq_x\, dq_y = 2\pi q\, dq$ giving for spherically symmetric V

$$\sigma_{m\to n} = (2\pi)^{-1}(\hbar v)^{-2}\int_{q_z}^{\infty} q\, dq\, |\mathcal{V}(q) F_{nm}(q)|^2 \qquad (19\text{-}32)$$

From the Born approximation, we have shown in (17-21b) that

$$\sigma_{m\rightarrow n} = \frac{1}{(2\pi)(\hbar v_m)^2} \int_{q_{min}}^{q_{max}} q\, dq \, |T_{nm}(q)|^2 \qquad (19\text{-}33)$$

$$T_{nm} = \mathcal{V}(q) F_{nm}(q) \qquad (19\text{-}34)$$

with $\mathcal{V}$ and F_{nm} defined in (19-20) and (19-23). This is identical with (19-32).

In the classical approximation, we have never assumed that $\hbar v > Ze^2$, in other words, we have *not* assumed the validity of the Born approximation for the motion of the scattered particle in the field of the atom. In particular, it is perfectly permissible for the phase shifts of the scattered particle to be large. We have shown that nevertheless the Born approximation gives the correct total cross section for any inelastic transition $m \rightarrow n$ if only the conditions for the *classical* approximation, (19-19), are satisfied. The first of these, $ka \gg 1$ is much weaker than the condition for the Born approximation if the scattered particle is heavy. Thus our theory is particularly useful for the inelastic scattering of heavy particles. We have shown that the *integrated cross section for the inelastic scattering can be calculated by the Born approximation* even if the velocity of the particle is comparable to, or smaller than, that of the electrons in the atom, $v \leq Ze^2/\hbar$. This result has been extensively used in accurate calculations of the stopping power.[4]

We still have to discuss the second condition (19-19). The potential near the nucleus diverges as $1/r$ so that for small enough impact parameter b, this second condition is always violated. However, if the particle energy $E \gg Ry$ and $m \gg m_e$, then this violation occurs only for $b \ll a$. Then the contribution of these small impact parameters is a negligible fraction of the total inelastic cross section.

For electron scattering, our theory is much less useful. Even here, however, if we assume the atomic radius a to be of the order of the Bohr radius a_0, the condition $ka \gg 1$ only means

$$\hbar v \gg e^2 \qquad (19\text{-}35)$$

It can then be argued that at least for the excitation of the *outer* atomic electrons the Born approximation should give a good approximation for the integrated cross section, even though the condition $\hbar v > Ze^2$ is not fulfilled. For the excitation of K electrons by electrons, $\hbar v > Ze^2$ is necessary just for energetic reasons. More generally, Mott and Massey have given the empirical rule

$$E > 7I \qquad (19\text{-}36)$$

for the validity of the Born approximation for the total cross section for excitation of an electron of ionization energy I by an itinerant electron o energy E.

All of our calculations in this and the preceding section are of course based on the assumption that we may use first-order, time-dependent pe turbation theory for the calculation of inelastic transitions, as we did in deriving (19-6). This is justified provided $\Sigma_n |T_{mn}|^2 \ll 1$ and can be verified in most practical cases. Sometimes the condition $\Sigma_n |T_{mn}|^2 \ll$ is not necessary, but two successive excitations of two different atomic electrons by the passage of the same particle can be permitted without invalidating the theory.

It need hardly be pointed out that the equivalence of the semiclassical theory with the Born approximation holds only for the integrated cross section, not for the angular distribution. The latter, as we discussed in Chapter 18, exhibits characteristic diffraction phenomena which are explainable neither by classical mechanics nor by the Born approximation.

PROBLEMS

1. Verify the integration (19-12) and the approximations (19-13).
2. Justify the neglect of the correlation term in (19-15) in the Hartree approximation.
3. Derive equations (19-6) and (19-7).
4. Derive equations (19-22) and (19-24).
5. Justify (19-29).

Chapter 20

CLASSICAL LIMIT OF QUANTUM MECHANICAL SCATTERING; SUMMARY

It is of interest to exhibit the relationship between the quantum mechanical and classical descriptions of the elastic scattering process, and investigate the conditions under which the two descriptions are equivalent. We begin with the classical description.

Recall that a classical particle in a central potential V(r) moves in a plane, which may be described by circular coordinates r and φ. The orbital equation[1]

$$\varphi = \int_{r_i}^{r} \frac{j \, d\tau'}{r'^2 \sqrt{2m(E - V(r')) - j^2/r'^2}} + \varphi_i \tag{20-1}$$

gives the connection between φ and r, and thus defines the classical trajectory. Here j is the angular momentum and E the energy of the particle. The initial conditions are specified by r_i and φ_i. (20-1) can also be written as[2]

$$\varphi = -\int_{r_i}^{r} dr' \frac{\partial}{\partial j} \sqrt{2m(E - V(r')) - j^2/r'^2} + \varphi_i \tag{20-2}$$

The scattering angle θ is then given by

---

[1]For example, H. Goldstein, *Classical Mechanics*, 2nd ed., Addison-Wesley, Reading MA., 1980, p. 105.

[2]We concern ourselves only with the case that (20-2) defines a unique relation between φ and j. More general cases are discussed by N. F. Mott and H. S. W. Massey, *The Theory of Atomic Collisions*, 3rd ed., Oxford University Press, Oxford, 1965.

$$\mp\theta = \pi + 2\int_{r_0}^{\infty} dr' \frac{\partial}{\partial j} \sqrt{2m(E - V(r') - j^2/r'^2} \tag{20-3}$$

where r_0 is the largest root of $2m(E - V) - j^2/r^2$. The upper (lower) sign holds for an attractive (repulsive) field. The differential cross section is[2]

$$\frac{d\sigma}{d\Omega} = -\frac{b}{\sin\theta}\frac{db}{d\theta} \tag{20-4}$$

where b is the impact parameter, related to the angular momentum by $b = j/mv_0 = j/p$. (Note that $d\theta/db < 0$.) Thus

$$\frac{d\sigma}{d\Omega} = -\frac{j}{p^2 \sin\theta}\frac{\partial j}{\partial\theta} \tag{20-5}$$

where the relationship between j and θ is given by (20-3).

The quantum mechanical description consists in solving

$$\frac{d^2u}{dr^2} + \left[\frac{2m}{\hbar^2}(E - V) - \frac{\ell(\ell+1)}{r^2}\right] u = 0 \tag{20-6}$$

leading to the asymptotic behavior

$$u \xrightarrow[r\to\infty]{} \sin(kr - \tfrac{1}{2}\ell\pi + \delta_\ell)$$

$$\hbar k = p = \sqrt{2mE} \tag{20-7}$$

Then the scattering amplitude is as usual

$$f(\theta) = \frac{1}{2ik}\sum_{\ell}(2\ell + 1)(e^{2i\delta_\ell} - 1)P_\ell(\cos\theta) \tag{20-8}$$

and the differential scattering cross section

$$\frac{d\sigma}{d\Omega} = |f(\theta)|^2 \tag{20-9}$$

In examining the classical limit of these expressions it is natural to make use of the semiclassical WKB method in solving (20-6). The WKB phase shifts were given before, (1-25b).

$$\delta_\ell^{WKB} = \lim_{r\to\infty}\left[\int_a^r \Phi^{1/2} dr' - kr + \left(\ell + \frac{1}{2}\right)\frac{\pi}{2}\right] \tag{1-25b}$$

$$\Phi(r) = \frac{2m}{\hbar^2}(E - V) - \frac{(\ell + \frac{1}{2})^2}{r^2}$$

a is the largest root of Φ.

The WKB phase shifts are now to be substituted into (20-8). If we restrict ourselves to nonforward scattering, we can simplify (20-8) by noting that[3]

$$\sum_{\ell=0}^{\infty} (2\ell + 1) P_\ell(u) = 2\,\delta(u - 1) \tag{20-10}$$

Hence[4]

$$f(\theta)_{\theta \neq 0} = \frac{1}{2ik} \sum_{\ell=0}^{\infty} (2\ell + 1)\, e^{2i\delta_\ell} P_\ell(\cos\theta) \tag{20-11}$$

Since WKB phase shifts are in general large for all ℓ up to some large ℓ, we may replace the sum over ℓ by an integral, and use the WKB approximation for P_ℓ

$$P_\ell(\cos\theta) \approx \frac{-i}{\sqrt{2\pi\ell \sin\theta}} \left\{ \exp i\left[(\ell + \tfrac{1}{2})\theta + \frac{\pi}{4}\right] - \exp -i\left[(\ell + \tfrac{1}{2})\theta + \frac{\pi}{4}\right] \right\} \tag{20-12}$$

Inserting these in (20-11)

$$f(\theta) \approx -\frac{1}{k\sqrt{2\pi \sin\theta}} \int_0^\infty d\ell\, \sqrt{\ell} \left\{ \exp i\left[2\delta_\ell + (\ell + \tfrac{1}{2})\theta + \frac{\pi}{4}\right] - \exp i\left[2\delta_\ell - (\ell + \tfrac{1}{2})\theta - \frac{\pi}{4}\right] \right\} \tag{20-13}$$

The exponentials oscillate rapidly due to the large magnitude of δ_ℓ except near the extremum of the exponents as a function of ℓ. The main contribution will therefore come from the neighborhood of this extremum (stationary phase approximation). The extremum will occur when one of the following two equations is fulfilled

---

[3]Problem 1 concerns itself with deriving this result.

[4]Strictly speaking, the sum in (20-11) does not converge for large ℓ while that in (20-8) does. We may insert in both formulas a convergence factor $e^{-\alpha\ell}$ and then go to the limit $\alpha = 0$; this will not make any difference in the following.

$$\frac{\partial}{\partial \ell}\left[2\delta_\ell + (\ell + \tfrac{1}{2})\theta + \frac{\pi}{4}\right] = 2\frac{\partial \delta_\ell}{\partial \ell} + \theta = 0 \qquad \text{(20-14a)}$$

$$\frac{\partial}{\partial \ell}\left[2\delta_\ell - (\ell + \tfrac{1}{2})\theta - \frac{\pi}{4}\right] = 2\frac{\partial \delta_\ell}{\partial \ell} - \theta = 0 \qquad \text{(20-14b)}$$

or

$$2\hbar \frac{\partial \delta_\ell}{\partial j} = \mp\theta \qquad \text{(20-14c)}$$

where $j = \hbar\ell$. Evaluating this from the WKB formula (1-25b)

$$2\frac{\partial \delta_\ell}{\partial j} = 2 \lim_{r\to\infty}\left[\int_a^r \hbar \frac{\partial \Phi^{1/2}(r')}{\partial j}\, dr' + \frac{\pi}{2}\right]$$

$$= 2\int_a^\infty \frac{\partial}{\partial j}\left[2m(E - V) - \frac{j^2}{r'^2}\right]^{1/2} dr' + \pi = \mp\theta \qquad \text{(20-15)}$$

The quantum mechanical scattering is seen to be sizeable only at angles satisfying (20-15); at other angles it is negligible by destructive interference. But (20-15) is identical with the classical equation determining the scattering angle (20-3). Thus we see that under the assumptions implicit in the above, viz., validity of WKB and nonforward scattering, the quantum mechanical and classical scattering angles are the same for a given angular momentum j.

The differential cross sections are also the same. To see this we assume that (20-14a) holds at a value of $\ell = \ell_0$, i.e.,

$$2\left.\frac{\partial \delta_\ell}{\partial \ell}\right|_{\ell=\ell_0} = -\theta$$

Then

$$2\delta_\ell + (\ell + \tfrac{1}{2})\theta + \frac{\pi}{4} \approx 2\delta_{\ell_0} + (\ell_0 + \tfrac{1}{2})\theta + \frac{\pi}{4}$$

$$+ \frac{\partial^2 \delta_{\ell_0}}{\partial \ell_0^2}(\ell - \ell_0)^2 \qquad \text{(20-16)}$$

Setting

$$A(\ell_0) \equiv 2\delta_{\ell_0} + (\ell_0 + \tfrac{1}{2})\theta + \frac{\pi}{4}$$

$$\frac{\partial^2 \delta_{\ell_0}}{\partial \ell_0^2} = -\frac{\hbar}{2}\frac{\partial\theta}{\partial j} \equiv \alpha \qquad (20\text{-}17)$$

then

$$f(\theta) = -\frac{1}{k\sqrt{2\pi \sin\theta}}\int_0^\infty d\ell \sqrt{\ell}\, e^{iA(\ell_0)}\, e^{i\alpha(\ell-\ell_0)^2}$$

$$\approx -e^{iA(\ell_0)}\sqrt{\frac{\ell_0}{2\pi k^2 \sin\theta}}\int_{-\infty}^{\infty} d\ell\, e^{i\alpha(\ell-\ell_0)^2}$$

$$= -e^{iA(\ell_0)+i\pi/4}\sqrt{\frac{\ell_0}{2k^2\alpha\sin\theta}}$$

$$= e^{i[A(\ell_0)-3\pi/4]}\sqrt{\frac{-j}{p^2\sin\theta}\frac{\partial j}{\partial\theta}} \qquad (20\text{-}18)$$

The absolute square of this is seen to agree with the classical cross section (20-5). Thus the quantum mechanical description of a single scattering, in the classical limit, is seen to coincide with the classical description. However even in the classical limit, the quantum mechanical result contains a phase factor. If there are several scatterers, the waves scattered from these will interfere, and then the phase factor will be important.

The classical limit, as stated above requires four assumptions, viz., (1) validity of the WKB, (2) $ka \gg 1$, (3) phase shift $\delta \gg 1$, and (4) nonforward scattering. The first condition is obvious; the second is necessary so that many ℓ contribute; the third is obviously the opposite of the Born approximation. For atoms it requires

$$\hbar v \ll Z e^2 \ln ka \qquad (20\text{-}19)$$

where a is the atomic radius. To fulfill this and yet to have $ka \gg 1$ we must consider heavy particles (i.e., not electrons) of moderately low velocity.

We should actually add a fifth condition: The calculation will not hold unless the value of ℓ_0 which is deduced from (20-14c) is substantial, let us say

$$\ell_0 \gtrsim 10 \qquad (20\text{-}20)$$

Only if this is fulfilled can (20-13) be evaluated by the method of stationary phase. If we have just a Coulomb field, the *classical* deflection is about

$$\theta = \frac{2Ze^2}{vbp} = \frac{2Ze^2}{v\hbar\ell_0} \tag{20-21}$$

where b is the impact parameter. Therefore condition (20-20) means that the classical approximation will only be good for

$$\theta < \theta_1 \simeq \frac{Ze^2}{5\hbar v} \tag{20-22}$$

For greater angles, direct summation of the partial waves will be required. However, in most cases of scattering of heavy particles, greater angles will mean impact parameters b well inside the Fermi radius of the atom, $a_0 Z^{-1/3}$ and therefore the phase shifts will be well approximated by those in the pure Coulomb field Ze^2/r plus a constant. We then get simple Coulomb scattering which is accidentally equal to the classical scattering in the same potential.

Finally, condition (4) is understood from elementary diffraction theory: A scatterer of radius a will diffract the waves through a "diffraction angle"

$$\theta_0 \approx \frac{1}{ka} \tag{20-23a}$$

Inside this angle, "geometrical optics," i.e., classical mechanics, is not valid. Thus our result is restricted to the interval

$$\theta_0 < \theta < \theta_1 \tag{20-23b}$$

FORWARD SCATTERING

We complete our study of the classical limit of scattering by examining the behavior of the cross section for forward scattering. We assume that the phase shifts are large (>1) for all ℓ up to some maximum $\ell = \ell_1$ and small (<1) for $\ell > \ell_1$. If we assume that the interacting region is bounded by a sphere of radius a, we have ℓ_1 = ka. Under these conditions the phase shift sum can be approximated by

$$f(\theta) \approx \frac{1}{2ik} \int_0^{\ell_1} 2\ell\,(e^{2i\delta_\ell} - 1)\, P_\ell(\cos\theta)\, d\ell \tag{20-24}$$

At small angles, P_ℓ does not vary rapidly with ℓ; then destructive

Table 20-1
Summary of Calculational Methods

Velocity range	Other conditions	Method
A. $v \gg Ze^2/\hbar$	None	Born approximation
B. $v \geq 2$ to $4e^2/\hbar$	None	Elastic: sum Rayleigh series Inelastic: Distorted waves
C. $mv \gg \hbar/a$	None	WKB may be used for calculating phase shifts and wave functions.
	(1) $\delta \gg 1$	Elastic: angular distribution = Classical limit (Chapter 20)
	(2) $\frac{1}{2}mv^2 \gg E_n - E_0$	Inelastic: total cross section for excitation of any given level (discrete or continuous) is equal to Born approximation (Chapter 19).
D. $v < 4e^2/\hbar$	None	Absorption and polarization of atom must be included. In some cases, repeated inelastic scattering must be considered. For electrons, exchange must be included.

interference eliminates the contribution from the exponential $e^{2i\delta_\ell}$. Explicitly, for small θ we may replace $P_\ell(\cos\theta)$ by $J_0(\ell\theta)$ which gives

$$f(\theta) = \frac{i}{k}\int_0^{\ell_1} \ell J_0(\ell\theta)\, d\ell = \frac{i}{k\theta}\,\ell_1 J_1(\ell_1\theta)$$

$$= \frac{ia}{\theta} J_1(ka\theta) \tag{20-25}$$

The cross section for forward scattering becomes

$$\frac{d\sigma}{d\Omega} = \frac{a^2}{\theta^2} J_1^2(ka\theta) \tag{20-26}$$

This is a completely nonclassical result, arising from the wave properties of the incident particle. Indeed (20-26) is identical with the expression for the intensity of light undergoing Fraunhofer diffraction around a black sphere of radius a.

SUMMARY OF METHODS

In Table 20-1, we give a summary of the methods we have discussed.

The calculation is simple in cases A and C(1). In the condition for C, a is the atomic radius. The condition C(2) is automatically fulfilled for electrons if condition C is satisfied. Cases C(1) and C(2) are very useful for heavy particles but not for electrons.

Mott and Massey[2] (pp. 323-448) give some further approximate methods for dealing with special cases in the difficult category D.

PROBLEMS

1. Derive (20-10c).
2. Verify the sign convention in equation (20-3).
3. Using the WKB approximation on the differential equation defining the Legendre polynomials, derive (20-12).
4. Verify equation (20-18).
5. Verify equation (20-21).

Part IV
RELATIVISTIC EQUATIONS

THIS part is devoted to an introduction to relativistic quantum mechanics, that is to generalizations of Schrödinger's equation which include relativistic effects. As we shall see, these extensions are only partially successful. Although well defined predictions, for example, of the energy levels of the hydrogen atom or of scattering, can be given, and they agree well with experiment, the theory is not really self-contained in that the solutions of what is hoped are one-particle equations make reference to many particles. This state of affairs is a reflection of the fact that a definite number of particles is not a relativistically invariant concept. The proper way to take into account this possibility of particle creation and annihilation is through relativistic quantum field theory, which however is beyond the scope of this book.

THIS part is devoted to an introduction to relativistic quantum mechanics, that is to generalizations of Schrödinger's equation which include relativistic effects. As we shall see, these extensions are only partially successful. Although well defined predictions, for example, of the energy levels of the hydrogen atom or of scattering, can be given, and they agree well with experiment, the theory is not really self-contained in that the solutions of what is hoped are one-particle equations make reference to many particles. This state of affairs is a reflection of the fact that a definite number of particles is not a relativistically invariant concept. The proper way to take into account this possibility of particle creation and annihilation is through relativistic quantum field theory, which however is beyond the scope of this book.

Chapter 21

KLEIN-GORDON EQUATION

The nonrelativistic Schrödinger equation for a free particle can be obtained by replacing E by $i\hbar(\partial/\partial t)$ and $\mathbf{p}$ by $(\hbar/i)\nabla$ in the nonrelativistic equation

$$\frac{p^2}{2m} = E \tag{21-1}$$

If the same substitution is performed in the relativistic equation,

$$c^2 p^2 + m^2 c^4 = E^2 \tag{21-2}$$

the *Klein-Gordon* equation is obtained:

$$-\hbar^2 \frac{\partial^2}{\partial t^2} \psi = -\hbar^2 c^2 \nabla^2 \psi + m^2 c^4 \psi \tag{21-3}$$

This equation was also derived by Schrödinger.

We wish to obtain quantities ρ and $\mathbf{j}$, analogous to the probability density and current of the Schrödinger equation. These quantities must satisfy the following criteria: ρ is real; $\int \rho \, d\tau$ is time-independent and transforms as a scalar under Lorenz transformations; and a continuity equation of the form

$$\nabla \cdot \mathbf{j} + \frac{\partial \rho}{\partial t} = 0 \tag{21-4}$$

is satisfied. To achieve this we define

$$\rho = \frac{i\hbar}{2mc^2}\left(\psi^* \frac{\partial\psi}{\partial t} - \psi \frac{\partial\psi^*}{\partial t}\right) = \frac{\hbar}{mc^2} \operatorname{Im}\psi \frac{\partial\psi^*}{\partial t} \qquad (21\text{-}5a)$$

$$\mathbf{j} = \frac{\hbar}{2im}(\psi^* \nabla\psi - \psi\nabla\psi^*) \qquad (21\text{-}5b)$$

It is clear that ρ is real; it can be verified that (21-4) is satisfied. Considering the four-vector ($c\rho$, $\mathbf{j}$) and requiring that ψ fall off sufficiently rapidly for large r, (21-4) implies that $\int\rho\, d\tau$ is Lorentz invariant and time independent.

Equation (21-3) has the solution

$$\psi = e^{i(\mathbf{k}\cdot\mathbf{r}-\omega t)} \qquad (21\text{-}6)$$

where

$$\hbar^2\omega^2 = (\hbar^2 c^2 k^2 + m^2 c^4) \qquad (21\text{-}7)$$

To regain (21-2) we require

$$E = \hbar\omega \qquad \mathbf{p} = \hbar\mathbf{k} \qquad (21\text{-}8)$$

then

$$E = \pm c\sqrt{p^2 + m^2c^2} \qquad (21\text{-}9)$$

It is seen that the Klein-Gordon equation admits solutions with negative energy.

PHYSICAL INTERPRETATION OF THE KLEIN-GORDON EQUATION

The Klein-Gordon equation can be written in an invariant-looking form:

$$\left(\Box + \frac{m^2c^2}{\hbar^2}\right)\psi = 0$$

$$\Box \equiv \frac{1}{c^2}\frac{\partial^2}{\partial t^2} - \nabla^2 \qquad (21\text{-}10)$$

Since ψ has only one component, it must transform as a scalar under Lorentz transformations.

$$\psi'(x') = \psi(x) \qquad (21\text{-}11)$$

Here x' is the Lorentz-transformed space-time point. This also implies that the particles described by ψ must have no degrees of freedom other than translations in space-time. In particular the Klein-Gordon equation (without the addition of subsidiary equations) can only describe particles of zero spin such as π or K mesons.

Replacing $i\hbar(\partial\psi/\partial t)$ by $E\psi$ in (21-5a) we obtain

$$\rho = \frac{E}{mc^2}|\psi|^2 \tag{21-12}$$

In the nonrelativistic limit, $E \approx mc^2$; ρ and j reduce to the nonrelativistic expression for position probability density and probability current. However ρ cannot be interpreted as a probability density nor can ψ be considered a position probability amplitude. This is because (21-5a) can be either positive or negative, while a probability density must be non-negative. The indefinite sign of (21-5a) arises from the fact that ψ *and* $\partial\psi/\partial t$ can (and must) be specified arbitrarily at an initial time t = 0 in order to determine the solution $\psi(\mathbf{r},t)$, since the Klein-Gordon equation is of second order in t, rather than of first order as is the nonrelativistic Schrödinger equation. Thus if for some value of $\partial\psi/\partial t$ the expression (21-5a) is positive, then since the time derivative is arbitrary, $-\partial\psi/\partial t$ is also an acceptable value, and for this choice ρ becomes negative. Since (21-3) is second order in the time derivative, ρ must contain the first derivative in order to satisfy the continuity equation (21-4). From (21-12) it is seen that it is the existence of negative-energy solutions (21-9) which permits ρ to become negative. A further consequence of these properties of the Klein-Gordon equation is that the quantum mechanical postulate that ψ is determined by its initial value at one time is no longer true.

Owing to these difficulties, the Klein-Gordon equation was at first discarded. It was Pauli and Weisskopf who provided the valid interpretation of the Klein-Gordon equation by considering it as a classical field equation (analogous to the electromagnetic field equations) and then quantizing it. It then becomes reasonable to interpret ρ and j as the charge and current densities of the particles of this field. Moreover, the energy of the quantized field theory is always positive.[2]

For a free particle the negative energy solutions need not be considered, for if the free particle has energy greater than mc^2 it will always have this energy and never will undergo a transition to a state with energy less than $-mc^2$. In this case ρ remains positive definite, and it might be asked whether or not a probabilistic interpretation is possible. This problem has been investigated by Newton and Wigner, who give an affirmative answer but find that the position operator x no longer has as its eigenfunctions $\delta(x - x')$ but instead a fairly complicated

---

[2]See for example J. D. Bjorken and S. D. Drell, *Relativistic Quantum Fields*, McGraw-Hill, New York, 1965.

function in which the particle is not localized at a point but is spread out over a region of order ħ/mc.

In the presence of external fields, transitions to states of negative E can occur. Such states, according to the theory of Pauli and Weisskopf, should be interpreted as particles of negative charge (if positive E corresponds to positive charge) but of positive energy. Transitions from a state of positive to one of negative E are interpreted as the production (or annihilation) of a pair of particles of opposite charge. This will be discussed in more detail in Chapter 23 in connection with the Dirac equation.

INTERACTION WITH EXTERNAL ELECTROMAGNETIC FIELD

To include the effects of an external electromagnetic field described by potentials A, we make the usual minimal replacement

$$\mathbf{p} \to \mathbf{p} - \frac{e}{c}\mathbf{A}$$

$$E \to E - e\varphi \tag{21-13}$$

and obtain

$$\left(i\hbar\frac{\partial}{\partial t} - e\varphi\right)^2 \psi = \left(\frac{c\hbar}{i}\nabla - e\mathbf{A}\right)^2 \psi + m^2 c^4 \psi \tag{21-14}$$

Obviously we are taking the particle described by the wave function ψ to have charge e.

With the substitution

$$\psi(\mathbf{r}, t) = \psi'(\mathbf{r}, t)\, e^{-(imc^2 t/\hbar)} \tag{21-15}$$

equation (21-14) reduces to

$$-\hbar^2\frac{\partial^2\psi'}{\partial t^2} + 2i\hbar[mc^2 - e\varphi]\frac{\partial\psi'}{\partial t}$$

$$-e\varphi\left[2mc^2 - e\varphi + i\hbar\frac{\partial \log\varphi}{\partial t}\right]\psi'$$

$$= [-\hbar^2 c^2 \nabla^2 + 2ie\hbar c\mathbf{A}\cdot\nabla + ie\hbar c(\nabla\cdot\mathbf{A}) + e^2 A^2]\psi' \tag{21-16a}$$

Assuming

$$\left|i\hbar\frac{\partial\psi'}{\partial t}\right| = O(e\varphi\psi')$$

and neglecting $e\varphi$ with comparison to mc^2 equation (21-16a) becomes

$$i\hbar \frac{\partial \psi'}{\partial t} = \left[\frac{-\hbar^2}{2m}\nabla^2 + \frac{ie\hbar}{mc}\mathbf{A}\cdot\nabla + \frac{ie\hbar}{2mc}\nabla\cdot\mathbf{A} + \frac{e^2 A^2}{2mc^2} + e\varphi\right]\psi' \tag{21-16b}$$

which is the nonrelativistic electromagnetic Schrödinger equation.
It can be shown that under gauge transformation

$$\mathbf{A} \to \mathbf{A}' = \mathbf{A} + \nabla\chi$$

$$\varphi \to \varphi' = \varphi - \frac{1}{c}\frac{\partial \chi}{\partial t}$$

ψ changes only by a phase factor $\exp(ie\chi/\hbar c)$, where χ is the gauge function.

If further potentials are to be included in the Klein-Gordon equation their Lorentz transformation properties must be studied. If they transform as four-vectors they are to be added to $(c\varphi, \mathbf{A})$. If they transform as relativistic scalars, they can be added to mc^2.

COULOMB FIELD

We now solve the Klein-Gordon equation where the external potential is Coulombic. Setting

$$\mathbf{A} = 0 \qquad e\varphi(\mathbf{r}) = -\frac{Ze^2}{r} \qquad \psi(\mathbf{r}, t) = R(r)Y_{\ell m}(\theta, \varphi)e^{-iEt/\hbar}$$

$$\rho = \alpha r \qquad \alpha^2 = \frac{4(m^2c^4 - E^2)}{\hbar^2 c^2} \qquad \lambda = \frac{2E\gamma}{\hbar c\alpha}$$

$$\gamma = \frac{Ze^2}{\hbar c} \qquad E = mc^2\left(1 + \frac{\gamma^2}{\lambda^2}\right)^{-1/2} \tag{21-17}$$

we obtain from (21-14)

$$\frac{1}{\rho^2}\frac{d}{d\rho}\left(\rho^2\frac{dR}{d\rho}\right) + \left(\frac{\lambda}{\rho} - \frac{1}{4} - \frac{\ell(\ell+1) - \gamma^2}{\rho^2}\right)R = 0 \tag{21-18}$$

This describes a spinless particle in a Coulomb field. It is not a description of the hydrogen atom, since electrons have spin $\frac{1}{2}$.

Equation (21-18) is like the Schrödinger equation for the hydrogen atom, except that the relativistic correction term γ^2 is subtracted from $\ell(\ell + 1)$. We notice $\lim_{c \to \infty} \gamma^2 = 0$. However, λ is *not* small, because γ is multiplied by the large factor E/c.

For small ρ, $R \propto \rho^s$, where

$$s_\pm = -\tfrac{1}{2} \pm \sqrt{(\ell + \tfrac{1}{2})^2 - \gamma^2} \tag{21-19}$$

In the nonrelativistic theory, a unique solution is obtained b requiring that ψ must be everywhere finite; viz., the nonrelativistic analogue of (21-19) gives $s_+ = \ell$ and $s_- = -\ell - 1$. The latter value is unacceptable because the wave function diverges at the origin. In the Klein-Gordon theory we cannot require that ψ be everywhere finite, since for $\ell = 0$ and $\gamma \neq 0$ $s_\pm = -\frac{1}{2} \pm \sqrt{\frac{1}{4} - \gamma^2} < 0$, and R always divergences at the origin.

It is reasonable to require that the wave function be normalizable, i.e., that $\int \psi^2 d\tau$ be finite. This implies that $s > -3/2$; which is satisfied for s_+. Unfortunately it is also satisfied for s_-, with $\ell = 0$ (also with higher ℓ for sufficiently large γ, but these large γ are unphysical, as we shall see below).

A sufficiently restrictive condition can be obtained by requiring that the matrix elements of the kinetic energy be finite. From (21-18) it is seen that this implies

$$\int \left(\frac{dR}{dr}\right)^2 r^2\, dr < \infty \tag{21-20}$$

i.e., $s > -\frac{1}{2}$, which is only satisfied by s_+. Condition (21-20) holds also in the nonrelativistic theory.

In this way, we have retained one and only one radial solution of (21-18) for each ℓ. This is the necessary and sufficient condition for the solutions R to form a complete, but not an over-complete, set. It is of course gratifying that s_+ in the nonrelativistic limit, takes on the value ℓ which is the value for exponent obtained in the nonrelativisti theory.

Equation (21-19) further shows that $s_\pm$ becomes complex for $\gamma > \frac{1}{2}$ and $\ell = 0$, viz., $s_\pm = -\frac{1}{2} \pm is'$, $s' = \sqrt{\gamma^2 - \frac{1}{4}}$. Such a complex value is not acceptable because then the wave function behaves at small r as

$$r^{-1/2}\, e^{\pm is' \log r}$$

which means it has infinitely many oscillations. Also (21-20) for complex $s_\pm$ leads to the condition Re $s > -\frac{1}{2}$ which is not fulfilled by either s_+ or s_-.

According to(21-17), $\gamma = Z/137$, so $\gamma > \frac{1}{2}$ for many existing nuclei. Further we know that the Klein-Gordon equation is supposed to hold for π mesons, and charged π mesons have a Coulomb interaction with nuclei. Thus it appears that the singularity at the origin prevents using the Klein-Gordon equation to obtain the bound states of π mesons in the electromagnetic field of heavy nuclei. Physically however, nuclei have finite size; in fact, the large nuclei with $Z > 137/2$ have radii several times the Compton wavelength of a π meson, $\hbar/m_\pi c = 1.4 \times 10^{-13}$ cm. Thus the electromagnetic interaction at small distances between these heavy nuclei and π mesons is no longer given by the Coulomb potential $-Ze^2/r$ and the problem does not arise.

The subsequent analysis of the differential equation (21-18) proceeds by the usual series method and is formally identical to the analysis of the Schrödinger equation in a Coulomb field, except that angular momentum ℓ in the Schrödinger equation is now replaced by $-\frac{1}{2} + \sqrt{(\ell + \frac{1}{2})^2 - \gamma^2}$. It is required that the series terminate; otherwise the wave function diverges exponentially at large r, which violates (21-20). The series terminates if and only if

$$\lambda = n' + \tfrac{1}{2} + [(\ell + \tfrac{1}{2})^2 - \gamma^2]^{1/2} \qquad (21\text{-}21)$$

and n' is a nonnegative integer. From (21-17) this means

$$E = \frac{mc^2}{\sqrt{1 + \dfrac{\gamma^2}{[n' + \frac{1}{2} + \sqrt{(\ell + \frac{1}{2})^2 - \gamma^2}\,]^2}}} \qquad (21\text{-}22)$$

or setting $n = n' + \ell + 1$ and expanding in powers of γ^2, we get

$$E = mc^2\left[1 - \frac{\gamma^2}{2n^2} - \frac{\gamma^4}{2n^4}\left(\frac{n}{\ell + \frac{1}{2}} - \frac{3}{4}\right)\right] + O(\gamma^6)$$

$$= mc^2 - \frac{Ry}{n^2} - \frac{Ry\,\gamma^2}{n^3}\left(\frac{1}{\ell + \frac{1}{2}} - \frac{3}{4n}\right) + O(Ry\,\gamma^4) \qquad (21\text{-}23)$$

The first term is the rest energy, the second the nonrelativistic Rydberg formula for the energy. The third term is the relativistic correction and is seen to remove the ℓ degeneracy. The total spread of the fine-structure levels is from (21-23)

$$\frac{4\,Ry\,\gamma^2}{n^3}\left[\frac{n-1}{2n-1}\right] \qquad (21\text{-}24)$$

The spread observed experimentally in the spectrum of hydrogen is only about half as large. The above should be valid for π-mesonic atoms (apart from the effect of the size of the nucleus), but no experimental verification has been obtained.

PROBLEMS

1. Verify that ρ and $\mathbf{j}$, defined in (21-5) satisfy the continuity equation (21-4); and that $\int \rho \, d\tau$ is a Lorentz scalar and is time independent.
2. What are the forms of the charge and current density, satisfying the continuity equation, for the Klein-Gordon equation in an external electromagnetic field.
3. Prove that gauge transformations on the electromagnetic potentials, occurring in the Klein-Gordon equation in the presence of external fields, induce a phase transformation of the Klein-Gordon function ψ.
4. Consider the Klein-Gordon equation in an attractive square well potential which is the fourth component of a four-vector. Solve the equation after determining the continuity conditions satisfied by the Klein-Gordon function ψ at the edge of the square well.
5. Obtain solutions of the Klein-Gordon equation in (a) a homogenous constant magnetic field, (b) a homogenous constant electric field.
6. Repeat the analysis of Problem 4, but now assume the square well to be a scalar potential, i.e., an addition to the mass term.
7. Verify the steps leading to equation (21-16).
8. Verify the steps leading to equation (21-18).

Chapter 22
DIRAC EQUATION, FORMAL THEORY

In what is undoubtedly one of the great papers in physics of this century, Dirac set up a relativistic wave equation which avoids the difficulties of negative probability density of the Klein-Gordon equation, and describes naturally the spin of the electron. Until Pauli and Weisskopf reinterpreted the Klein-Gordon equation, it was believed that this Dirac equation was the only valid relativistic equation. It is now recognized that the Dirac equation and the Klein-Gordon equation are equally valid; the Dirac equation governs particles of spin ½, the Klein-Gordon equation those of spin zero. Between them they describe most of the known elementary particles (although the proper definition of "elementary particle" is unclear). Formally one can extend the ideas of the Dirac theory to particles with nonzero rest mass with higher spin, but these theories have not proved to be successful, in that their interaction with the electromagnetic field leads to uncorrectable divergences. We shall not discuss these extensions, nor shall we discuss the successful Weyl equations, which describe relativistic massless particles of spin ½ and 1. The former, which describes the neutrino, can be considered a natural simplification of the Dirac equation.

DERIVATION OF THE DIRAC EQUATION

To prevent the occurrence of negative probability densities, we must avoid time derivatives in the expression for ρ. Therefore, we must admit time derivatives no higher than first order in the wave equation. Since in the theory of relativity x, y, z, and ct are treated symmetrically, the Dirac wave function ψ must satisfy a first-order differential equation in all four coordinates. Furthermore, the equation must be linear so that the superposition principle of quantum mechanics holds. The correspondence principle also requires that the Klein-Gordon equation be satisfied, since the latter merely implies that (21-2) is valid; i.e., that in the limit of large quantum numbers classical relativity holds.

A similar situation obtains in electrodynamics. Maxwell's equations are first order in both space and time. On the other hand, each field component satisfies a second-order wave equation similar to (21-3) with zero rest mass. These two requirements are reconciled by the fact that each of Maxwell's equations connects different field components. This structure may be used as the guiding principle to obtain the Dirac equation.

Assuming that ψ consists of N components ψ_ℓ, where the number N is yet unspecified, the most general equation which satisfies the above requirements is

$$\frac{1}{c}\frac{\partial\psi_\ell}{\partial t} + \sum_{k=1}^{3}\sum_{n=1}^{N}\alpha^k_{\ell n}\frac{\partial\psi_n}{\partial x^k} + \frac{imc}{\hbar}\sum_{n=1}^{N}\beta_{\ell n}\psi_n = 0 \qquad (22\text{-}1)$$

$$\ell = 1, 2, \ldots, N \qquad x^k = x, y, z \qquad \text{for } k = 1, 2, 3$$

For a free particle, all points in space-time are equivalent (homogeneity of space-time). Hence the $\alpha^k_{\ell n}$ and $\beta_{\ell n}$ must be dimensionless constants independent of and commuting with $\mathbf{r}$ and $\mathbf{p}$.

The N equations (22-1) can be written more compactly by introducing the one-column matrix

$$\psi = \begin{pmatrix}\psi_1\\ \psi_2\\ \vdots\\ \psi_N\end{pmatrix}$$

and the $N \times N$ matrices α^k and β. Further, we may define the vector matrix

$$\boldsymbol{\alpha} = \alpha^1\mathbf{i} + \alpha^2\mathbf{j} + \alpha^3\mathbf{k}$$

Then (22-1) becomes

$$\frac{1}{c}\frac{\partial\psi}{\partial t} + \boldsymbol{\alpha}\cdot\nabla\psi + \frac{imc}{\hbar}\beta\psi = 0 \tag{22-2}$$

The N components of ψ describe a new degree of freedom of the particle just as the components of the Maxwell field describe the polarization of the light quantum. We shall see below that this new degree of freedom is the spin of the particle, and ψ is called a *spinor*.

We next seek an expression for probability density ρ and current $\mathbf{j}$ which satisfies the conditions given before (see p. 341) together with the requirement that ρ be positive definite. We define

$$\rho = \psi^\dagger \psi \tag{22-3}$$

where $\psi^\dagger$ is the Hermitian adjoint of ψ, hence a one-row matrix with N columns, whose components are the complex conjugate of the corresponding components of ψ. Clearly, $\psi^\dagger$ satisfies

$$\frac{1}{c}\frac{\partial\psi^\dagger}{\partial t} + \nabla\psi^\dagger\cdot\boldsymbol{\alpha}^\dagger - \frac{imc}{\hbar}\psi^\dagger\beta^\dagger = 0 \tag{22-4}$$

where the interchange of ψ with $\boldsymbol{\alpha}$ and β is necessary, since $\psi^\dagger$ is a row matrix.

To arrive at $\mathbf{j}$ we write the continuity equation

$$\frac{\partial}{\partial t}(\psi^\dagger\psi) + \nabla\cdot\mathbf{j} = 0 \tag{22-5}$$

Multiplying (22-2) on the left by $\psi^\dagger$, (22-4) on the right by ψ, and adding, we obtain

$$\frac{1}{c}\left(\psi^\dagger\frac{\partial}{\partial t}\psi + \frac{\partial\psi^\dagger}{\partial t}\psi\right) + \nabla\psi^\dagger\cdot\boldsymbol{\alpha}^\dagger\psi + \psi^\dagger\boldsymbol{\alpha}\cdot\nabla\psi$$
$$+ \frac{imc}{\hbar}(\psi^\dagger\beta\psi - \psi^\dagger\beta^\dagger\psi) = 0 \tag{22-6}$$

To identify this with (22-5) we require

$$\beta^\dagger = \beta$$

$$\boldsymbol{\alpha}^\dagger = \boldsymbol{\alpha} \tag{22-7}$$

$$\mathbf{j} = c\psi^\dagger\boldsymbol{\alpha}\psi \tag{22-8}$$

Equation (22-7) expresses the very reasonable condition that the Dirac matrices be Hermitian. That (22-7) is indeed necessary follows from the fact that (22-2) can be written

$$i\hbar \frac{\partial \psi}{\partial t} = H\psi \tag{22-9}$$

$$H = \left(c\boldsymbol{\alpha} \cdot \frac{\hbar}{i} \nabla + \beta mc^2 \right) \tag{22-10}$$

It is clear that if H is to be Hermitian $\boldsymbol{\alpha}$ and β must also be. Equations (22-3) and (22-8) are the very simple expressions for probability density and current which satisfy the continuity equation.

DIRAC MATRICES I

To derive further properties of the Dirac matrices, we operate on (22-2) with

$$\frac{1}{c} \frac{\partial}{\partial t} - \boldsymbol{\alpha} \cdot \nabla - \frac{imc}{\hbar} \beta$$

and obtain

$$\frac{1}{c^2} \frac{\partial^2 \psi}{\partial t^2} = \frac{1}{2} \sum_k \sum_\ell (\alpha^k \alpha^\ell + \alpha^\ell \alpha^k) \frac{\partial^2 \psi}{\partial x^k \partial x^\ell}$$

$$- \frac{m^2 c^2}{\hbar^2} \beta^2 \psi + \frac{imc}{\hbar} \sum_k (\alpha^k \beta + \beta \alpha^k) \frac{\partial \psi}{\partial x^k} \tag{22-11}$$

We have symmetrized the $\alpha^k \alpha^\ell$ term, which is permissible, since $\partial/\partial x^k$ and $\partial/\partial x^\ell$ commute. If (22-11) is to agree with the Klein-Gordon equation for each component, the Dirac matrices must satisfy

$$\tfrac{1}{2}(\alpha^k \alpha^\ell + \alpha^\ell \alpha^k) = \delta^{k\ell} I$$

$$\alpha^k \beta + \beta \alpha^k = 0 \tag{22-12}$$

$$(\alpha^k)^2 = \beta^2 = I$$

where I is the unit matrix, and $\delta^{k\ell}$ is the Kronecker delta.

We now prove two important theorems about the Dirac matrices. From (22-12) it follows that

$$\beta \alpha^k = -\alpha^k \beta = (-I) \alpha^k \beta \tag{22-13a}$$

Taking determinants, we obtain

$$(\det \beta)(\det \alpha^k) = (-1)^N (\det \alpha^k)(\det \beta) \qquad (22\text{-}13b)$$

Since (22-12) indicates that α^k and β have inverses, none of the determinants vanish, and

$$(-1)^N = 1 \qquad (22\text{-}13c)$$

Thus N, the dimension of the matrices, is even.

Since $\beta^2 = (\alpha^k)^2 = 1$, each matrix has itself for an inverse. Therefore (22-13a) can be rewritten as

$$(\alpha^k)^{-1} \beta \alpha^k = -\beta \qquad (22\text{-}14a)$$

Taking the trace of both sides we obtain

$$\mathrm{Tr}\,[(\alpha^k)^{-1} \beta \alpha^k] = \mathrm{Tr}\,[\alpha^k (\alpha^k)^{-1} \beta] = \mathrm{Tr}(\beta) = \mathrm{Tr}(-\beta) \qquad (22\text{-}14b)$$

Therefore,

$$\mathrm{Tr}(\beta) = 0 \qquad (22\text{-}14c)$$

The same holds for the other matrices:

$$\mathrm{Tr}(\alpha^k) = 0 \qquad (22\text{-}14d)$$

COVARIANT FORM OF THE DIRAC EQUATION

To write the Dirac equation in covariant form we define

$$\beta = \gamma^0$$

$$\beta \boldsymbol{\alpha} = \boldsymbol{\gamma}$$

$$\gamma^\mu = (\gamma^0, \boldsymbol{\gamma})$$

$$x^\mu = (ct, \mathbf{r}) \qquad (22\text{-}15)$$

γ^0 is Hermitian, the other γ's are anti-Hermitian. These Hermiticity relations can be summarized compactly by

$$(\gamma^\mu)^\dagger = \gamma^0 \gamma^\mu \gamma^0 \qquad (22\text{-}16)$$

The metric tensor that we shall use is $g_{\mu\nu}$:

$$g_{00} = 1 \qquad g_{kk} = -1,\ k = 1, 2, 3 \qquad g_{\mu\nu} = 0,\ \mu \neq \nu$$

We shall raise and lower indices on the γ's even though they are *not* components of a four-vector:

$$\gamma_\mu = g_{\mu\nu}\gamma^\nu \tag{22-17}$$

(summation over repeated Greek dummy indices implied). The commutation relations (22-12) can be summarized:

$$\gamma^\mu\gamma^\nu + \gamma^\nu\gamma^\mu = 2g^{\mu\nu}$$

$$\gamma_\mu\gamma_\nu + \gamma_\nu\gamma_\mu = 2g_{\mu\nu} \tag{22-18}$$

Multiplying (22-2) by $i\hbar\beta$, we obtain

$$i\hbar\gamma^\mu \frac{\partial\psi}{\partial x^\mu} - mc\psi = 0 \tag{22-19}$$

Defining a (Dirac) adjoint spinor $\bar{\psi}$ by

$$\bar{\psi} = \psi^\dagger\gamma^0 \tag{22-20}$$

and multiplying (22-4) by $i\hbar$ one obtains

$$i\hbar \frac{\partial}{\partial x^\mu} \bar{\psi}\gamma^\mu + mc\bar{\psi} = 0 \tag{22-21}$$

The four-current vector is defined

$$\frac{j^\mu}{c} = \bar{\psi}\gamma^\mu\psi = \left(\rho, \frac{\mathbf{j}}{c}\right) \tag{22-22}$$

and the continuity equation (22-5) can be written in covariant form.

$$\frac{\partial}{\partial x^\mu} j^\mu = 0 \tag{22-23}$$

DIRAC MATRICES II

We now obtain several theorems about the Dirac matrices. To study these matrices it is not necessary to assume Hermiticity and we shall not do so. The fundamental relationship is of course (22-18), which when satisfied is said to define a *Clifford algebra*.

We may form new matrices from the four γ's by multiplying two or more together. Since the square of each γ equals $\pm I$ we need to consider only products whose factors are different. Order is immaterial, since different matrices commute or anticommute. This means there are $2^4 - 1$ different products of the four γ's. (The number of ways of making different combinations from n objects is $2^n - 1$.) If we also include I we can enumerate 16 different matrices

$$I$$

$$\gamma^0 \quad i\gamma^1 \quad i\gamma^2 \quad i\gamma^3$$

$$\gamma^0\gamma^1 \qquad \gamma^0\gamma^2 \qquad \gamma^0\gamma^3 \qquad i\gamma^2\gamma^3 \qquad i\gamma^3\gamma^1 \qquad i\gamma^1\gamma^2$$

$$i\gamma^0\gamma^2\gamma^3 \qquad i\gamma^0\gamma^3\gamma^1 \qquad i\gamma^0\gamma^1\gamma^2 \qquad \gamma^1\gamma^2\gamma^3$$

$$i\gamma^0\gamma^1\gamma^2\gamma^3 \equiv i\gamma_5 \tag{22-24}$$

Denoting elements in the above array by Γ_ℓ, $\ell = 1, 2, \ldots, 16$, the following properties can be verified (no summation over repeated Latin indices implied):

$$\Gamma_\ell \Gamma_m = a_{\ell m} \Gamma_n \qquad a_{\ell m} = \pm 1 \text{ or } \pm i \tag{22-25}$$

$$\Gamma_\ell \Gamma_m = I \qquad \text{if and only if } \ell = m \tag{22-26}$$

$$\Gamma_\ell \Gamma_m = \pm \Gamma_m \Gamma_\ell \tag{22-27}$$

If $\Gamma_\ell \neq I$, there always exists a Γ_k such that

$$\Gamma_k \Gamma_\ell \Gamma_k = -\Gamma_\ell \tag{22-28}$$

We now establish certain important properties of the Γ_ℓ by proving the following theorems.

Theorem 1

$$\mathrm{Tr}(\Gamma_\ell) = 0 \qquad (\Gamma_\ell \neq I) \tag{22-29}$$

Proof
Choose a k such that by applying (22-26) and (22-28) we have

$$\mathrm{Tr}(\Gamma_k \Gamma_\ell \Gamma_k) = \mathrm{Tr}(\Gamma_\ell \Gamma_k \Gamma_k) = \mathrm{Tr}(\Gamma_\ell) = -\mathrm{Tr}(\Gamma_\ell)$$

Theorem 2

$$\sum_{k=1}^{16} x_k \Gamma_k = 0 \qquad \text{only if } x_k = 0;\ k = 1, 2, \ldots, 16 \qquad (22\text{-}30)$$

Proof

Multiply (22-30) by Γ_m to get

$$x_m I + \sum_{k \neq m} x_k \Gamma_k \Gamma_m = x_m I + \sum_{k \neq m} x_k a_{km} \Gamma_n = 0$$

with $\Gamma_n \neq I$, since $k \neq m$. Taking the trace:

$$x_m \operatorname{Tr} I = -\sum_{k \neq m} x_k a_{km} \operatorname{Tr}(\Gamma_n) = 0 \qquad [\text{by}(22\text{-}29)]$$

$$x_m = 0$$

Theorem 2 gives us the following important result: The Γ_k's cannot be represented by matrices whose dimension is less than 4×4, since it is impossible to construct 16 linearly independent matrices from such matrices. Furthermore there exist precisely 16 linearly independent 4×4 matrices which we can use to represent the Γ_k's. We shall assume from now on that the γ's are in fact 4×4 matrices. It should be emphasized that the identity between the dimensionality of the Dirac matrices and the dimensionality of space-time is accidental.

A corollary to Theorem 2 is that any 4×4 matrix X can be written uniquely as a linear combination of the Γ_k's.

$$X = \sum_{k=1}^{16} x_k \Gamma_k \qquad (22\text{-}31a)$$

Multiplying by Γ_m and taking the trace,

$$\operatorname{Tr}(X\Gamma_m) = x_m \operatorname{Tr}(\Gamma_m \Gamma_m) + \sum_{k \neq m} x_k \operatorname{Tr}(\Gamma_k \Gamma_m)$$

$$= x_m \operatorname{Tr}(I)$$

$$x_m = \tfrac{1}{4} \operatorname{Tr}(X\Gamma_m) \qquad (22\text{-}31b)$$

A further corollary to Theorem 2 is a stronger statement of

(22-25). We can now say that $\Gamma_\ell \Gamma_m = a_{\ell m} \Gamma_n$, where Γ_n is a different Γ_n for each m, for fixed ℓ. For suppose

$$\Gamma_\ell \Gamma_m = a_{\ell m} \Gamma_n$$
$$\Gamma_\ell \Gamma_{m'} = a_{\ell m'} \Gamma_n \qquad \Gamma_m \neq \Gamma_{m'} \tag{22-32a}$$

Then

$$\Gamma_m = a_{\ell m} \Gamma_\ell \Gamma_n$$
$$\Gamma_{m'} = a_{\ell m'} \Gamma_\ell \Gamma_n \tag{22-32b}$$

and

$$\Gamma_m = \frac{a_{\ell m}}{a_{\ell m'}} \Gamma_{m'} \tag{22-32c}$$

which contradicts the fact that the Γ_k's are linearly independent.

Theorem 3

Any matrix X that commutes with γ^μ (for all μ) is a multiple of the identity.

Proof

Assume X is not a multiple of the identity. If X commutes with all the γ's, it commutes with all the Γ_i's: $X = \Gamma_i X \Gamma_i$. From (22-31) we can write

$$X = x_m \Gamma_m + \sum_{k \neq m} x_k \Gamma_k \qquad \Gamma_m \neq I \tag{22-33a}$$

From (22-28) there exists a Γ_i such that $\Gamma_i \Gamma_m \Gamma_i = -\Gamma_m$ and by hypothesis X commutes with this Γ_i. Therefore,

$$X = x_m \Gamma_m + \sum_{k \neq m} x_k \Gamma_k = \Gamma_i X \Gamma_i$$
$$= x_m \Gamma_i \Gamma_m \Gamma_i + \sum_{k \neq m} x_k \Gamma_i \Gamma_k \Gamma_i$$
$$= -x_m \Gamma_m + \sum_{k \neq m} x_k (\pm 1) \Gamma_k \tag{22-33b}$$

Since the expansion is unique, we have $x_m = -x_m = 0$. Since Γ_m was arbitrary except that $\Gamma_m \neq I$, we have proved

$$[X, \gamma^{\mu}] = 0 \text{ for all } \mu, \text{ implies } X = aI \tag{22-33c}$$

Theorem 4

The above allows us to derive another important result, called *Pauli's fundamental theorem*, which states: Given two sets of 4 × 4 matrices γ^{μ} and γ'^{μ}, both of which satisfy (22-18), there exists a nonsingular matrix S such that

$$\gamma'^{\mu} = S\gamma^{\mu}S^{-1} \tag{22-34}$$

Proof

Set

$$S = \sum_{i=1}^{16} \Gamma_i' F \Gamma_i \tag{22-35}$$

where F is as yet an arbitrary 4 × 4 matrix and the Γ_i''s are the set of 16 products formed from the γ'^{μ}'s. These are constructed in the same fashion as the Γ_i's are constructed from the γ^{μ}'s. From (22-25) $\Gamma_i \Gamma_j = a_{ij} \Gamma_k$; then $\Gamma_i \Gamma_j \Gamma_i \Gamma_j = a_{ij}^2 \Gamma_k^2 = a_{ij}^2$; therefore

$$\Gamma_j \Gamma_i = a_{ij}^2 \Gamma_i \Gamma_j = a_{ij}^3 \Gamma_k$$

We also have

$$\Gamma_i' \Gamma_j' = a_{ij} \Gamma_k'$$

since in the primed system the Γ_i''s are constructed in the same fashion as in the unprimed system. Then, for any i,

$$\Gamma_i' S \Gamma_i = \sum_j \Gamma_i' \Gamma_j' F \Gamma_j \Gamma_i = \sum_j a_{ij}^4 \Gamma_k' F \Gamma_k \tag{22-36}$$

Since $a_{ij}^4 = 1$ and since the sum over j ranges through all 16 elements, it can be replaced by a sum over k = k (j) to give, from (22-35),

$$\Gamma_i' S \Gamma_i = S \tag{22-37}$$

To show that S is nonsingular, we define a quantity S′:

$$S' = \sum_{i=1}^{16} \Gamma_i G \Gamma_i' \tag{22-38}$$

where G is as yet arbitrary. By symmetry

$$\Gamma_i S' \Gamma_i' = S' \tag{22-39a}$$

for any i. From (22-37),

$$S'S = \Gamma_i S' \Gamma_i' \Gamma_i' S \Gamma_i = \Gamma_i S' S \Gamma_i \tag{22-39b}$$

This by (22-33) implies that $S'S = aI$. Since F and G occurring in (22-35) and (22-38) are arbitrary, they can be so chosen that $a \neq 0$. (It is easy to show by choosing for F and G matrices with only one nonzero element that $a = 0$ for all F and G would contradict the linear independence of the Γ_k's.) Hence S is nonsingular and

$$\gamma^{\mu'} = S\gamma^{\mu} S^{-1} \tag{22-34}$$

This proves the theorem. S is unique up to a constant, for suppose $S_1\gamma^{\mu} S_1^{-1} = S_2\gamma^{\mu} S_2^{-1}$. Then $S_2^{-1} S_1 \gamma^{\mu} = \gamma^{\mu} S_2^{-1} S$, which means that $S_2^{-1} S_1 = aI$ or $S_1 = aS_2$. Further, given four γ^{μ}'s that satisfy (22-18) and defining $\gamma^{\mu'} = S\gamma^{\mu} S^{-1}$, it is clear that the $\gamma^{\mu'}$ will also satisfy (22-18).

We conclude by noting that γ_5 defined in (22-24) anticommutes with γ^{μ} for any μ and $(\gamma_5)^2 = -I$. We shall not raise the subscript 5.

EXPLICIT FORM OF THE DIRAC MATRICES

We now exhibit a nonunique matrix representation for the Dirac matrices. It should be clear that relations (22-7) and (22-12) or, alternatively, (22-16) and (22-18), do not specify the matrices uniquely. Thus it is usually best not to express the matrices explicitly when working problems.

We have seen that the Dirac matrices must have at least four rows and columns, and we shall restrict ourselves to 4 × 4 matrices. We have further seen that the trace of β and α^k must be zero. For most problems involving particles of moderate speed, including atomic problems, the term mc^2 in the Hamiltonian dominates; therefore it is convenient to represent β by a diagonal matrix. This, together with $\mathrm{Tr}\,\beta = 0$ and $\beta^2 = I$, leads to the choice

$$\beta = \begin{pmatrix} I & 0 \\ 0 & -I \end{pmatrix} \tag{22-40a}$$

where I is the 2 × 2 identity matrix.

The three α matrices, in order to anticommute with β and to be Hermitian, must have the form

$$\alpha^k = \begin{pmatrix} 0 & A^k \\ A^{k\dagger} & 0 \end{pmatrix} \tag{22-40b}$$

where A^k is a 2×2 matrix, not necessarily Hermitian. They must also anticommute with each other, and their squares must be unity. We remember here the Pauli spin matrices σ, which have just this property. Clearly, all relations (22-12) are satisfied if we put

$$\alpha = \begin{pmatrix} 0 & \sigma \\ \sigma & 0 \end{pmatrix} \tag{22-40c}$$

Then (22-15) and (22-23) give

$$\gamma = \begin{pmatrix} 0 & \sigma \\ -\sigma & 0 \end{pmatrix} \qquad \gamma_5 = -i\begin{pmatrix} 0 & I \\ I & 0 \end{pmatrix} \tag{22-40d}$$

We shall find the choice (22-40a) and (22-40c) useful in discussion of spin. In the extreme relativistic case, it is usually more convenient to diagonalize γ_5. Of course all physical consequences must be independent of the representation.

The Dirac adjoint $\overline{M}$ of a matrix M is defined in anology with (22-20).

$$\overline{M} = \gamma^0 M^\dagger \gamma^0 \tag{22-41}$$

With these notations it is true that bilinears $M \equiv \overline{\psi} M \psi$ are real for matrices that are Dirac self-adjoint $\overline{M} = M$. Note that in the representation (22-40) the γ^μ's and γ_5 are Dirac self-adjoint.

RELATIVISTIC INVARIANCE OF THE DIRAC EQUATION

Before obtaining solutions and physical consequences from the Dirac equation we shall show that the physical results are independent of the Lorentz frame used to derive them. If the Dirac equation is solved in two different frames the solutions must describe the same physical state. This does not mean that the components of ψ are the same in the two Lorentz frames. This is analogous to the electromagnetic field tensor, where the components of $\mathcal{E}$ and $\mathcal{H}$ transform, but the form of the Maxwell equation remains invariant. So here also we shall see that ψ transforms but that the form of the Dirac equation remains the same.

The most general homogenous Lorentz transformation (i.e., omitting space-time translations) between two coordinate systems may be written

$$x'^{\mu} = a^{\mu}{}_{\nu} x^{\nu} \tag{22-42a}$$

with

$$a_{\mu}{}^{\nu} a^{\mu}{}_{\lambda} = a^{\mu\nu} a_{\mu\lambda} = a^{\nu\mu} a_{\lambda\mu} = \delta^{\nu}{}_{\lambda} \qquad a_{\mu}{}^{\nu} \text{ real} \tag{22-42b}$$

Equation (22-41b) follows from the requirement that the real quadratic form $x^{\mu} x_{\mu}$ is an invariant.

In the special case of the standard Lorentz transformation (motion of two coordinate systems along their mutual x axes with relative velocity v; origins coincident at t = 0),

$$a^{\mu}{}_{\nu} = \begin{pmatrix} \gamma & -\beta\gamma & 0 & 0 \\ -\beta\gamma & \gamma & 0 & 0 \\ 0 & 0 & 1 & 0 \\ 0 & 0 & 0 & 1 \end{pmatrix} \quad (\mu \downarrow,\ \nu \rightarrow).$$

$$\beta = v/c \qquad \gamma = (1 - \beta^2)^{-1/2}$$

Using

$$\frac{\partial}{\partial x^{\mu}} = \frac{\partial x'^{\lambda}}{\partial x^{\mu}} \frac{\partial}{\partial x'^{\lambda}} = a^{\lambda}{}_{\mu} \frac{\partial}{\partial x'^{\lambda}} \tag{22-43}$$

which follows from (22-42a), we obtain from (22-19),

$$\frac{\hbar}{i} a^{\lambda}{}_{\mu} \gamma^{\mu} \frac{\partial \psi}{\partial x'^{\lambda}} + mc\psi = 0 \tag{22-44}$$

Defining $\gamma'^{\lambda} = a^{\lambda}{}_{\mu}\gamma^{\mu}$, we can verify with the help of (22-42b) that γ'^{λ} satisfies (22-18). Hence by Pauli's fundamental theorem, there exists a (unique up to multiplicative constant) S such that

$$a^{\lambda}{}_{\mu} \gamma^{\mu} = S^{-1} \gamma^{\lambda} S \tag{22-45}$$

Substituting this quantity in (22-44) and premultiplying by S gives

$$\frac{\hbar}{i} \gamma^{\mu} \frac{\partial S\psi}{\partial x'^{\mu}} + mcS\psi = 0 \tag{22-46}$$

Defining the Lorentz transformation law for spinors to be

$$\psi'(x') = S\psi(x) \tag{22-47}$$

(22-46) gives

$$\frac{\hbar}{i}\gamma^{\mu}\frac{\partial\psi'}{\partial x'^{\mu}} + mc\psi' = 0 \tag{22-48}$$

This is of the same form as (22-19). As we predicted above, the ψ transforms but the γ^{μ}'s remain the same. Thus if we can show that $S\psi = \psi'$ has the same physical significance in the primed system as ψ has in the unprimed, we shall have demonstrated the covariance of the theory. To do this we derive some further properties of S.

From the Hermiticity relations (22-16), from (22-42b), and from (22-45) we have

$$a^{\lambda}{}_{\mu}\gamma^{\mu} = \gamma^0\left(a^{\lambda}{}_{\mu}\gamma^{\mu}\right)^{\dagger}\gamma^0 = \gamma^0\left(S^{-1}\gamma^{\lambda}S\right)^{\dagger}\gamma^0 \tag{22-49a}$$

$$a^{\lambda}{}_{\mu}\gamma^{\mu} = (\gamma^0 S^{\dagger}\gamma^0)\gamma^{\lambda}(\gamma^0 S^{\dagger}\gamma^0)^{-1} \tag{22-49b}$$

Substituting in (22-49b) $S^{-1}\gamma^{\lambda}S$ for $a^{\lambda}{}_{\mu}\gamma^{\mu}$, we obtain

$$(\gamma^0 S^{\dagger}\gamma^0)\gamma^{\lambda}(\gamma^0 S^{\dagger}\gamma^0)^{-1} = S^{-1}\gamma^{\lambda}S \tag{22-50a}$$

$$(S\gamma^0 S^{\dagger}\gamma^0)\gamma^{\lambda} = \gamma^{\lambda}(S\gamma^0 S^{\dagger}\gamma^0) \tag{22-50b}$$

By Theorem 3 and (22-33) we conclude that

$$(S\gamma^0 S^{\dagger}\gamma^0) = bI \tag{22-51a}$$

$$S\gamma^0 S^{\dagger} = b\gamma^0 \tag{22-51b}$$

Taking Hermitian conjugates of (22-51b), we find that b is real. Prescribing a normalization on S by det S = 1 gives from (22-51b) $b^4 = 1$, and since b is real,

$$b = \pm 1 \tag{22-52}$$

To examine the physical significance of (22-52) we consider

$$S^{\dagger}S = S^{\dagger}\gamma^0\gamma^0 S = b\gamma^0 S^{-1}\gamma^0 S = b\gamma^0 a^0{}_{\nu}\gamma^{\nu}$$

$$= ba^0{}_0 I - \sum_{k=1}^{3} ba^0{}_k \gamma^0 \gamma^k \tag{22-53}$$

where (22-45) and (22-51) have been used. Since $S^\dagger S$ has positive definite eigenvalues, taking the trace of (22-53) we get

$$0 < \mathrm{Tr}(S^\dagger S) = 4ba^0{}_0 \qquad ba^0{}_0 > 0 \tag{22-54}$$

i.e.,

$$a^0{}_0 < 0 \qquad b = -1 \tag{22-55a}$$

$$a^0{}_0 > 0 \qquad b = 1 \tag{22-55b}$$

The former case, $a^0{}_0 < 0$, corresponds to time reversal.

Next we consider the transformation properties of the adjoint function $\overline{\psi} = \psi^\dagger \gamma^0$.

$$\psi' = S\psi \tag{22-47}$$

$$(\psi')^\dagger = \psi^\dagger S^\dagger$$

$$\overline{\psi'} = (\psi')^\dagger \gamma^0 = \psi^\dagger S^\dagger \gamma^0$$

$$= b\psi^\dagger \gamma^0 S^{-1} = b\overline{\psi} S^{-1} \tag{22-56}$$

$$\overline{\psi}'(x') = b\overline{\psi}(x) S^{-1}$$

We are now ready to answer the question posed above; i.e., does ψ' describe the same physical situation in the primed system as ψ does in the unprimed? This will be so if $(\psi')^\dagger \psi'$ gives the probability density in the primed system. We consider the transformation properties of the current j^μ.

$$\frac{j^\mu}{c} = \overline{\psi}\gamma^\mu \psi \tag{22-22}$$

$$\frac{j'^\mu}{c} = \overline{\psi}'\gamma^\mu \psi' = b\overline{\psi} S^{-1}\gamma^\mu S \psi = ba^\mu{}_\lambda \overline{\psi}\gamma^\lambda \psi$$

$$= ba^\mu{}_\lambda \frac{j^\lambda}{c} \tag{22-57}$$

Time refersal will not be discussed here, nor will the related transformation of charge conjugation, which interchanges particles with anti-particles. While both of these operations are symmetries of the Dirac theory, they are discussed most easily in the field-theoretical context, which we have not developed here.

Hence if we restrict ourselves to Lorentz transformations which do not include time reversal, j^μ transforms as a four-vector, which gives the proper transformation law for $\psi^\dagger\psi = j^0/c$.

EXPLICIT TRANSFORMATION MATRIX

Finally we exhibit the actual S corresponding to a definite Lorentz transformation $(a^\mu{}_\nu)$. Since (22-42) implies that $\det^2(a^\mu{}_\nu) = 1$, Lorentz transformations may be classified according to whether $\det(a^\mu{}_\nu) = \pm 1$ and $a^0_0 \gtrless 0$. Remaining only with those Lorentz transformations that do not reverse time, there are still two kinds--the proper, with determinant equal to 1 and the improper, with determinant equal to -1. The identity transformation is of the former kind, while the latter can always be thought of as a proper Lorentz transformation, followed by a standard improper one, for example a parity transformation in which the spatial coordinates change sign

$$\mathbf{x}' = -\mathbf{x} \tag{22-58}$$

Thus we discuss separately proper Lorentz transformations, and the parity transformation (22-58).

Proper Lorentz transformations are continuous, i.e., they are parameterized by a continuously varying parameter, e.g., the relative velocity between the two frames. As this parameter ranges from zero to a definite finite value, the transformations evolve from the identity to a definite Lorentz transform. Because these transformations form a group it is suffient to deal only with infinitesimal transformations, sinc repeated iteration of the infinitesimal will result in a finit transformation. For an infinitesimal transformation,

$$a^\mu{}_\nu = \delta^\mu{}_\nu + \lambda\epsilon^\mu{}_\nu \tag{22-59a}$$

λ is the smallness parameter and in the following only terms linear in λ are kept. From (22-42b) it follows that

$$\epsilon^{\mu\nu} = -\epsilon^{\nu\mu} \tag{22-59b}$$

We write

$$S = 1 + \lambda T$$

$$S^{-1} = 1 - \lambda T \tag{22-60}$$

$$a^{\mu}{}_{\nu}\gamma^{\nu} = S^{-1}\gamma^{\mu}S = (1 - \lambda T)\gamma^{\mu}(1 + \lambda T)$$
$$= \gamma^{\mu} + \lambda(\gamma^{\mu}T - T\gamma^{\mu}) \tag{22-61a}$$

$$a^{\mu}{}_{\nu}\gamma^{\nu} = \gamma^{\mu} + \lambda\epsilon^{\mu}{}_{\nu}\gamma^{\nu} \tag{22-61b}$$

hence

$$\epsilon^{\mu}{}_{\nu}\gamma^{\nu} = \gamma^{\mu}T - T\gamma^{\mu} \tag{22-61c}$$

T is uniquely defined up to the addition of a constant multiple of the identity. For if there were two such T's, from (22-61) their difference would commute with the γ^{μ}'s and would be a constant multiple of the identity. The normalization requirement on S removes this ambiguity, since

$$1 = \det S = \det(1 + \lambda T)$$
$$= \det 1 + \lambda \operatorname{Tr} T \qquad \operatorname{Tr} T = 0 \tag{22-62}$$

One readily establishes that

$$T = \frac{1}{8}\varepsilon^{\mu\nu}(\gamma_{\mu}\gamma_{\nu} - \gamma_{\nu}\gamma_{\mu}) = \frac{1}{4i}\varepsilon^{\mu\nu}\sigma_{\mu\nu} \tag{22-63}$$

where

$$\sigma_{\mu\nu} \equiv \frac{i}{2}(\gamma_{\mu}\gamma_{\nu} - \gamma_{\nu}\gamma_{\mu}),\ \gamma_{\mu}\gamma_{\nu} = g_{\mu\nu} - i\sigma_{\mu\nu} \tag{22-64}$$

satisfies (22-61) and (22-62). Therefore T is the required transformation matrix. The finite transformation matrix is $S = e^{\lambda T}$.

For parity transformations, we see from (22-44) and (22-57) that

$$S = \gamma^{0}$$

is the correct choice. (22-65)

TRACES OF GAMMA MATRICES

Since the trace of a matrix is insensitive to a similarity transformation, traces of products of gamma matrices do not depend on the specific representation used. Therefore, physical results determined by the Dirac theory should be expressed in terms of these traces. Some examples are

$$\operatorname{Tr}\gamma^{\mu}\gamma^{\nu} = 4g^{\mu\nu} \tag{22-66}$$

This follows from (22-18) and from the cyclicity of the trace. The trace of three, indeed of any odd number, of gamma matrices vanishes, as is seen from the following sequence of equations.

$$\begin{aligned}\mathrm{Tr}\,\gamma^{\mu_1}\ldots\gamma^{\mu_{2n+1}} &= -\mathrm{Tr}\,\gamma_5^2\,\gamma^{\mu_1}\ldots\gamma^{\mu_{2n+1}}\\ &= -\mathrm{Tr}\,\gamma_5\gamma^{\mu_1}\ldots\gamma^{\mu_{2n+1}}\,\gamma_5 = \mathrm{Tr}\,\gamma_5\gamma_5\,\gamma^{\mu_1}\ldots\gamma^{\mu_{2n+1}}\\ &= -\mathrm{Tr}\,\gamma^{\mu_1}\ldots\,\gamma^{\mu_{2n+1}}\end{aligned} \tag{22-67}$$

We have used the fact that the γ_5 matrix, with square equal to -I, anti-commutes with a product of odd gamma matrices. To calculate the trace of four or more gamma matrices, we make use of the identity

$$\gamma^\alpha\gamma^\beta\gamma^\gamma = g^{\alpha\beta}\gamma^\gamma + g^{\beta\gamma}\gamma^\alpha - g^{\alpha\gamma}\gamma^\beta - \varepsilon^{\alpha\beta\gamma\delta}\gamma_\delta\gamma_5 \tag{22-68}$$

where the ε symbol is the four dimensional version of three-index symbol, defined with the normalization

$$\varepsilon^{0123} = 1 \tag{22-69}$$

Applications are in Problem 10.

PROBLEMS

1. **Verify that the probability density and current defined in (22-3) and (22-8) satisfy the continuity equation. Show also that $\int \rho\, d\tau$ is a time independent, Lorentz scalar.**
2. **Verify (22-14d).**
3. **Verify (22-62).**
4. **Since the transpose of γ^μ satisfies the defining relation (22-18) if γ^μ does, Pauli's fundamental theorem states that there exists a nonsingular matrix S such that: Transpose $(\gamma^\mu) = S\gamma^\mu S^{-1}$. Find S in the representation in which γ^μ is given by (22-40).**
5. **Find the representation of the γ^μ's if γ_5 is chosen to be diagonal.**
6. The Majorana representation is a representation in which the γ_μ matrices are all imaginary and the $\boldsymbol{\alpha}$ matrices symmetric. Exhibit the explicit form of the Dirac matrices in this representation.
7. Determine the Lorentz transformation properties of $\bar\psi\psi$, $\bar\psi\gamma_5\psi$, $\bar\psi i\gamma^\mu\gamma_5\psi$ and $\bar\psi\sigma^{\mu\nu}\psi$.
8. (a) Verify (22-23) directly from (22-19), (22-21) and (22-22).
   (b) Consider the Dirac equation with zero mass m = 0.

Show that in addition to the vector current $j^\mu = c\bar{\psi}\gamma^\mu\psi$, the axial vector current $j^\mu_5 = \bar{\psi}i\gamma^\mu\gamma_5\psi$ is conserved.

(c) If the mass is not zero, what is $\partial_\mu j^\mu_5$?

9. The matrices $\gamma'^\mu \equiv (a - ib\gamma_5)\,\gamma^\mu$, with $a^2 - b^2 = 1$, obey the anticommutation rules $\{\gamma'^\mu, \gamma'^\nu\} = 2g^{\mu\nu}$. Obtain the explicit representation of the similarity transformations S with relates γ'^μ and γ^μ:

$$\gamma'^\mu = S^{-1}\gamma^\mu S$$

10. (a) Compute the following quantities:

$$\gamma_\mu\gamma^\mu \qquad \gamma_\mu\gamma_\alpha\gamma^\mu \qquad \gamma_\mu\gamma_\alpha\gamma_\beta\gamma^\mu$$

(b) It is customary to write $\not{p}$ for the product $\gamma_\mu p^\mu$ (summation over repeated index implied). Reduce the following products of Dirac matrices to fewer matrices: $\not{p}\not{p}$, $\not{p}\gamma_\mu\not{p}$, $\not{p}\gamma_\mu\gamma_\nu\not{p}$

(c) Compute: Trace $\gamma^\mu\gamma^\nu$, Trace $\gamma^\mu\gamma^\nu\gamma^\alpha\gamma^\beta$,

Trace $(\not{p}+m)\gamma_5(\not{p}+\not{q}+m)\gamma_5(\not{p}+m)\gamma_5(\not{p}+\not{q}+m)\gamma_5$

where p^μ and q^μ are two distinct 4-vectors with $p_\mu p^\mu = m^2$ and $q_\mu q^\mu = M^2$.

11. Show that in a three-dimensional space-time (physics on a plane) Dirac matrices can be 2 × 2, and find an explicit form. What is the situation in two-dimensional space-time (physics on a line) and five-dimensional space-time?

Chapter 23

SOLUTIONS OF THE DIRAC EQUATION

FREE-PARTICLE SOLUTION

The Dirac equation possesses plane-wave solutions, i.e., solutions describing a single particle in the absence of interaction. We write

$$\psi(\mathbf{r}, t) = u \exp\left\{\frac{-ip_\nu x^\nu}{\hbar}\right\} \tag{23-1}$$

where u is a four-component column matrix which satisfies, according to (22-19),

$$(\gamma^\mu p_\mu - mcI)u = 0 \tag{23-2}$$

This is a system of four simultaneous homogenous algebraic (not differential) equations for the four components of u. It has a nontrivial solution if and only if the matrix

$$\gamma^\mu p_\mu - mcI \tag{23-3}$$

has no inverse. Since the formal inverse of (23-3) is

$$(p_\mu p^\mu - m^2c^2)^{-1}(\gamma^\mu p_\mu + mcI) \tag{23-4}$$

(23-2) has a solution if and only if

$$p_\mu p^\mu - m^2c^2 = 0$$

$$E^2 = c^2p^2 + m^2c^4 \tag{23-5a}$$

This shows that the relativistic relation between energy and momentum has been preserved.

For a given p, (23-5a) yields two solutions; one with

$$E = +\sqrt{c^2p^2 + m^2c^4} \equiv E_+ \tag{23-5b}$$

the other with

$$E = -\sqrt{c^2p^2 + m^2c^2} \equiv E_- \tag{23-5c}$$

Thus we see that the Dirac equation leads to positive energy and negative energy solutions just as the Klein-Gordon equation did.

To obtain a specific expression for u we must represent the Dirac matrices in some definite fashion. It is convenient to choose the representation given by (22-40). Clearly there will be four linearly independent solutions, two belonging to E_+ and two to E_-. These can be shown to be

$$\text{(A)}\quad u_1 = 1 \quad u_2 = 0 \quad u_3 = \frac{cp_z}{E_+ + mc^2} \quad u_4 = \frac{c(p_x + ip_y)}{E_+ + mc^2} \tag{23-6a}$$

$$\text{(B)}\quad u_1 = 0 \quad u_2 = 1 \quad u_3 = \frac{c(p_x - ip_y)}{E_+ + mc^2} \quad u_4 = \frac{-cp_z}{E_+ + mc^2} \tag{23-6b}$$

for E_+; and

$$\text{(C)}\quad u_1 = \frac{-cp_z}{-E_- + mc^2} \quad u_2 = \frac{-c(p_x + ip_y)}{-E_- + mc^2} \quad u_3 = 1 \quad u_4 = 0 \tag{23-7a}$$

$$\text{(D)}\quad u_1 = \frac{-c(p_x - ip_y)}{-E_- + mc^2} \quad u_2 = \frac{cp_z}{-E_- + mc^2} \quad u_3 = 0 \quad u_4 = 1 \tag{23-7b}$$

for $E_- = -E_+$. Here u is taken to be

$$u = \begin{pmatrix} u_1 \\ u_2 \\ u_3 \\ u_4 \end{pmatrix}$$

To normalize u, i.e., to get $\Sigma_{n=1}^{4} |u_n|^2 = 1$, each u_i must be multiplied by

$$\left\{1 + \frac{c^2p^2}{(E_+ + mc^2)^2}\right\}^{-1/2} = \left\{\frac{E_+ + mc^2}{2E_+}\right\}^{1/2} \tag{23-8}$$

We see that each solution has two components which, in the nonrelativistic limit, $E_+ \approx mc^2$, are of the order v/c. These are called *small components* and the other two *large components*. For the positive energy solutions, u_1 and u_2 are the large components. In the nonrelativistic limit we expect the large components to correspond to solutions of the Schrödinger free-particle equation. Examining (23-1) we see that this is indeed the case.

Further insight can be obtained by considering the Pauli matrices in the 4 × 4 representation

$$\Sigma = \begin{pmatrix} \sigma & 0 \\ 0 & \sigma \end{pmatrix}, \quad \sigma^{ab} = \varepsilon^{abc} \Sigma^c \tag{23-9}$$

where in the second equation, a, b, c are cyclic permutations of x, y, z and σ^{ab} is the spatial part of the matrix defined in (22-64). In the limit when the small components can be neglected, (23-6a) and (23-7a) are eigenfunctions of Σ^z with eigenvalue +1; (23-6b) and (23-7b) are eigenfunctions of Σ^z with eigenvalue -1. We shall see below that $\frac{1}{2}\hbar\boldsymbol{\Sigma}$ is the spin angular momentum. Hence the four solutions of the free-particle Dirac equation correspond to positive energy, spin $\pm \frac{1}{2}$; negative energy, spin $\pm \frac{1}{2}$.

PHYSICAL INTERPRETATION OF DIRAC MATRICES

The matrix α occurs in the expression for the probability current, $\psi^+ c\alpha\psi$. We thus expect that $c\alpha$ should be interpreted as velocity: $\dot{\mathbf{r}}$. That this is indeed so follows from the Heisenberg expression for the time derivative of an operator

$$\dot{A} = \frac{1}{i\hbar} [A, H] \tag{23-10}$$

which holds in the Dirac theory because of (22-9). Then

$$\dot{x} = \frac{1}{i\hbar} [x, H] = c\alpha_x$$

$$\dot{\mathbf{r}} = c\alpha \tag{23-11}$$

The meaning of (23-11) is, of course, that any matrix elements of the two sides are equal; i.e.,

$$\frac{d}{dt} \int \varphi^\dagger x \psi \, d\tau = c \int \varphi^\dagger \alpha_x \psi \, d\tau$$

The eigenvalues of α_k are ±1, since $\alpha_k^2 = 1$. Thus the eigenvalues of velocity are ±c. This is a very remarkable result, since we know in classical relativity theory that a particle of finite mass can never attain the velocity of light. Moreover, since the components of α do not commute, when the velocity in any one direction is measured, the velocity in the other two directions is entirely undefined. This seems to deny the possibility of velocity measurements.

Because of these difficulties it has been suggested that the position operator should be redefined. One can go into a representation of the Dirac matrices which does not connect positive-energy states with negative ones. This is the so-called *Foldy-Wouthuysen* representation. The position operator x in this representation differs from that in the usual Dirac representation, which can be obtained from it by a unitary similarity transformation. The details of this theory can be found in Bjorken and Drell[1], pp. 45-52.

A consequence of (23-11) is that the relativistic spin $\frac{1}{2}$ particle executes a complicated motion which is an average translation and a superimposed erratic motion called *Zitterbewegung*. To see this we consider

$$\dot{\alpha}_x = \frac{1}{i\hbar}[\alpha_x, H] = \frac{2}{i\hbar}(\alpha_x H - cp_x)$$

$$= \frac{-2}{i\hbar}(H\alpha_x - cp_x) \tag{23-12}$$

H and p_x are constant; therefore,

$$\ddot{\alpha}_x = \frac{2}{i\hbar}\dot{\alpha}_x H = \frac{-2}{i\hbar} H\dot{\alpha}_x \tag{23-13}$$

Integrating,

$$\dot{\alpha}_x = \dot{\alpha}_x^0 e^{-2iHt/\hbar} = e^{2iHt/\hbar}\dot{\alpha}_x^0 \tag{23-14}$$

where $\dot{\alpha}_x^0$ is the value of $\dot{\alpha}_x$ at t = 0. From (23-12),

$$\alpha_x H = \frac{i\hbar}{2}\dot{\alpha}_x^0 e^{-2iHt/\hbar} + cp_x$$

$$\dot{x} = \frac{i\hbar c}{2}\dot{\alpha}_x^0 e^{-2iHt/\hbar} H^{-1} + c^2 p_x H^{-1} \tag{23-15}$$

---

[1] J. D. Bjorken and S. D. Drell, *Relativistic Quantum Mechanics*, McGraw-Hill, New York, 1964.

Noting that $H^{-1} = H/E^2$, we may rewrite (23-15) as

$$\dot{x} = \frac{i\hbar c}{2}\dot{\alpha}_x^0 e^{-2iHt/\hbar}\frac{H}{E^2} + \frac{c^2 p_x H}{E^2} \tag{23-16}$$

Thus we see that the *Zitterbewegung* is an oscillatory motion with frequency $2H/\hbar$, which is at least $2mc^2/\hbar$ and is thus very high. Of course, no practical experiment can observe this. Had we used the position operator which was discussed above in connection with the Foldy-Wouthuysen representation, no such *Zitterbewegung* would arise.

No physical interpretation is given to the β matrix but the following relations can be verified:

$$\frac{d}{dt}\left[\mathbf{r} + \frac{\hbar i}{2mc}\beta\boldsymbol{\alpha}\right] = \frac{\beta\mathbf{p}}{m}$$

$$\frac{d}{dt}\left[t + \frac{\hbar i}{2mc^2}\beta\right] = \frac{\beta}{mc^2}H$$

(23-17)

$$i\hbar\frac{d}{dt}[\alpha_x\alpha_y\alpha_z] = -i\hbar\frac{d}{dt}\gamma_5 = -2mc^2\beta\alpha_x\alpha_y\alpha_z = 2mc^2\gamma^0\gamma_5$$

SPIN

We consider the angular momentum operator $\mathbf{L}$ and inquire about $[L_x, H]$

$$\frac{1}{i\hbar}[L_x, H] = \frac{1}{i\hbar}[(yp_z - zp_y), (c\boldsymbol{\alpha}\cdot\mathbf{p} + \beta mc^2)]$$

$$= c(\alpha_y p_z - \alpha_z p_y) \tag{23-18a}$$

or

$$\frac{d\mathbf{L}}{dt} = c\boldsymbol{\alpha}\times\mathbf{p} \tag{23-18b}$$

Hence the angular momentum is no longer a constant of the motion. On the other hand, the existence of two linearly independent solutions corresponding to a given value of the energy indicates that an operator that commutes with the Hamiltonian must exist. We shall show that this operator is

$$\mathbf{J} = \mathbf{L} + \frac{1}{2}\hbar\boldsymbol{\Sigma} \tag{23-19}$$

where Σ is, in the special representation (22-40a) and (22-40c). Remembering the commutation relations of the Pauli matrices σ, we find

$$[\Sigma_x, \alpha_y] = 2i\alpha_z \tag{23-20}$$

and cyclic permutations. Then

$$\frac{1}{i\hbar}\frac{\hbar}{2}[\Sigma_x, H] = c(\alpha_z p_y - \alpha_y p_z) \tag{23-21a}$$

and therefore

$$[\mathbf{L} + \frac{1}{2}\hbar\mathbf{\Sigma}, H] = 0 \tag{23-21b}$$

Thus (23-19) is indeed a constant of motion. It is the total angular momentum J and is the vector sum of the orbital momentum L and a second term, which has the eigenvalues $\pm\frac{1}{2}\hbar$. We have thus shown that a particle obeying the Dirac equation has a spin of $\frac{1}{2}$. All our previous results about the compounding of orbital and spin angular momentum can be used.

The correct total angular momentum operator has become $\mathbf{J} = \mathbf{L} + \frac{1}{2}\hbar\mathbf{\Sigma}$. This is an example of the fact that the elementary rules for constructing operators from classical dynamical variables [e.g., in the $\mathbf{r}$ representation, replacing $\mathbf{r}$ by the operator $\mathbf{r}$ and $\mathbf{p}$ by $(\hbar/i)\nabla$] are not general enough. Instead we must allow for the possibility of adding further terms which disappear in the limit $\hbar \to 0$. Thus to obtain the proper angular momentum operator from $\mathbf{r} \times \mathbf{p}$, for particles which satisfy the Dirac equation, we must add $\frac{1}{2}\hbar\,\mathbf{\Sigma}$ to the operator $\mathbf{r} \times \mathbf{p}$.

If we do not wish to specify the representation of the Dirac matrices we may put

$$\mathbf{\Sigma} = i\gamma_5\boldsymbol{\alpha} \tag{23-23a}$$

Using the expression (22-23) for γ_5, and noting that γ_5 commutes with any α_k we regain (23-20). From the definitions (22-63), (23-9) and (23-22a) we obtain

$$\Sigma^a = i\gamma^b\gamma^c \tag{23-22b}$$

where a, b, c are x, y, z or follow cyclically.

DIRAC EQUATION IN EXTERNAL FIELD

As in the Klein-Gordon case, the inclusion of an electromagnetic field can be achieved by the gauge-invariant, Lorentz-covariant *minimal* replacement

$$p_\mu \to p_\mu - \frac{e}{c}A_\mu \tag{23-23}$$

i.e.,

$$E \to E - e\varphi \qquad \mathbf{p} \to \mathbf{p} - \frac{e}{c}\mathbf{A}$$

This leads to the equation

$$\{(i\hbar\frac{\partial}{\partial t} - e\varphi) - \boldsymbol{\alpha}\cdot(\frac{c\hbar}{i}\nabla - e\mathbf{A}) - \beta mc^2\}\psi = 0 \tag{23-24a}$$

$$\left(i\hbar\gamma^\mu\left(\frac{\partial}{\partial x^\mu} + \frac{ie}{\hbar c}A_\mu\right) - mc\right)\psi = 0 \tag{23-24b}$$

Evidently we are assuming that the Dirac particle carries charge e. Other fields can be taken into account by including the corresponding potentials with mc^2 if the potentials are relativistic scalars, in A_μ if they are four-vectors.

The substitution (23-23) is not completely general; a possible additional term will be given in (23-28).

To obtain a second-order equation similar in form to the Klein-Gordon equation we premultiply (23-24b) by $\gamma^\nu(i\hbar\,\partial/\partial x^\nu - (e/c)A_\nu)$ and obtain

$$\gamma^\mu\gamma^\nu\left(i\hbar\frac{\partial}{\partial x^\mu} - \frac{e}{c}A_\mu\right)\left(i\hbar\frac{\partial}{\partial x^\nu} - \frac{e}{c}A_\nu\right)\psi = m^2c^2\psi \tag{23-25a}$$

and use (22-64) to rewrite this as

$$(g^{\mu\nu} - i\sigma^{\mu\nu})\left(i\hbar\frac{\partial}{\partial x^\mu} - \frac{e}{c}A_\mu\right)\left(i\hbar\frac{\partial}{\partial x^\nu} - \frac{e}{c}A_\nu\right)$$

$$= \left(i\hbar\frac{\partial}{\partial x^\mu} - \frac{e}{c}A_\mu\right)\left(i\hbar\frac{\partial}{\partial x_\mu} - \frac{e}{c}A^\mu\right)$$

$$- \frac{i}{2}(\sigma^{\mu\nu} - \sigma^{\nu\mu})\left(i\hbar\frac{\partial}{\partial x^\mu} - \frac{e}{c}A_\mu\right)\left(i\hbar\frac{\partial}{\partial x^\nu} - \frac{e}{c}A_\nu\right)$$

$$= \left(i\hbar\frac{\partial}{\partial x^\mu} - \frac{e}{c}A_\mu\right)\left(i\hbar\frac{\partial}{\partial x_\mu} - \frac{e}{c}A^\mu\right)$$

$$- \frac{i}{2}\sigma^{\mu\nu}\left[\left(i\hbar\frac{\partial}{\partial x^\mu} - \frac{e}{c}A_\mu\right), \left(i\hbar\frac{\partial}{\partial x^\nu} - \frac{e}{c}A_\nu\right)\right]$$

The commutator in the last line yields

$$- \frac{ie\hbar}{c}\left[\frac{\partial A_\nu}{\partial x^\mu} - \frac{\partial A_\mu}{\partial x^\nu}\right] = - \frac{ie\hbar}{c}F_{\mu\nu}$$

where $F_{\mu\nu}$ is the electromagnetic field tensor. We finally obtain the equation

$$\left[\left(i\hbar\frac{\partial}{\partial x^\mu} - \frac{e}{c}A_\mu\right)\left(i\hbar\frac{\partial}{\partial x_\mu} - \frac{e}{c}A^\mu\right) - \frac{e\hbar}{2c}\sigma^{\mu\nu}F_{\mu\nu}\right]\psi \qquad \text{(23-25b)}$$

$$= (mc)^2\psi$$

Using the definition of Σ, we can write the above as

$$\left[\left(i\hbar\frac{\partial}{\partial t} - e\varphi\right)^2 - \left(\frac{c\hbar}{i}\nabla - e\mathbf{A}\right)^2 + e\hbar c\left(\Sigma\cdot\mathcal{H} - i\alpha\cdot\mathcal{E}\right)\right]\psi \qquad \text{(23-26)}$$

$$= m^2c^4\psi$$

where

$$F^{oa} = -\mathcal{E}^a, \; F^{ab} = -\varepsilon^{abc}\mathcal{H}^c \qquad \text{(23-27)}$$

The first two terms on the left-hand side occur in the Klein-Gordon equation. The other two terms arise only in the Dirac theory and are seen to vanish as $\hbar \to 0$. The last term is by itself relativistically invariant. It is in principle possible to multiply it by an arbitrary factor $1 + K$, i.e., to add to the equation a term

$$K\frac{e\hbar}{2c}\sigma^{\mu\nu}F_{\mu\nu}\psi \qquad \text{(23-28)}$$

This is known as a *Pauli moment* term, and means of course a corresponding modification of the original Dirac equation. Such a modification is in principle permissible because it tends to zero as $\hbar \to 0$, but it would complicate the theory. Indeed one may insert a Pauli moment already in the first order Dirac equation (23-24b) without violating the principles of Lorentz and gauge invariance. Such additional electromagnetic coupling between the Dirac particles and the electromagnetic field is called *non-minimal*. We shall not consider this more complicated possibility, but it will be mentioned again below.

NONRELATIVISTIC LIMIT

To discuss the non-relativistic limit, we use the explicit form for the Dirac matrices (22-40), for which βmc^2 is diagonal and write in (23-24a)

$$\psi = \begin{pmatrix} \psi_A \\ \psi_B \end{pmatrix} \tag{23-29}$$

where ψ_A and ψ_B are still two-component functions. In the Dirac equation (23-24a) time is separated in the usual way, since we assume the external potentials to be static. Thus we arrive at

$$\left\{ \begin{bmatrix} 0 & I \\ I & 0 \end{bmatrix} \boldsymbol{\sigma} \cdot (c\mathbf{p} - e\mathbf{A}) + \begin{bmatrix} I & 0 \\ 0 & -I \end{bmatrix} mc^2 \right\} \begin{pmatrix} \psi_A \\ \psi_B \end{pmatrix}$$

$$= (E - e\varphi) \begin{pmatrix} \psi_A \\ \psi_B \end{pmatrix} \tag{23-30}$$

where $\mathbf{p}$ stands for the gradient operator $\frac{\hbar}{i} \nabla$. Equation (23-30) is equivalent to two coupled equations

$$\boldsymbol{\sigma} \cdot (c\mathbf{p} - e\mathbf{A}) \psi_B + mc^2 \psi_A = (E - e\varphi) \psi_A \tag{23-31a}$$

$$\boldsymbol{\sigma} \cdot (c\mathbf{p} - e\mathbf{A}) \psi_A - mc^2 \psi_B = (E - e\varphi) \psi_B \tag{23-31b}$$

Setting $E = E' + mc^2$, we obtain from (23-31b)

$$\psi_B = (E' - e\varphi + 2mc^2)^{-1} \boldsymbol{\sigma} \cdot (c\mathbf{p} - e\mathbf{A}) \psi_A \tag{23-32a}$$

which is then inserted into (23-31a) to give

$$\frac{1}{2m}\,\sigma\cdot(p - \frac{e}{c}A)(1 + \frac{E'-e\varphi}{2mc^2})^{-1}\ \sigma\cdot(p - \frac{e}{c}A)\psi_A = (E' - e\varphi)\psi_A \tag{23-32b}$$

Equations (23-32) are exact; we now approximate them in the nonrelativistic regime, where

$$E' \ll mc^2,\ e\varphi \ll mc^2$$

and the eigenvalues of **p** are of order mv, and much less than mc. Hence the lower components are

$$\psi_B = O(\frac{v}{c})\psi_A \tag{23-33}$$

The four-component solution has two large components ψ_A and two small components ψ_B, just as in the free-particle case.

If terms of order v^2/c^2 are ignored, the equation for ψ_A, (23-32b) reduces to

$$\frac{1}{2m}[\sigma\cdot(p - \frac{e}{c}A)]^2\psi_A + e\varphi\,\psi_A = E'\psi_A \tag{23-34a}$$

Working out the product of two Pauli matrices in the above leaves

$$\left[\frac{1}{2m}(p - \frac{e}{c}A)^2 + e\varphi - \frac{e}{mc}\frac{\hbar}{2}\,\sigma\cdot\mathcal{H}\right]\psi_A = E'\psi_A \tag{23-34b}$$

This is the *Pauli equation*, introduced by him to take electron spin into account nonrelativistically. The term involving the magnetic field has the form of magnetic dipole interaction energy $-\frac{e}{mc}\,S\cdot\mathcal{H}$, with gyromagnetic ratio $2 \times \frac{e}{2mc}$. This verifies of course the Goudsmit-Uhlenbeck hypothesis that the electron has the magnetic moment $(e\hbar/2mc)\sigma$, and gives the correct gyromagnetic ratio.

It is beyond the scope of the present discussion to treat the corrections that quantum electrodynamics requires. Suffice it to say that the interaction of a charged particle with its own field leads to a correction factor to the magnetic moment which is calculated to be

$$1 + \frac{e^2}{2\pi\hbar c} + O(\frac{e^2}{\hbar c})^2 = 1.00116$$

This value of the magnetic moment has been verified experimentally. Another observable consequence of the self-interaction of the electron is the Lamb shift (see below and Chapter 8, Problem 8).

The Dirac theory does not predict the correct value for the magnetic moment of the proton or neutron. One could obtain the required result by adding to the first-order Dirac equation the so-called Pauli term,

$$K(\boldsymbol{\Sigma}\cdot\boldsymbol{\mathcal{H}} - i\boldsymbol{\alpha}\cdot\boldsymbol{\mathcal{E}})$$

which would still lead to a relativistically invariant equation. (The term in $\boldsymbol{\mathcal{E}}$ is necessary to preserve Lorentz covariance, but does not contribute in the non-relativistic limit since $\boldsymbol{\alpha}\cdot\boldsymbol{\mathcal{E}}$ is $O(\frac{v}{c})$ relative to $\boldsymbol{\sigma}\cdot\boldsymbol{\mathcal{H}}$.) The constant K is then adjusted to give the correct answer. The arbitrary nature of this device makes it rather unsatisfactory. Physically, it is believed that the proton's and neutron's magnetic moments are due to interaction with a meson field, but attempts at a quantitative theory have been only partially successful.

When terms of order v^2/c^2 are kept, we set $\mathbf{A}$ to zero for simplicity, define $e\varphi$ to be V and obtain for (23-32b)

$$\frac{1}{2m}\,\boldsymbol{\sigma}\cdot\mathbf{p}\left(1 + \frac{E'-V}{2mc^2}\right)^{-1}\boldsymbol{\sigma}\cdot\mathbf{p}\,\psi_A + V\psi_A = E'\psi_A \tag{23-35}$$

The following relations are true.

$$\left[1 + \frac{E' - V}{2mc^2}\right]^{-1} \approx 1 - \frac{E' - V}{2mc^2}$$

$$\mathbf{p}V = V\mathbf{p} - i\hbar\nabla V$$

$$(\boldsymbol{\sigma}\cdot\nabla V)(\boldsymbol{\sigma}\cdot\mathbf{p}) = \nabla V\cdot\mathbf{p} + i\boldsymbol{\sigma}\cdot[\nabla V \times \mathbf{p}]$$

Taking V to be spherically symmetric, we may reduce (23-35) to

$$\left[\left(1 - \frac{E' - V}{2mc^2}\right)\frac{p^2}{2m} + V\right]\psi_A - \frac{\hbar^2}{4m^2c^2}\,\frac{dV}{dr}\,\frac{\partial\psi_A}{\partial r}$$

$$+ \frac{1}{2m^2c^2}\,\frac{1}{r}\,\frac{dV}{dr}\,\mathbf{S}\cdot\mathbf{L}\,\psi_A = E'\psi_A \tag{23-36a}$$

with

$$\mathbf{S} = \tfrac{1}{2}\hbar\boldsymbol{\sigma} \qquad \mathbf{L} = \mathbf{r}\times\mathbf{p}$$

Finally, setting in the correction term $E' - V \approx p^2/2m$, we obtain

$$\left[\frac{p^2}{2m} + V - \frac{p^4}{8m^3c^2} - \frac{\hbar^2}{4m^2c^2}\frac{dV}{dr}\frac{\partial}{\partial r} + \frac{1}{2m^2c^2}\frac{1}{r}\frac{dV}{dr}\mathbf{S}\cdot\mathbf{L}\right]\psi_A = E'\psi_A \tag{23-36b}$$

The first two terms on the left side of (23-36b) are recognized from the nonrelativistic Schrödinger equation. The third term is the second contribution to the expansion of E' in terms of p^2, i.e.,

$$E' = E - mc^2 = mc^2\left(1 + \frac{p^2}{m^2c^2}\right)^{1/2} - mc^2 \approx \frac{p^2}{2m} - \frac{p^4}{8m^3c^2}$$

The next term has no classical analogue. Finally the last term is the spin-orbit coupling energy including the Thomas factor of $\frac{1}{2}$ (see p. 153).

The procedure of solution for (23-36b) is to solve the nonrelativistic Schrödinger equation for the two components of ψ_A; then to form linear combinations such that J^2, J_z, L^2, S^2 are diagonal; and then to consider

$$\frac{-p^4}{8m^3c^2} - \left(\frac{\hbar}{2mc}\right)^2\frac{dV}{dr}\frac{\partial}{\partial r}$$

and the spin-orbit interaction as a perturbation.

EXACT SOLUTION OF DIRAC EQUATION FOR COULOMB POTENTIAL

We solve the Dirac differential equation for a Coulomb field. We use the explicit Dirac representation; i.e., we solve the four equations, (23-31) with $e\varphi = -Ze^2/r$, $\mathbf{A} = 0$. We write

$$\psi_A = \begin{pmatrix} u_1 \\ u_2 \end{pmatrix} \qquad \psi_B = \begin{pmatrix} u_3 \\ u_4 \end{pmatrix}$$

and get

$$-\frac{i}{\hbar c}\left[E + \frac{Ze^2}{r} - mc^2\right]u_1 + \frac{\partial u_4}{\partial x} - i\frac{\partial u_4}{\partial y} + \frac{\partial u_3}{\partial z} = 0$$

$$-\frac{i}{\hbar c}\left[E + \frac{Ze^2}{r} - mc^2\right]u_2 + \frac{\partial u_3}{\partial x} + i\frac{\partial u_3}{\partial y} - \frac{\partial u_4}{\partial z} = 0$$

$$-\frac{i}{\hbar c}\left[E + \frac{Ze^2}{r} + mc^2\right]u_3 + \frac{\partial u_2}{\partial x} - i\frac{\partial u_2}{\partial y} + \frac{\partial u_1}{\partial z} = 0$$

$$-\frac{i}{\hbar c}\left[E + \frac{Ze^2}{r} + mc^2\right]u_4 + \frac{\partial u_1}{\partial x} + i\frac{\partial u_1}{\partial y} - \frac{\partial u_2}{\partial y} = 0 \qquad (23\text{-}37)$$

To find a solution we now make use of the fact that if we consider only large components, i.e., set the small components equal to zero, then $[\boldsymbol{\ell}, H]$, which is proportional to $\boldsymbol{\alpha} \times \mathbf{p}$, will be 0, since α connects the large and small components. Thus ψ_A will be an eigenfunction of $\boldsymbol{\ell}$. Furthermore it must contain one component with spin up and another with spin down. Of course j and j_z are constants of the motion. Hence for $j = \ell + \frac{1}{2}$ we set

$$u_1 = g(r)\sqrt{\frac{\ell + m + \frac{1}{2}}{2\ell + 1}}\; Y_{\ell,m-\frac{1}{2}}(\Omega)$$

$$u_2 = -g(r)\sqrt{\frac{\ell - m + \frac{1}{2}}{2\ell + 1}}\; Y_{\ell,m+\frac{1}{2}}(\Omega) \qquad (23\text{-}38a)$$

This differs from the Pauli nonrelativistic treatment in that g(r) is an as-yet arbitrary radial function and not the solution of the radial nonrelativistic Schrödinger equation.

To obtain the small components, we note from (23-32a) that the operator which gives the small component from the large component has odd parity (**p** is odd, **A** is zero, everything else is even) and commutes with **j**. Hence ψ_B must belong to the same j value as ψ_A but must have a different ℓ. Corresponding to $j = \ell + \frac{1}{2}$ the only other possible value of the orbital momentum is $\ell' = \ell + 1$. Hence we set, remembering the Clebsch-Gordan coefficients,

$$u_3 = -if(r)\sqrt{\frac{\ell - m + \frac{3}{2}}{2\ell + 3}}\; Y_{\ell+1,m-\frac{1}{2}}(\Omega)$$

$$u_4 = -if(r)\sqrt{\frac{\ell + m + \frac{3}{2}}{2\ell + 3}}\; Y_{\ell+1,m+\frac{1}{2}}(\Omega) \qquad (23\text{-}38b)$$

where f(r) is some radial function. Inserting (23-38) in (23-37), we find that for $j = \ell + \frac{1}{2}$ the connection between f and g is given by

$$\frac{1}{\hbar c}\left(E + \frac{Ze^2}{r} + mc^2\right) f = \frac{dg}{dr} - \ell \frac{g}{r}$$

$$\frac{1}{\hbar c}\left(E + \frac{Ze^2}{r} - mc^2\right) g = -\frac{df}{dr} - (\ell + 2)\frac{f}{r} \tag{23-39}$$

Arguing completely analogously one finds that for $j = \ell - \frac{1}{2}$

$$u_1 = g(r)\sqrt{\frac{\ell - m + \frac{1}{2}}{2\ell + 1}}\; Y_{\ell, m-\frac{1}{2}}(\Omega)$$

$$u_2 = g(r)\sqrt{\frac{\ell + m + \frac{1}{2}}{2\ell + 1}}\; Y_{\ell, m+\frac{1}{2}}(\Omega)$$

$$u_3 = -if(r)\sqrt{\frac{\ell + m - \frac{1}{2}}{2\ell - 1}}\; Y_{\ell-1, m-\frac{1}{2}}(\Omega)$$

$$u_4 = if(r)\sqrt{\frac{\ell - m - \frac{1}{2}}{2\ell - 1}}\; Y_{\ell-1, m+\frac{1}{2}}(\Omega) \tag{23-40}$$

and

$$\frac{1}{\hbar c}\left(E + \frac{Ze^2}{r} + mc^2\right) f = \frac{dg}{dr} + (\ell + 1)\frac{g}{r}$$

$$\frac{1}{\hbar c}\left(E + \frac{Ze^2}{r} - mc^2\right) g = -\frac{df}{dr} + (\ell - 1)\frac{f}{r} \tag{23-41}$$

We define

$$k = -(\ell + 1) \qquad \text{if } j = \ell + \tfrac{1}{2}$$

$$k = \ell \qquad \text{if } j = \ell - \tfrac{1}{2}$$

i.e.,

$$k = 1, 2, \ldots \qquad \text{for } j = \ell - \tfrac{1}{2}$$

$$k = -1, -2, \ldots \qquad \text{for } j = \ell + \tfrac{1}{2} \tag{23-42}$$

Equations (23-39) and (23-41) can be combined as

$$\frac{1}{\hbar c}\left(E + \frac{Ze^2}{r} + mc^2\right) f - \left(\frac{dg}{dr} + (1 + k)\frac{g}{r}\right) = 0$$

$$\frac{1}{\hbar c}\left(E + \frac{Ze^2}{r} - mc^2\right) g + \left(\frac{df}{dr} + (1 - k)\frac{f}{r}\right) = 0 \qquad (23\text{-}43)$$

Setting

$$F = rf \qquad G = rg$$

$$\alpha_1 = \frac{mc^2 + E}{\hbar c} \qquad \alpha_2 = \frac{mc^2 - E}{\hbar c}$$

$$\alpha = (\alpha_1 \alpha_2)^{1/2} \qquad \gamma = \frac{Ze^2}{\hbar c} \qquad \rho = \alpha r \qquad (23\text{-}44)$$

we obtain

$$\left(\frac{d}{d\rho} + \frac{k}{\rho}\right) G - \left(\frac{\alpha_1}{\alpha} + \frac{\gamma}{\rho}\right) F = 0$$

$$\left(\frac{d}{d\rho} - \frac{k}{\rho}\right) F - \left(\frac{\alpha_2}{\alpha} - \frac{\gamma}{\rho}\right) G = 0 \qquad (23\text{-}45)$$

We use the time-honored series method to obtain the positive energy bound state solutions. First we substitute

$$F = \varphi(\rho)e^{-\rho} \qquad G = \chi(\rho)e^{-\rho} \qquad (23\text{-}46)$$

and obtain

$$\chi' - \chi + \frac{k\chi}{\rho} - \left(\frac{\alpha_1}{\alpha} + \frac{\gamma}{\rho}\right) \varphi = 0$$

$$\varphi' - \varphi - \frac{k\varphi}{\rho} - \left(\frac{\alpha_2}{\alpha} - \frac{\gamma}{\rho}\right) \chi = 0 \qquad (23\text{-}47)$$

We then substitute the power series

$$\varphi = \rho^s \sum_{m=0}^{\infty} a_m \rho^m \qquad a_0 \neq 0$$

$$\chi = \rho^s \sum_{m=0}^{\infty} b_m \rho^m \qquad b_0 \neq 0 \qquad (23\text{-}48)$$

The requirement that f, g be everywhere finite will be seen to be impossible to satisfy. We therefore require that the integrated probability density be finite, i.e., that

$$\int_0^\infty \{|F(\rho)|^2 + |G(\rho)|^2\}\, d\rho < \infty \tag{23-49}$$

This assures us that $s \neq -\infty$. Substituting the series into (23-47) and equating coefficients of the same power of ρ gives

$$(s + v + k)b_v - b_{v-1} - \gamma a_v - \frac{\alpha_1}{\alpha} a_{v-1} = 0$$

$$(s + v - k)a_v - a_{v-1} + \gamma b_v - \frac{\alpha_2}{\alpha} b_{v-1} = 0 \tag{23-50}$$

For $v = 0$ we get

$$(s + k)b_0 - \gamma a_0 = 0$$

$$(s - k)a_0 + \gamma b_0 = 0 \tag{23-51}$$

This will have a nontrivial solution if and only if

$$s = \pm(k^2 - \gamma^2)^{1/2} \tag{23-52a}$$

We first examine the negative root. For small ρ the integrand of (23-49) is $\sim\rho^{2s}$ and we must have $2s > -1$, $(k^2 - \gamma^2)^{1/2} < \frac{1}{2}$. The minimum s occurs for $k^2 = 1$; this would mean $Z \geq 109$. For $k^2 > 1$, no value of Z will permit the negative root. Restricting ourselves to $Z < 109$, we choose the positive root

$$s = (k^2 - \gamma^2)^{1/2} \tag{23-52b}$$

For $k = 1$, $s < 1$, and f, g diverge at the origin; however (23-49) converges.

The recursion relations (23-50) can easily be seen to lead to functions of the order $e^{2\rho}$; hence we must require the series to terminate for (23-49) to hold. Suppose this happens for $v = n'$, i.e., $a_{n'+1} = b_{n'+1} = 0$. Then from (23-50) we get

$$\alpha_1 a_{n'} = -\alpha b_{n'} \qquad n' = 0, 1, \ldots \tag{23-53a}$$

Multiply the first of (23-50) by α, the second by α_1, and subtract

$$b_v[\alpha(s + v + k) - \alpha_1\gamma] = a_v[\alpha_1(s + v - k) + \alpha\gamma] \tag{23-53b}$$

then insert $v = n'$ and use (23-53).

$$2\alpha(s + n') = \gamma(\alpha_1 - \alpha_2) = \frac{2E\gamma}{\hbar c} \tag{23-53c}$$

All this leads to

$$E = mc^2\left[1 + \frac{\gamma^2}{(s + n')^2}\right]^{-1/2} \tag{23-54a}$$

$$E = mc^2\left[1 + \frac{\gamma^2}{(n' + \sqrt{(k^2 - \gamma^2)})^2}\right]^{-1/2} \tag{23-54b}$$

Since $|k| = j + \frac{1}{2}$,

$$E = mc^2\left[1 + \frac{\gamma^2}{(n' + \sqrt{(j + \frac{1}{2})^2 - \gamma^2})^2}\right]^{-1/2} \tag{23-54c}$$

$$n' = 0, 1, \ldots; \qquad j + \tfrac{1}{2} = 1, 2, \ldots$$

From (23-54b) it is seen that negative values of $k = -1, -2, -3, \ldots$ are also acceptable. However, for $n' = 0$ (23-51) and (23-53) give

$$\frac{b_0}{a_0} = \frac{\gamma}{s + k} = -\frac{\alpha_1}{\alpha} \tag{23-55}$$

Since $s < |k|$, the first expression for b_0/a_0 will be positive or negative, depending on whether k is positive or negative. The second will always be negative. Hence for $n' = 0$, k can only be negative, i.e., $j = \ell + \frac{1}{2}$. Equation (23-54c) can be expanded in powers of γ^2:

$$E = mc^2\left[1 - \frac{\gamma^2}{2n^2} - \frac{\gamma^4}{2n^3}\left(\frac{1}{|k|} - \frac{3}{4n}\right)\right] \tag{23-56}$$

where $n = n' + |k|$.

It is seen that the Dirac theory leads to accidental degeneracy in ℓ; i.e., states with the same j but different ℓ have the same energy. This degeneracy is removed by the Lamb shift, which is due to the interaction of the electron with its own field. For $j = \frac{1}{2}$, the effect is one order of magnitude smaller than the fine-structure splitting; for $j \geq \frac{3}{2}$, two orders of magnitude. For example, the Dirac theory predicts, for $n = 2$, two states with the same energy $s_{1/2}$, $p_{1/2}$ and a fine-structure splitting of $\sim 0.36\ \text{cm}^{-1}$ between these and the $p_{3/2}$ state. The Lamb shift raises the $s_{1/2}$ state by about $0.035\ \text{cm}^{-1}$ above the $p_{1/2}$. The binding energy is $27{,}000\ \text{cm}^{-1}$, so we are really dealing with a "fine" structure.

NEGATIVE ENERGY SOLUTIONS

We have seen that both the Klein-Gordon and the Dirac theories lead to positive energy states $\gtrsim mc^2$ and negative energy states $\lesssim -mc^2$. In classical theory negative energy solutions exist also, but they can be excluded by physical continuity: it is impossible for a classical particle to pass from positive energy states to negative energy without going through energy states in between. Hence the removal of the negative energy states is equivalent to the boundary condition that "in the beginning" all particles had positive energy.

A completely free, single quantum mechanical particle also will not make transitions. However, no particle is completely free and transitions can always occur, by radiation if by no other means; e.g., for an electron bound in a hydrogen atom, one can calculate that the transition to the negative energy states by radiation will take place in about 10^{-10} sec. Once an electron has made the transition it would quickly "fall" toward negative infinite energy. This is clearly contrary to observation.

Dirac proposed that the negative energy states be regarded as full. Then the Pauli exclusion principle will prevent transitions into such occupied states. It is assumed that there are no gravitational or electromagnetic effects of the electrons occupying the negative states. That is to say, according to Dirac the vacuum is the condition when all the negative energy states are filled. Occasionally one or more of the negative energy states can become empty. The absence of a negative energy electron would manifest itself as the presence of a positively charged electron, i.e., of a positron. When Pauli in 1932 considered (in his article in the *Handbuch der Physik*) this interpretation of the negative energy states, he rejected it because at the time there was no experimental evidence for the existence of such positrons. However, by the time the article had appeared in print in 1933, the theory was vindicated by Anderson's discovery of the positron.

Using the concept of the "sea of negative energy electrons" one can calculate the probability of pair production in the electric field of a nucleus by considering the probability of raising an electron from a negative to a positive state.

It should be pointed out that similar techniques cannot be applied to the Klein-Gordon equation, because particles of zero spin do not obey an exclusion principle. Also there remain difficulties of principle with this approach to the Dirac equation. If the vacuum consists of an infinity of filled states, one wonders about the total inertia, energy and charge of such a state. These problems can be avoided by viewing the Dirac equation as a classical field equation and then quantizing it, just as is done for Klein-Gordon equation. In this approach $e\rho$ and ej become the charge and current density, and the quantized field energy is always positive. The parameter E in the wave equation is positive for positively charged particles and negative for negatively charged particles, and so is the charge density.[2]

PERTURBATION THEORY

It is clear from the general structure of the Dirac theory that time-independent and time-dependent perturbation theory is formally the same as for the nonrelativistic Schrödinger theory, except that the matrix elements are now calculated between spinors and not just between one-component (scalar) wave functions. We shall examine the results for scattering of free particles by a potential V, which is the fourth component of a four-vector.

The transition probability per unit time is given by Fermi's golden rule

$$w = \frac{2\pi}{\hbar} |V_{fi}|^2 \rho(E) \tag{1-15}$$

$$\rho(E) = \frac{d\Omega\, p^2\, dp}{(2\pi\hbar)^3\, dE} = \frac{d\Omega\, p^2}{(2\pi\hbar)^3 v} \tag{1-18}$$

For free particles the initial wave function is

$$\psi_i = u_i e^{i\mathbf{k}_i \cdot \mathbf{r}} \tag{23-57}$$

where u_i is a four-spinor independent of $\mathbf{r}$. The final states are

$$\psi_f = u_f e^{i\mathbf{k}_f \cdot \mathbf{r}} \tag{23-58}$$

where u_f is again independent of $\mathbf{r}$.

$$V_{fi} = \int V e^{i\mathbf{q}\cdot\mathbf{r}}\, d\tau\, (u_f^\dagger u_i)$$

$$\mathbf{q} = \mathbf{k}_i - \mathbf{k}_f \tag{23-59}$$

This is exactly the same as the nonrelativistic Born approximation result except for the new factor $(u_f^\dagger u_i)$.

---

[2] See for example J. D. Bjorken and S. D. Drell, *Relativistic Quantum Fields*, McGraw-Hill, New York, 1965.

We consider the quantity $|(u_f^\dagger u_i)|^2$. Usually we are not interested in the specific final spin, hence we must sum this over all final spin states. Furthermore we may wish to take one-half of the sum over initial spin states, which amounts to averaging over the initial spins —a necessary procedure if the initial states are unpolarized.

A simple method for evaluating such sums is in terms of *Casimir projection operators*. It should be noted that dealing with elastic scattering we cannot use closure, since sums do not extend over all possible energy states. Specifically, both initially and finally energy must be positive. The projection operator is defined by

$$P = \frac{1}{2}\left[1 + \frac{c}{|E|}(\alpha\cdot p + \beta mc)\right] = \frac{1}{2}(1 + \alpha\cdot\frac{v}{c} + \beta\mu)$$

$$\mu = \frac{mc^2}{|E|} = (1 - \frac{v^2}{c^2})^{1/2} \qquad (23\text{-}60)$$

We note that

$$Pu = \frac{1}{2}\left[u + \frac{c}{|E|}(\alpha\cdot p + \beta mc)u\right]$$

$$= \frac{1}{2}[u \pm u] = u \text{ or } 0 \qquad (23\text{-}61)$$

depending on whether u is a positive energy state or a negative energy state. Therefore,

$$\frac{1}{2}\sum_{i}\sum_{\substack{f\\E>0}}(u_i^\dagger u_f)(u_f^\dagger u_i)$$

$$= \frac{1}{2}\sum_{i}\sum_{\substack{f\\\text{all } E}}(u_i^\dagger P_f u_f)(u_f^\dagger P_i u_i) \qquad (23\text{-}62)$$

The two sums are the same since the projection operators annihilate the states with negative energy. We further note that

$$P_f = \frac{1}{2}(1 + \alpha\cdot\frac{v_f}{c} + \beta\mu)$$

$$P_i = \frac{1}{2}(1 + \alpha\cdot\frac{v_i}{c} + \beta\mu) \qquad (23\text{-}63)$$

The μ's in the two operators are the same because the scattering is elastic. Using closure in the sum over the final states in (23-62) we get

$$\frac{1}{8}\sum_i u_i^\dagger\left(1 + \boldsymbol{\alpha}\cdot\frac{\mathbf{v}_f}{c} + \beta\mu\right)\left(1 + \boldsymbol{\alpha}\cdot\frac{\mathbf{v}_i}{c} + \beta\mu\right)u_i \tag{23-64a}$$

Since the set of spinors u_i is also complete, the above is

$$\frac{1}{8}\operatorname{Tr}\left(1 + \boldsymbol{\alpha}\cdot\frac{\mathbf{v}_f}{c} + \beta\mu\right)\left(1 + \boldsymbol{\alpha}\cdot\frac{\mathbf{v}_i}{c} + \beta\mu\right) \tag{23-64b}$$

Expanding the product and making use of the results that

$$\operatorname{Tr}\boldsymbol{\alpha} = \operatorname{Tr}\beta = \operatorname{Tr}\boldsymbol{\alpha}\beta = 0$$

we finally get for the sum over initial and final states

$$\frac{1}{2}\left(1 + \frac{\mathbf{v}_i\cdot\mathbf{v}_f}{c^2} + \mu^2\right) = \frac{1}{2}\left(2 - \frac{v^2}{c^2} + \frac{v^2}{c^2}\cos\theta\right) \tag{23-64c}$$

where

$$\frac{\mathbf{v}_i\cdot\mathbf{v}_f}{v^2} = \cos\theta$$

We conclude that the scattering cross section of Dirac particles in a potential which is the fourth component of a four-vector is the same as the nonrelativistic result times a factor

$$1 - \frac{v^2}{c^2}\sin^2\left(\frac{1}{2}\theta\right) \tag{23-65}$$

PROBLEMS

1. **A representation for the free particle solutions u, which is independent of the choice of the γ^μ is $(p^\mu\gamma_\mu + mcI)\chi_i$; where χ_i is any one of the four linearly independent column vectors**

$$\begin{pmatrix}1\\0\\0\\0\end{pmatrix}\quad\begin{pmatrix}0\\1\\0\\0\end{pmatrix}\quad\begin{pmatrix}0\\0\\1\\0\end{pmatrix}\quad\begin{pmatrix}0\\0\\0\\1\end{pmatrix}$$

$(p^\mu\gamma_\mu + mcI)\chi_i$ evidently is a solution of $(p^\mu\gamma_\mu - mcI)u = 0$ if we set $p^2 = m^2c^2$ or $E = \pm\sqrt{p^2c^2 + m^2c^4}$. It appears therefore that eight solutions are obtained; four for $E > 0$ and four for $E < 0$. By using our specific representation of the γ^μ, show that in fact only four independent solutions are obtained and relate these to the free solutions obtained in the text.

2. Show that gauge transformations of the electromagnetic potential A_μ induce phase transformations of the Dirac wave function.
3. Solve the Dirac equation for a particle in a square well after determining the continuity condition on the Dirac wave function at the edge of the well. Assume that the potential is the time component of a Lorentz four-vector.
4. Repeat the analysis of Problem 3, but now assume the square well to be a scalar potential; i.e., an addition to the mass.
5. Show that the vector current remains conserved in the presence of electromagnetic interactions. What is the divergence of the axial vector current?
6. Solve the Dirac equation in an external constant magnetic field along with z direction.
7. Verify (23-17)
8. Verify that (23-39) and (23-41) follow from (23-37), (23-38) and (23-39).

INDEX